Y0-BWC-298

NONLINEAR SYSTEM ANALYSIS

ELECTRICAL SCIENCE

A Series of Monographs and Texts

Edited by

Henry G. Booker
UNIVERSITY OF CALIFORNIA AT SAN DIEGO
LA JOLLA, CALIFORNIA

Nicholas DeClaris
CORNELL UNIVERSITY
ITHACA, NEW YORK

JOSEPH E. ROWE. Nonlinear Electron-Wave Interaction Phenomena. 1965

MAX J. O. STRUTT. Semiconductor Devices: Volume I. Semiconductors and Semiconductor Diodes. 1966

AUSTIN BLAQUIÈRE. Nonlinear System Analysis. 1966

In preparation:

CHARLES BIRDSALL AND WILLIAM BRIDGES. Electron Dynamics of Diode Regions

CHARLES COOK AND MARVIN BERNFELD. Radar Signals: An Introduction to Theory and Application.

A. D. KUZ'MIN AND A. E. SALOMONOVICH. Radioastronomical Methods of Antenna Measurements

VICTOR RUMSEY. Frequency Independent Antennas

NONLINEAR SYSTEM ANALYSIS

AUSTIN BLAQUIÈRE

FACULTY OF SCIENCES, PARIS
UNIVERSITY OF PARIS, FRANCE

1966

ACADEMIC PRESS New York and London

ACADEMIC PRESS INC.
111 Fifth Avenue, New York, New York 10003

United Kingdom Edition published by
ACADEMIC PRESS INC. (LONDON) LTD.
Berkeley Square House, London W.1

LIBRARY OF CONGRESS CATALOG CARD NUMBER: 65-26390

PRINTED IN THE UNITED STATES OF AMERICA

Foreword

Nonlinear aspects of classical physics were first viewed, in the 19th century, as a set of phenomena which occur at the limit of validity of linear laws and linear models. By the end of the 19th century, however, accurate observations in celestial mechanics had been accumulated over a long enough period (some 100 years) to enable one to recognize as well as to measure the effects of the basic nonlinearity of Newton's law of attraction. Herein lay the source of the fundamental work of the great pioneers in nonlinear theory, such as Poincaré, Linstedt, Hill and Lyapunov. With great imagination, they went beyond the bounds of the particular astronomical problem to build the powerful general theories which still encompass most of our present knowledge in the field. Their line of reasoning was further extended in succeeding years, mainly in the USSR and Japan. The work of applied mathematicians and theoretical physicists like Krylov, Bogolioubov, Andronov, Chaikin, Hayashi and Minorski developed, and continue to develop, the mathematical basis of the methods which are the object of a good part of this book.

Other nonlinear phenomena, also discovered during the 19th century, failed to arouse the interest they deserved; they remained half understood curiosities among the wealth of orderly linear phenomena. Such, for example, was the case with Melde's famous experiment which though examined by generations of students, was not widely recognized to be an application of "parametric excitation" until very recent years.

Lack of appreciation of the nonlinear domain reached an end with the appearance of "electronic amplifiers" and "electronic oscillators" just prior to the 1920's because the flexibility and accuracy of these electronic devices offered a wealth of results which demanded analytic interpretation. The multivibrator of the World War I and the neon lamp oscillator were soon explained by the clear, elegant theory of Balth Van der Pol. Varying only one parameter in a single second order differential equation, any graduate student could understand the transition from the linear to the completely nonlinear domain, and soon "relaxation" oscillations were accepted on the same footing as sinusoïdal ones. Shortly before World War II, a large domain of applications of nonlinear theory appeared with the development of oscillography, television, and pulse communications.

The tremendous scientific revolution which accompanied World War II gave new impetus to research in applied electronics. It gave birth to such important domains as nonlinear servomechanisms, discontinuous servos, and still more recently, parametric amplification. While each of these appears today as a well-defined and developed specialized domain, they nevertheless merit special mention in this short historical review because they helped to underscore the importance of noise in nonlinear phenomena. Thus, they all lead to an intimate union of two hitherto separate disciplines: nonlinear theory and stochastic methods. This necessary fusion of two separate, but equally difficult, disciplines typifies the nature of the pedagogical problems which confront teachers and research workers today. The present book aims at solving this type of pedagogical problem for the nonlinear and stochastic fusion. It leans on previous efforts at a more elementary level, as represented by the books of Hayashi[†] and Stern[‡] but extends this work sufficiently to enable the graduate student to initiate his own research. Although the scope is broad, the book evolves in a very orderly manner.

The attack starts with the presentation of slightly nonlinear cases, and then proceeds with the development of all the general methods in roughly the same order in which they were discovered. In each step, the physical problem is clearly enunciated first, and then the mathematical methods of solution are developed completely, but with the utmost economy of mathematics. Numerous modern examples are thoroughly and clearly treated. The book considers, for example, the theories of the line width of maser clocks, the synchronization of laser modes, and the stability of betatron oscillations in the synchrotron. The latter is a particularly instructive example, because this oscillator appears to be devoid of damping. Thus, slight perturbations have

[†] C. Hayashi. "Nonlinear Oscillations in Physical Systems." McGraw-Hill, New York, 1964.

[‡] T. E. Stern. "Nonlinear Networks and Systems." Addison-Wesley, Reading, Massachusetts, 1965.

sufficient time to develop large effects. The betatron oscillations are also interesting because they involve two variables instead of one as in the classical oscillator. This characteristic was enough, in the early days of the big European accelerator in Geneva, to render inefficient any brute force analytic methods, including the use of giant calculators. Step-by-step computations of the divergence of the beam proved nearly too expensive, even for the "international budget" of CERN. Although the modern theory employs the calculator also, it is put to a much more efficient use, and so achieves an otherwise unobtainable complete solution.

It is hoped that this form of teaching, by consideration of representative examples which are rather thoroughly treated, will be well received by the reader as appropriate for such a broad new field. My further hope, which is also shared by the author, is that numerous readers of this book will soon try their own hand at the teaching and/or research effort. Many new and promising research applications appear all the time. For example, it is becoming increasingly apparent that there are links between classical nonlinear mechanisms and quantum mechanics. These begin to take on practical importance, in such areas as quantum electronics, as in Lamb's[†] theory of the laser, and in nonlinear optics, as recently presented by Bloembergen.[‡] Furthermore, it now appears that the links between classical and quantum nonlinear aspects will become of increasing theoretical importance in the future.

If the present book succeeds in helping to pave the way to further research and teaching in nonlinear theory, it will have fulfilled its aim, and this writer and the author will be more than pleased.

Faculty of Sciences, Orsay
University of Paris, France
April 1966

PIERRE GRIVET

[†] W. E. Lamb, Jr. Theory of optical maser oscillators, *in* "Quantum Electronics and Coherent Light" (C. H. Townes, ed.) pp. 78-110. Proceedings of the International School of Physics "Enrico Fermi" Course 31. Academic Press, New York, 1964.

[‡] N. Bloembergen. "Nonlinear Optics." Benjamin, New York, 1965.

Preface

This book is devoted to the study of systems whose behavior is governed by nonlinear differential equations. As the subject is a broad one, I have chosen to simplify it and to focus attention on a few problems which play a central role in engineering and physics: illustrative examples are discussed, with applications to particle accelerators, frequency measurement, and masers; and important practical problems, such as synchronization, stability of systems with periodic coefficients, and effect of random disturbances, are analyzed from different viewpoints.

One of the purposes of this monograph is to provide engineers and physicists with basic knowledge concerning typically nonlinear problems. While the main stress is laid on the theory of oscillations, the subject is carefully limited as the book does not exhaustively cover the problems occurring in this field. Because there are many good introductory books to which the reader may refer, a few developments concerning general features of nonlinear systems have been deliberately shortened or omitted, allowing a more detailed discussion of the selected examples.

On the other hand, I have tried to gain perspective of the domain of nonlinear theory by considering, whenever possible, applications of different methods to the same problems, and by examining the chronological and logical connections between classical methods and more recent ones. This is emphasized by the bibliography at the end of each chapter, which is arranged in chronological order. The approach to problems in many chapters is original, being based on some of my previously published works in France and elsewhere. Other approaches have been developed in collaboration with Professor P. Grivet.

A part of this book was the material of a graduate course in nonlinear oscillations, at the University of California, Berkeley. The present volume might serve as a text for a course in nonlinear vibrations or nonlinear circuit analysis, given to graduate students or to advanced seniors in electrical and mechanical engineering. Graduate students in physics will also be interested in the treatment of the up-to-date examples mentioned above.

The first chapter is an introduction to quantitative and qualitative methods, with applications to the simple pendulum. This simple example is of particular interest since it can be analyzed very precisely and allows comparison between approximate and exact solutions. Moreover, its equation provides a good model for describing the behavior of many other nonlinear systems.

The methods described in Chapter I are elaborated on in the second chapter which deals with self-oscillatory systems, and in the third chapter which gives a brief classification of singular points in the phase plane.

Whereas the first three chapters analyze the motion of oscillators with one degree of freedom, governed by a differential equation of the second order, Chapter IV considers more complicated devices in which a number of oscillators of this kind are linked together. The aim of this chapter is to show how the theory of oscillations can be extended to such systems having several degrees of freedom. It starts with systems having two degrees of freedom and, according to the earlier classification, it separately considers conservative oscillators and self-oscillatory systems. The theory is extended to systems with n degrees of freedom.

Chapters V and VI extend the concept of frequency response to nonlinear systems. They describe a heuristic method which is based on the principle of *equivalent linearization* first introduced by Krylov and Bogoliubov. As this method is suitable for many nonlinear problems of practical interest, it has been extensively studied, developed, and applied to numerous engineering problems during the last decade. It has become known as the "describing function" method. A less classical version of this method leads to a matrix representation of the transfer function of nonlinear systems.

Chapter VII is devoted to nonlinear equations with periodic coefficients. The questions regarding the stability of their solutions play an important role in many fields. Here we have mostly considered the applications of the theory to particle accelerators, in connection with the construction of the alternating-gradient proton-synchrotron of CERN, at Geneva. A computational procedure which can also be applied to the artificial-satellite problem is described.

The last two chapters deal with system response to random inputs. In Chapter VIII methods are described for analyzing the fluctuations of nonlinear systems subjected to random forcing functions. Their starting point is the equation of Fokker-Planck-Kolmogorov and Campbell's theorems. Such investigations play an important role by precisely stating the limit which is prescribed for accuracy of the measurements. In Chapter IX the theory of noise in nonlinear systems is applied to the analysis of random fluctuations of self-oscillators. Such problems recently have attracted renewed interest with the discovery of atomic clocks and lasers, devices which produce an extremely pure sine wave.

The subject is a broad one and, therefore, many of its practical aspects are disregarded and the discussion concerns mainly the "line-width" problems in radioelectric oscillators and masers.

I wish to acknowledge Professor P. Grivet who revealed to me, many years ago, the wonderful roads of nonlinear theory. No doubt this book never would have appeared were it not for our fruitful discussions concerning the many problems encountered and for the works on which we collaborated. I am grateful to Professor F. Bertein who agreed to complement this book with up-to-date developments on synchronization phenomena in quantum oscillators.

I am indebted to the French Commissariat à l'Énergie Atomique and to Monsieur le Haut Commissaire, Professor F. Perrin, whose support enabled me to communicate with many specialists in this field, in France and abroad. I am also grateful to Dr. J. Debiesse, Director of the Institut National des Sciences et Techniques Nucléaires (Saclay).

I wish to acknowledge the University of California, Berkeley, for offering me the opportunity to spend several months on its campus. During that time I had many enlightening discussions concerning nonlinear problems with many of my American colleagues, among them, Professors Hsu, Leitmann, Rosenberg, and Zadeh, and with the students of the Department of Mechanical Engineering.

Last but not least, I am grateful to Academic Press Inc. who gave birth to the plan for this book, and who published it promptly and competently.

Paris, France AUSTIN BLAQUIÈRE
May 1966

Contents

CHAPTER I

Linearity and Nonlinearity

1. AN EXAMPLE OF A NONLINEAR SYSTEM: THE SIMPLE PENDULUM

One of the simplest nonlinear oscillators is the simple pendulum (Fig. 1), whose equation, when friction is neglected, is

$$ml\ddot{x} + mg \sin x = 0 \tag{1}$$

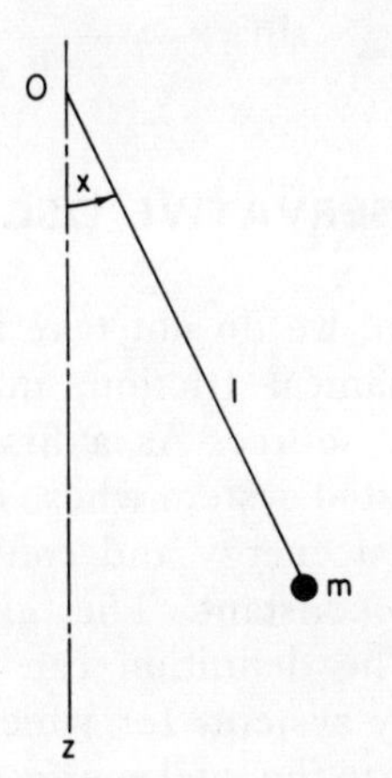

FIG. 1. Simple pendulum.

where m is the mass of the pendulum, l its length, and x the angle between the pendulum, at time t, and the vertical axis $0z$. By putting $\omega_0 = (g/l)^{1/2}$, (1) is rewritten

$$\ddot{x} + \omega_0^2 \sin x = 0 \tag{2}$$

When angle x is sufficiently small, $\sin x$ may be approximated by x, so that (2) is replaced by the linear equation

$$\ddot{x} + \omega_0^2 x = 0 \tag{3}$$

The solutions of (3) are sine functions,

$$x = a_0 \sin(\omega_0 t + \varphi)$$

whose angular frequency ω_0 does not depend on amplitude a_0.

When assuming that $x = 0$ at $t = 0$, we have

$$x = a_0 \sin \omega_0 t \tag{4}$$

Equation (4) describes approximately the oscillation of the pendulum when a_0 (and consequently x) is small. However, a better approximation may be obtained by using the nonlinear equation

$$\ddot{x} + \omega_0^2 \left(x - \frac{x^3}{6}\right) = 0 \tag{5}$$

instead of (3). Equation (5) is obtained from (2) by replacing $\sin x$ by the first two terms of its series expansion,

$$\sin x \sim x - \frac{x^3}{6}$$

2. CONSERVATIVE OSCILLATORS

In the discussion above we do not take into account the dissipation of energy due to mechanical friction, and the pendulum does not receive energy from any source. As a first approximation it may be considered to be an insulated system whose kinetic energy is periodically transformed into potential energy and conversely, in such a way that its total energy remains constant. This kind of oscillator is called a *conservative oscillator*. The definition can be made more general by including in this category systems for which there exists a function of the state variables, similar to the total energy function, which is a constant of the motion [13].

Next we shall take account of forces of friction which damp the oscillator, and we shall also study the case of regenerative oscillators which receive energy from an external source. Before discussing these more realistic oscillating systems, let us first describe briefly other conservative oscillators, whose equations are strongly similar to the one of the simple pendulum.

First consider the oscillation of a mass m hung from a spring (Fig. 2).

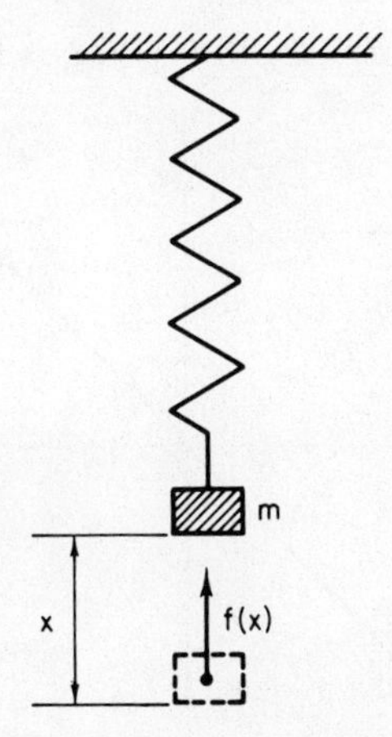

FIG. 2. Elongation of a spring.

Again neglect forces of friction, so that the total energy of the system remains constant, and let $f(x)$ be the relation between the restoring force and the elongation x of the spring. $f(x)$ is assumed to be a continuous function of x. Then the differential equation of motion is

$$m\ddot{x} + f(x) = 0 \tag{6}$$

Assuming that $f(x)$ is an odd function, which is not always true in practice, we are faced with several different situations:

(a) df/dx, which is the *rigidity* of the spring, may be a strictly increasing function of x, in which case the spring is said to be *hard* (Fig. 3).

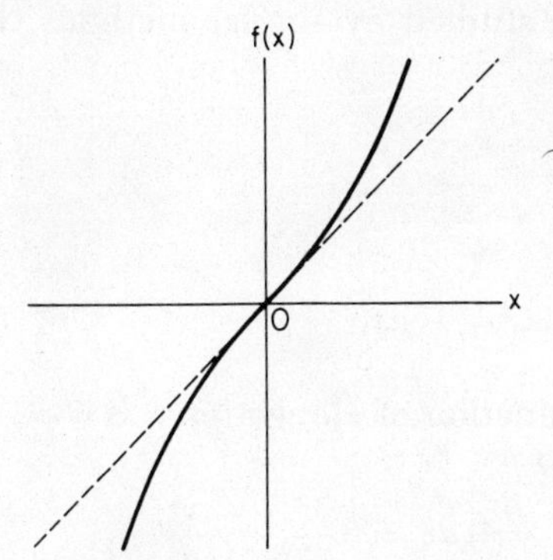

FIG. 3. Hard spring; the rigidity is an increasing function of x.

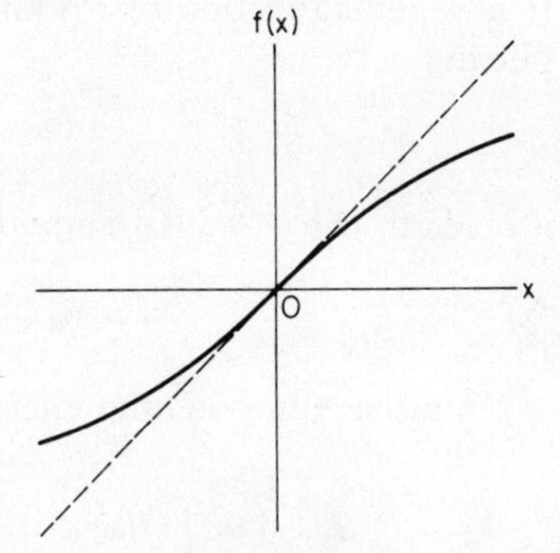

FIG. 4. Soft spring; the rigidity is a decreasing function of x.

(b) It may be that df/dx is a strictly decreasing function of x; then the spring is *soft* (Fig. 4).

(c) When df/dx is a constant, the spring is linear (Fig. 5).

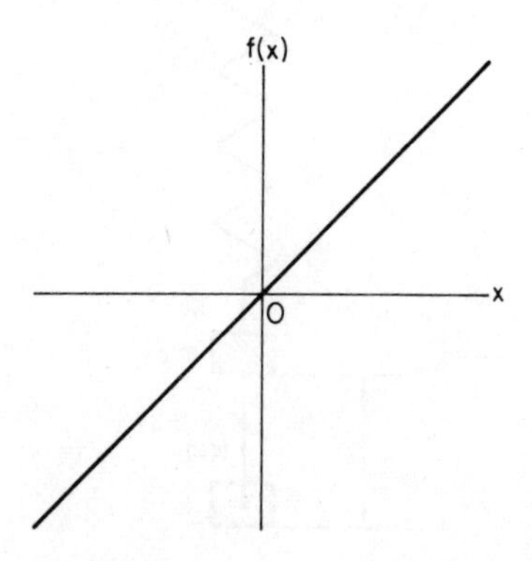

FIG. 5. Linear spring; the rigidity is a constant.

If, for example, the relation between the restoring force and the elongation is

$$f(x) = \beta x + \gamma x^3 \qquad \text{with} \quad \beta > 0 \tag{7}$$

the spring is hard if $\gamma > 0$; it is soft if $\gamma < 0$ and linear if $\gamma = 0$.

Equation (6) reduces to

$$\ddot{x} + \omega_0{}^2 x + \mu x^3 = 0 \qquad \omega_0{}^2 = \frac{\beta}{m} \quad \mu = \frac{\gamma}{m} \tag{8}$$

It is a generalization of (5) and can be studied by similar methods. By putting

$$x = x_1 \qquad \dot{x} = x_2$$

it leads to the so-called *normal set*

$$\dot{x}_1 = x_2 \qquad \dot{x}_2 = -\omega_0{}^2 x_1 - \mu x_1{}^3 \tag{9}$$

Note that the potential energy as a function of elongation x is

$$E_p(x) = \int_0^x f(x)\,dx = \int_0^x (\beta x + \gamma x^3)\,dx = \beta\frac{x^2}{2} + \gamma\frac{x^4}{4}$$

It is plotted in Fig. 6 for $\beta > 0$ and for $\gamma > 0$, $\gamma = 0$, and $\gamma < 0$.

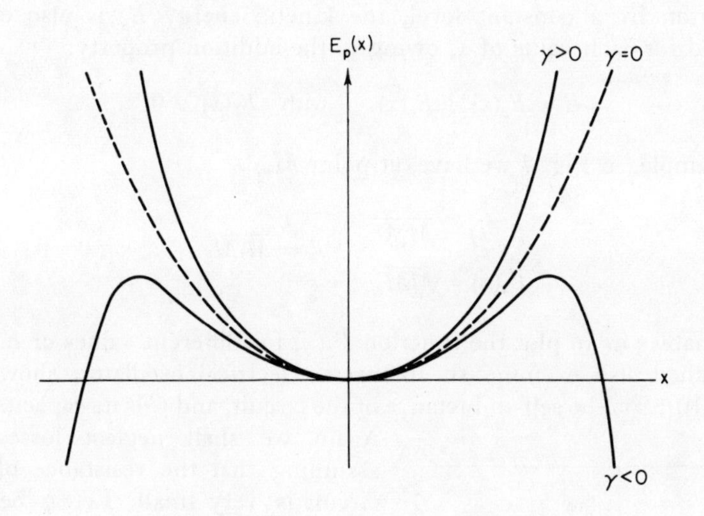

FIG. 6. Potential energy as a function of elongation.

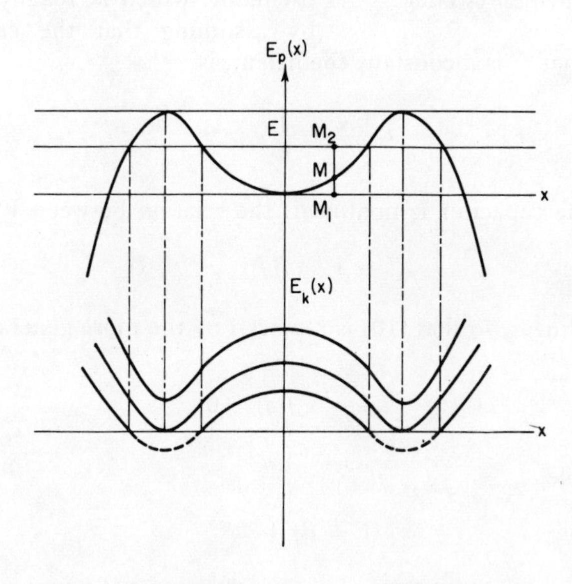

FIG. 7. Energy diagram for a nonlinear pendulum.

If the total energy E of the oscillator is represented in the same illustration by a constant level, the kinetic energy E_k is also easily obtained for each value of x, owing to the addition property:

$$E = E_p(x) + E_k(x) \qquad \text{with} \quad E_k(x) \geqslant 0$$

For example, in Fig. 7 we have, at point M,

$$E_p(x) = \overline{M_1M} \qquad E = \overline{M_1M_2}$$
$$E_k(x) = \overline{MM_2}$$

This enables us to plot the function $E_k(x)$ for different values of E.

We shall also examine an analogous electrical oscillator, shown in Fig. 8 [10]; L is the self-inductance of the circuit, and C is its capacitance. Again we shall neglect losses by assuming that the resistance of the circuit is very small. Let q be the charge of the capacitor at any time t and let V be the corresponding potential difference between its end points, related to q by $V = (1/C)q$. The differential equation of the oscillator, which is readily obtained by assuming that the capacitor is linear, i.e., that C is a constant coefficient, is

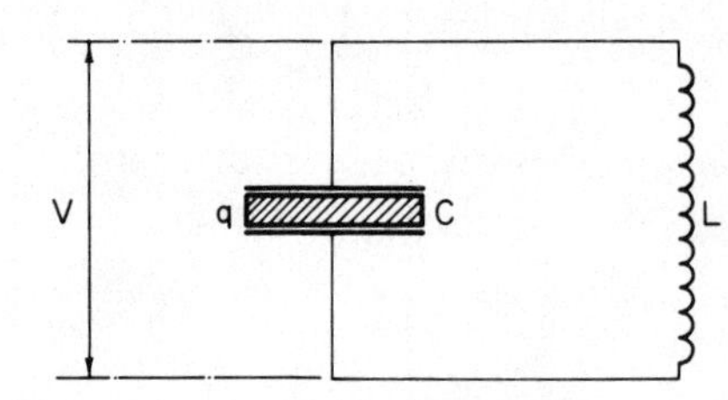

FIG. 8. Electrical oscillator.

$$L\frac{d^2q}{dt^2} + \frac{1}{C}q = 0 \tag{10}$$

Now, if the capacitor is nonlinear, the relation between V and q,

$$V = f(q)$$

is no longer linear, so that (10) is replaced by the more general equation

$$L\frac{d^2q}{dt^2} + f(q) = 0$$

If, for example,

$$V = \beta q + \gamma q^2$$

we find an equation which is similar to (8).

3. APPROXIMATE SOLUTIONS OF THE PENDULUM EQUATION

3.1. Secular Terms

Let us consider the equation

$$\ddot{x} + \omega_0^2 x + \mu x^3 = 0 \tag{11}$$

and assume for the time being that it is weakly nonlinear, i.e., that the coefficient μ is small.

First of all, following a heuristic assumption due to Poisson, we shall try to find a series expansion with respect to a small parameter μ which verifies (11), say

$$x(t) = x_0(t) + \mu x_1(t) + \mu^2 x_2(t) + \cdots + \mu^m x_m(t) + \cdots \tag{12}$$

where $x_0(t)$, $x_1(t)$, ..., $x_m(t)$ are functions to be determined.

By substituting (12) in (11) and equating the coefficients of the like powers of μ, one gets a set of recursive equations from which these functions can be deduced. If, for example, we consider terms of degree 0 and 1 with respect to μ, we get

$$\ddot{x}_0 + \omega_0^2 x_0 + \mu(\ddot{x}_1 + \omega_0^2 x_1 + x_0^3) + o(\mu) = 0$$

from which follows

$$\ddot{x}_0 + \omega_0^2 x_0 = 0 \tag{13}$$

$$\ddot{x}_1 + \omega_0^2 x_1 = -x_0^3 \tag{14}$$

From (13) we have

$$x_0 = a_0 \sin(\omega_0 t + \varphi)$$

Then, substituting in [14],

$$\ddot{x}_1 + \omega_0^2 x_1 = -\tfrac{3}{4} a_0^3 \sin(\omega_0 t + \varphi) + \tfrac{1}{4} a_0^3 \sin 3(\omega_0 t + \varphi) \tag{15}$$

The solution of (15) is

$$x_1 = \frac{3}{8\omega_0} t a_0^3 \cos(\omega_0 t + \varphi) - \frac{a_0^3}{32\omega_0^2} \sin 3(\omega_0 t + \varphi)$$

Finally, we obtain

$$x(t) = a_0 \sin(\omega_0 t + \varphi) + \mu \left[\frac{3}{8\omega_0} t a_0^3 \cos(\omega_0 t + \varphi) - \frac{a_0^3}{32\omega_0^2} \sin 3(\omega_0 t + \varphi)\right] + o(\mu) \tag{16}$$

We encounter a serious difficulty when we apply this small-parameter method, because the second term of the right side of (16) tends to infinity when t tends to infinity, whereas we shall see later that the exact solution of (11) is bounded whatever the value of t. This term is called a *secular term*. Indeed, the reason we are not able to discuss boundedness of $x(t)$ from such expansions is because we are considering only a small number of terms. However, in this example the infinite series is a bounded function of t because the process is a convergent one.

Another simple example of such a situation is the one given by Bogoliubov and Mitropolsky [7], namely the series expansion of $\sin(\omega_0 + \mu)t$:

$$\sin(\omega_0 + \mu)t = \sin \omega_0 t + \mu t \cos \omega_0 t - \frac{\mu^2 t^2}{2!} \sin \omega_0 t - \frac{\mu^3 t^3}{3!} \cos \omega_0 t + \cdots$$

where μ is a small parameter. This equation describes a periodic function which is bounded whatever the value of t; however, the expansion in the right side does not disclose such properties. This is a real difficulty, because the method does not bring to light important aspects of the solution, and in some problems the convergence of the process may be hard to prove.

3.2. Perturbation Method

We shall modify the above method according to a suggestion due to Lindstedt [1], which tends to eliminate secular terms in each step of the approximation. To circumvent the difficulty which is introduced by secular terms, let us rely on engineering practice by noting that the angular frequency of the oscillator is a function of amplitude a., and put

$$\omega^2 = \omega_0{}^2 + \mu\xi_1(a_0) + \mu^2\xi_2(a_0) + \cdots$$

If, as in the above derivation, we consider only linear terms in μ, we have

$$\omega_0{}^2 = \omega^2 - \mu\xi_1(a_0) + o(\mu) \tag{17}$$

which gives, by substituting (12) and (17) in (11),

$$\ddot{x}_0 + \omega^2 x_0 + \mu(\ddot{x}_1 + \omega^2 x_1 + x_0{}^3 - \xi_1 x_0) + o(\mu) = 0$$

from which follows

$$\ddot{x}_0 + \omega^2 x_0 = 0 \tag{18}$$

$$\ddot{x}_1 + \omega^2 x_1 = -x_0{}^3 + \xi_1 x_0 \tag{19}$$

From (18) we have

$$x_0 = a_0 \sin(\omega t + \varphi)$$

This particular zero-order solution is called the *generating solution*. Then (19) becomes

$$\ddot{x}_1 + \omega^2 x_1 = -\tfrac{3}{4}a_0^3 \sin(\omega t + \varphi) + \tfrac{1}{4}a_0^3 \sin 3(\omega t + \varphi) + \xi_1 a_0 \sin(\omega t + \varphi) \tag{20}$$

Since the secular term was generated by the first harmonic on the right side of (15), and ξ, is at our disposal, we can get rid of this secular term by putting

$$\xi_1 = \tfrac{3}{4}a_0^2 \tag{21}$$

Then the solution of (20) is

$$x_1 = A \sin(\omega t + \varphi_1) - \frac{a_0^3}{32\omega^2} \sin 3(\omega t + \varphi)$$

where A is a constant of integration which we can cancel to simplify the solution. Finally, we get

$$x(t) = a_0 \sin(\omega t + \varphi) - \mu \frac{a_0^3}{32\omega^2} \sin 3(\omega t + \varphi) + o(\mu) \tag{22}$$

with

$$\omega^2 = \omega_0^2 + \frac{3\mu}{4} a_0^2 + o(\mu) \tag{23}$$

A better approximation might be readily obtained by considering expansions to the order of μ^3:

$$x(t) = a_0 \sin(\omega t + \varphi) - \mu \frac{a_0^3}{32\omega^2} \sin 3(\omega t + \varphi) + \mu^2 \frac{a_0^5}{1024\omega^2} \sin 5(\omega t + \varphi) + o(\mu^2) \tag{24}$$

with

$$\omega^2 = \omega_0^2 + \frac{3\mu}{4} a_0^2 - \frac{3\mu^2}{128} \frac{a_0^4}{\omega^2} + o(\mu^2) \tag{25}$$

It is worthwhile to note that the nonlinear term in (11) raises odd harmonics, and introduces a relation between the frequency of the first harmonic and its amplitude. Then the angular frequency ω is different from ω_0, which corresponds to the linear approximation.

3.3. First-Harmonic Approximation[†]

Assume that a_0 is small and, as a first approximation, that the behavior of the pendulum, which is governed by (5), is described by

$$x = a_0 \sin \omega t \tag{26}$$

where ω is a function of amplitude a_0,

$$\omega = \omega(a_0)$$

which we wish to determine. Indeed, this function was determined by the above perturbation method, which justifies the present assumption; however, we shall now consider the problem from another viewpoint.

It may easily be seen that (26) is not an exact solution of (5). As a matter of fact, we have

$$x^3 = a_0^3 \sin^3 \omega t = a_0^3(\tfrac{3}{4} \sin \omega t - \tfrac{1}{4} \sin 3\omega t)$$

and substituting in the left side of (5) we obtain

$$\ddot{x} + \omega_0^2 \left(x - \frac{x^3}{6}\right) = \left(\omega_0^2 - \omega^2 - \omega_0^2 \frac{a_0^2}{8}\right) a_0 \sin \omega t + \omega_0^2 \frac{a_0^3}{24} \sin 3\omega t \tag{27}$$

Obviously this expression cannot vanish identically. It is a consequence of incompleteness of (26). As we have seen in the preceding section, together with the first harmonic (26), the oscillation of the pendulum exhibits higher-order harmonics which should be taken into account. For example, it would be better to assume that x has the form

$$x = a_0 \sin \omega t + \alpha_3 \sin 3\omega t + \alpha_5 \sin 5\omega t + \cdots$$

If so, higher-order harmonics appear in (27), the amplitudes of which can be put to zero by a proper choice of coefficients α_3, α_5, ... We shall not discuss here the details of such a calculus, which is a matter of engineering practice rather than a rigorous method of obtaining the solution. However, this practice is useful, as will be seen later.

If we disregard the third-order harmonic term in (27), according to the above arguments, we arrive at the condition

$$\left(\omega_0^2 - \omega^2 - \omega_0^2 \frac{a_0^2}{8}\right) a_0 \sin \omega t \equiv 0$$

which is verified if

$$\omega^2 = \omega_0^2 \left(1 - \frac{a_0^2}{8}\right)$$

That is formula (23) when $\mu = -\omega_0^2/6$.

[†] See [5].

Because a_0 is small, we can use the approximate formula

$$\omega = \omega_0 \left(1 - \frac{a_0^2}{16}\right) \tag{23*}$$

which is the well-known relation between amplitude and angular frequency of a simple pendulum.

3.4. Ritz-Galërkin Approximation

In general an oscillator is governed by an equation of the form

$$f(x, \dot{x}, \ddot{x}, ..., t) = 0 \tag{28}$$

where x is the state variable. It may be that t does not occur explicitly in the equation, in which case the system is said to be *autonomous*. In the other case the system is *nonautonomous*.

Later we shall see that it is possible to find an exact solution of the equation of the simple pendulum. However, it is a very special example, because in most practical problems only approximate solutions of (28) can be obtained.

In the Ritz-Galërkin method, the search for an approximate solution takes into account previous knowledge of its general form; i.e., starting with some empirical assumptions, the aim of this method is to improve the accuracy of a conveniently choosen approximate solution.

Let $\lambda_0 x_0(t)$ be this approximate solution, and write

$$x(t) = \lambda_0 x_0(t) + \lambda_1 x_1(t) + \cdots + \lambda_m x_m(t) \tag{29}$$

where $x(t)$ must verify some prescribed initial conditions. $x_1(t), ..., x_m(t)$ are correcting functions and $\lambda_0, \lambda_1, ..., \lambda_m$ are constant coefficients which will be computed so as to optimize $x(t)$.

The criterion of optimality is now as follows: Because $x(t)$ is not in general an exact solution of (28), by substituting (29) in (28) we get a function $\epsilon(t)$ which is a measurement of the accuracy of the approximation. Indeed, should $x(t)$ be an exact solution of (28), $\epsilon(t)$ would be identically zero.

Accordingly, a convenient way to ensure optimality of $x(t)$ is to minimize such an integral as

$$\int_a^b \epsilon^2(t)\, dt \tag{30}$$

where $[a, b]$ is a time interval which is defined by practical considerations. This leads to $m + 1$ equations,

$$\frac{\partial}{\partial \lambda_i} \int_a^b \epsilon^2 \, dt = 2 \int_a^b \epsilon \frac{\partial \epsilon}{\partial \lambda_i} \, dt = 0 \qquad (i = 0, 1, \dots, m)$$

from which $\lambda_0, \lambda_1, \dots, \lambda_m$ can be computed.

Let us apply this method to (5), by again using the approximate solution

$$x = a_0 \sin \omega t$$

In this example we have

$$\lambda_0 = a_0 \qquad x_0(t) = \sin \omega t$$

$$\lambda_1 = \lambda_2 = \cdots = \lambda_m = 0$$

By substituting in (5) we get

$$\epsilon = \left[\omega_0^2 - \omega^2 - \frac{\omega_0^2}{8} a_0^2\right] a_0 \sin \omega t + \frac{\omega_0^2}{24} a_0^3 \sin 3\omega t$$

Then, by choosing $a = 0$, $b = T$, with $T = 2\pi/\omega$;

$$\int_0^T \epsilon \frac{\partial \epsilon}{\partial a_0} \, dt \equiv \int_0^T \left[\left(\omega_0^2 - \omega^2 - \frac{\omega_0^2}{8} a_0^2\right) a_0 \sin \omega t + \frac{\omega_0^2}{24} a_0^3 \sin 3\omega t\right]$$

$$\times \left[\left(\omega_0^2 - \omega^2 - \frac{3\omega_0^2}{8} a_0^2\right) \sin \omega t + \frac{\omega_0^2}{8} a_0^2 \sin 3\omega t\right] dt = 0$$

which reduces, if we exclude the trivial solution $a_0 = 0$, to

$$\left(\omega_0^2 - \omega^2 - \frac{\omega_0^2}{8} a_0^2\right)\left(\omega_0^2 - \omega^2 - \frac{3\omega_0^2}{8} a_0^2\right) + \frac{\omega_0^2 a_0^2}{24} \frac{\omega_0^2 a_0^2}{8} = 0$$

or

$$\omega^4 - 2\left(1 - \frac{a_0^2}{4}\right) \omega_0^2 \omega^2 + \omega_0^4 \left(1 - \frac{a_0^2}{2} + \frac{5a_0^4}{96}\right) = 0$$

This equation has two roots,

$$\omega^2 = \omega_0^2(1 - 0.15 a_0^2) \tag{31}$$

$$\omega^2 = \omega_0^2(1 - 0.35 a_0^2) \tag{32}$$

and it may be shown that the first one minimizes $\int_0^T \epsilon^2 \, dt$ while the second one maximizes this quantity. Consequently, we shall keep approximation (31), which is slightly different from the above expression:

$$\omega^2 = \omega_0^2 \left(1 - \frac{a_0^2}{8}\right) = \omega_0^2 (1 - 0.125 a_0^2)$$

The result would be improved by starting with

$$x = a_0 \sin \omega t + \alpha_3 \sin 3\omega t$$

say

$$\lambda_0 = a_0 \qquad x_0(t) = \sin \omega t$$

$$\lambda_1 = \alpha_3 \qquad x_1(t) = \sin 3\omega t$$

$$\lambda_2 = \cdots = \lambda_m = 0$$

3.5. Optimal Linearization Method†

Now let us start with

$$\ddot{x} + \omega_0^2 \sin x = 0 \tag{33}$$

and try to *approach* this nonlinear equation by the linear one,

$$\ddot{x} + \lambda x = 0 \tag{34}$$

where λ is a constant coefficient which we wish to choose in an optimal way. We shall again specify what is meant by optimality concerning λ.

Consider the difference between the left sides of (33) and (34):

$$\epsilon(\lambda) = \lambda x - \omega_0^2 \sin x$$

and

$$\epsilon^2(\lambda) = \lambda^2 x^2 - 2\lambda\omega_0^2 x \sin x + \omega_0^4 \sin^2 x \tag{35}$$

ϵ^2 is a function of λ and x, in which we can substitute *any* function $x(t)$. Then ϵ^2 becomes a function of λ and t whose integral

$$\int_a^b \epsilon^2(\lambda, t)\, dt$$

provided that it is defined, will again be used as a criterion of the accuracy of the approximation, and the optimality of λ will be defined by the condition

$$\frac{\partial}{\partial \lambda} \int_a^b \epsilon^2(\lambda, t)\, dt = 0 \tag{36}$$

† See [11, 12].

where the time interval $[a, b]$ is again deduced from practical considerations.

Note that the value of λ thus obtained depends on the choice of $x(t)$. Accordingly we shall say that (34) is the optimal linear equation with constant coefficient, which approaches (33) for $x = x(t)$. Let us write

$$\overline{x^2} = \int_a^b x^2(t)\, dt \qquad \overline{x \sin x} = \int_a^b x(t) \sin[x(t)]\, dt$$

$$\overline{\sin^2 x} = \int_a^b \sin^2[x(t)]\, dt$$

We deduce from (36) that

$$\frac{\partial}{\partial \lambda} \int_a^b \epsilon^2\, dt = 2\lambda \overline{x^2} - 2\omega_0^2 \overline{x \sin x} = 0$$

from which follows

$$\lambda = \omega_0^2 \frac{\overline{x \sin x}}{\overline{x^2}} \tag{37}$$

Now, as in Ritz-Galërkin approximation, if we have some information concerning the general form of the solution of (33), we can compute an approximate value of λ. Assume, for example, that x is small and replace $\sin x$ by $x - (x^3/6)$; we get

$$\lambda = \omega_0^2 \left(1 - \frac{1}{6} \frac{\overline{x^4}}{\overline{x^2}}\right) \qquad \text{with} \quad \overline{x^4} = \int_a^b x^4(t)\, dt$$

Furthermore, replace x by

$$x = a_0 \sin \omega t$$

and put $a = 0$, $b = T = 2\pi/\omega$. Then

$$\frac{\overline{x^4}}{\overline{x^2}} = \tfrac{3}{4} a_0^2$$

from which follows

$$\lambda = \omega_0^2 \left(1 - \frac{a_0^2}{8}\right)$$

Finally, from (34) we obtain

$$\omega^2 = \omega_0^2 \left(1 - \frac{a_0^2}{8}\right)$$

If x is not necessarily small, we have

$$\sin x = \sin(a_0 \sin \omega t)$$
$$= 2 \sum_{n=0}^{\infty} J_{2n+1}(a_0) \sin(2n+1)\omega t$$

and

$$\lambda = \omega_0^2 \frac{2J_1(a_0)}{a_0}$$

It follows that

$$\omega^2 = \omega_0^2 \frac{2J_1(a_0)}{a_0}$$

3.6. Exact Solution of the Approximate Equation by Elliptic Integrals

Again consider (5) and let

$$x = x_1 \qquad \dot{x} = x_2$$

Equation (5) is thus decomposed into

$$\dot{x}_1 = x_2 \qquad \dot{x}_2 = -\omega_0^2 \left(x_1 - \frac{x_1^3}{6}\right) \tag{38}$$

We shall return to this decomposition later. For the time being let us note that

$$\dot{x}_2 = \frac{dx_2}{dt} = \frac{dx_2}{dx_1} \dot{x}_1 = \frac{dx_2}{dx_1} x_2$$

Then, substituting in the second equation of (38), we obtain

$$\frac{dx_2}{dx_1} x_2 = -\omega_0^2 \left(x_1 - \frac{x_1^3}{6}\right)$$

which may be easily integrated, and gives

$$x_2^2 - x_{20}^2 = -\omega_0^2 \left[\left(x_1^2 - \frac{x_1^4}{12}\right) - \left(x_{10}^2 - \frac{x_{10}^4}{12}\right)\right] \tag{39}$$

x_{10} and x_{20} are integration constants, and we know that when $x_1 = x_{10} = a_0$ (the maximum value of x), then $x_2 = x_{20} = 0$. Therefore (39) is rewritten

$$x_2^2 = \omega_0^2[(a_0^2 - x_1^2) - \tfrac{1}{12}(a_0^4 - x_1^4)] \tag{40}$$

or, more conveniently,

$$x_2^2 = \omega_0^2(a_0^2 - x_1^2)[1 - \tfrac{1}{12}(a_0^2 + x_1^2)] \tag{41}$$

Now introduce the angular variable ψ, such that

$$x_1 = a_0 \sin \psi \tag{42}$$

which is possible, since $| x_1 | \leqslant a_0$. We have

$$a_0^2 - x_1^2 = a_0^2 \cos^2 \psi$$

$$a_0^2 + x_1^2 = a_0^2(1 + \sin^2 \psi)$$

and

$$x_2 = \frac{dx_1}{dt} = a_0 \cos \psi \frac{d\psi}{dt}$$

Substituting in (41) we obtain, after a straightforward simplification,

$$\left(\frac{d\psi}{dt}\right)^2 = \omega_0^2 \left[1 - \frac{1}{12} a_0^2(1 + \sin^2 \psi)\right]$$

or

$$\left(\frac{d\psi}{dt}\right)^2 = \omega_0^2 \left(1 - \frac{a_0^2}{12}\right)(1 - k^2 \sin^2 \psi)$$

where

$$k^2 = \frac{a_0^2/12}{1 - (a_0^2/12)}$$

Finally, if we choose the positive sign in the expression of $d\psi/dt$, say

$$\frac{d\psi}{dt} = \omega_0 \left(1 - \frac{a_0^2}{12}\right)^{1/2} (1 - k^2 \sin^2 \psi)^{1/2}$$

we get by integration

$$t - t_0 = \frac{1}{\omega_0[1 - (a_0^2/12)]^{1/2}} \int_{\psi_0}^{\psi} \frac{d\psi}{(1 - k^2 \sin^2 \psi]^{1/2}}$$

Let the initial time $t_0 = 0$ be the one which corresponds to $\psi_0 = 0$; then

$$t = \frac{1}{\omega_0[1 - (a_0^2/12)]^{1/2}} F(k, \psi) \tag{43}$$

where $F(k, \psi)$ is an incomplete elliptic integral of the first kind, modulus k, which can be computed from the series expansion

$$F(k, \psi) = \int_0^{\psi} (1 + \tfrac{1}{2}k^2 \sin^2 \psi + \tfrac{3}{8}k^4 \sin^4 \psi + \tfrac{5}{16}k^6 \sin^6 \psi + \cdots)\, d\psi$$

The period $T = 2\pi/\omega$ of the oscillation is readily obtained, by noting that, starting from $x_1 = 0$ at time $t_0 = 0$, x_1 reaches its maximum value for the first time when $\psi = \pi/2$. The corresponding transit time is $T/4$; therefore

$$T = \frac{4}{\omega_0[1 - (a_0^2/12)]^{1/2}} F(k, \pi/2)$$

or

$$\omega = \omega_0 \left(1 - \frac{a_0^2}{12}\right)^{1/2} \frac{\pi}{2F(k, \pi/2)} \tag{44}$$

$F(k, \pi/2)$ is a complete elliptic integral, modulus k.

A strong difference between this method and the ones which we have discussed in the preceding sections lies in the fact that the present method is an *exact* method. Indeed, (5) is an approximate equation of the simple pendulum, so that, in this respect, solution (43) again describes approximately the behavior of the simple pendulum. However, we must keep in mind the fact that (5) can also be considered as the exact equation of a spring pendulum. From this viewpoint (43) is an exact solution.

Note that, when a_0 is sufficiently small,

$$k^2 \simeq \frac{a_0^2}{12}$$

and

$$F(k, \pi/2) = \int_0^{\pi/2} \frac{d\psi}{(1 - k^2 \sin^2 \psi)^{1/2}} \simeq \int_0^{\pi/2} \left(1 + \frac{k^2}{2} \sin^2 \psi\right) d\psi$$

$$= \frac{\pi}{2}\left(1 + \frac{k^2}{4}\right)$$

Then (44) is approximately

$$\omega \simeq \omega_0 \left(1 - \frac{a_0^2}{24}\right) \Big/ \left(1 + \frac{a_0^2}{48}\right)$$

$$\simeq \omega_0 \left(1 - \frac{a_0^2}{16}\right)$$

It reduces to formula (23*) (p. 11).

4. EXACT SOLUTION BY ELLIPTIC INTEGRAL

Now let us return to the exact equation (2) and derive its solution by elliptic integral (see also [8, 9]). By introducing variables x_1 and x_2,

$$x = x_1 \qquad \dot{x} = x_2$$

we get from (2) the new equation

$$\frac{dx_2}{dx_1} x_2 = -\omega_0^2 \sin x_1$$

which may be integrated and gives

$$x_2^2 - x_{20}^2 = 2\omega_0^2(\cos x_1 - \cos x_{10}) \tag{45}$$

or, with $x_{10} = a_0$ and $x_{20} = 0$, as in Section 3,

$$x_2^2 = 2\omega_0^2(\cos x_1 - \cos a_0) \tag{46}$$

Equation (46) may be easily rewritten in the more convenient form

$$x_2^2 = 4\omega_0^2 \left(\sin^2 \frac{a_0}{2} - \sin^2 \frac{x_1}{2}\right) \tag{47}$$

after which transformation we shall introduce a new angular variable ψ, such that

$$\sin \frac{x_1}{2} = \sin \frac{a_0}{2} \sin \psi \tag{48}$$

Obviously when a_0 and consequently x_1 $(|x_1| \leqslant a_0)$ are sufficiently small, (48) reduces to (42).

By differentiating (48) with respect to time t we obtain

$$x_2 \cos \frac{x_1}{2} = 2 \sin \frac{a_0}{2} \cos \psi \frac{d\psi}{dt}$$

Then, substituting in (47),

$$\frac{4 \sin^2(a_0/2) \cos^2 \psi}{\cos^2(x_1/2)} \left(\frac{d\psi}{dt}\right)^2 = 4\omega_0^2 \sin^2 \frac{a_0}{2} (1 - \sin^2 \psi)$$

from which we deduce

$$\left(\frac{d\psi}{dt}\right)^2 = \omega_0^2 \cos^2 \frac{x_1}{2} = \omega_0^2 \left(1 - \sin^2 \frac{a_0}{2} \sin^2 \psi\right)$$

Finally, if we choose the positive sign in the expression of $d\psi/dt$, we get by integration

$$t - t_0 = \frac{1}{\omega_0}\int_{\psi_0}^{\psi} \frac{d\psi}{(1 - K^2 \sin^2 \psi)^{1/2}}$$

with $K = \sin(a_0/2)$, or, letting $\psi_0 = 0$ correspond to $t_0 = 0$;

$$t = \frac{1}{\omega_0} F(K, \psi) \tag{49}$$

The period of the oscillation is readily deduced from this formula by the same argument as above. It is

$$T = \frac{4}{\omega_0} F(K, \pi/2)$$

5. REPRESENTATION IN A PHASE PLANE

The motion of an oscillator can also be studied from a geometric viewpoint by means of the variables x_1 and x_2, which we shall now consider to be the coordinates of a moving point P in a plane, the so-called *phase plane*.

This method is easily extended to more elaborate systems involving a higher number of variables, say $x_1, x_2, ..., x_n$. However, it is less beneficial in this case because then point P moves in an n-dimensional phase space ($n > 2$) which does not yield a fair representation (as is the case where the phase space reduces to a plane).

In this section we shall restrict ourselves to the case of a phase plane, and will see that an interesting feature of this geometric representation lies in the fact that it makes very clear the evolution of the behavior of the system as nonlinearity is increased.

For example, let us start with the normal equations (9). $\dot{x}_1$ and $\dot{x}_2$ are components of the velocity of point P. Because they only depend on x_1 and x_2, equations (9) define a vector field in the phase plane, namely a field of velocities; i.e., to each point of the plane is associated one and only one vector, and P moves in such a way that at each time its velocity is the vector which is associated to the point of the plane with which it is in coincidence at that time.

It may be that the velocity vector is zero at some point. For example, when $x_1 = x_2 = 0$ in (9), at the origin 0 of the phase plane, we have $\dot{x}_1 = \dot{x}_2 = 0$. Such points are called *equilibrium points*. What happens in the neighborhood of such points is a very important question, chiefly in connection with the notion of *stability*.

On the other hand, every trajectory of P in the phase plane is an integral curve of the differential equations that govern the system under consideration, equations (9) in our example. At each point of such a curve, the slope is defined by the ratio

$$\frac{dx_2}{dx_1} = \frac{\dot{x}_2}{\dot{x}_1}$$

Obviously, at an equilibrium point this ratio is undetermined. Such points are also designated *critical or singular points.*

For example, if we refer to (8) and to the normal set (9) associated to it, we get by a straightforward integration,

$$x_2^2 - x_{20}^2 = -\left[\left(\omega_0^2 x_1^2 + \mu \frac{x_1^4}{2}\right) - \left(\omega_0^2 x_{10}^2 + \mu \frac{x_{10}^4}{2}\right)\right] \tag{50}$$

which is the equation of the family of integral curves.

First of all, when x_1 is sufficiently small, so that x_1^3 can be neglected with respect to x_1 in (8), (50) reduces to

$$x_2^2 - x_{20}^2 = -\omega_0^2(x_1^2 - x_{10}^2)$$

or

$$\omega_0^2 x_1^2 + x_2^2 = C \qquad \text{with} \quad C = \omega_0^2 x_{10}^2 + x_{20}^2$$

Trajectories are ellipses centered at the origin 0. When large values of x_1 are allowed, nonlinearity of equations must be taken into account, and trajectories may have different shapes, as shown in Fig. 9 for the

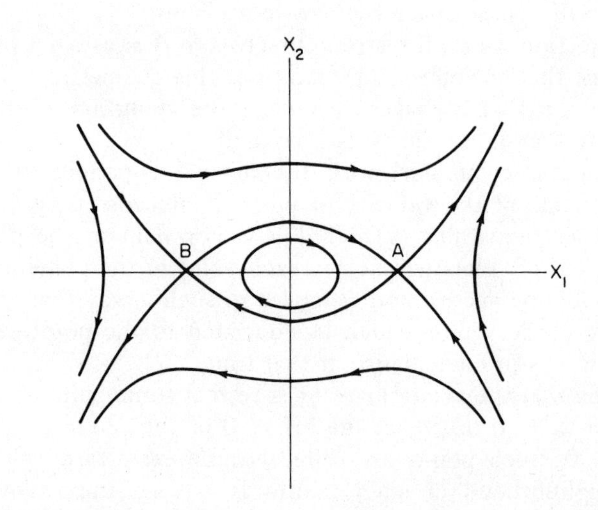

FIG. 9. Trajectories in a phase plane for Duffing's pendulum.

case $\mu < 0$ (soft spring). On each trajectory the arrows indicate in which direction point P moves as time increases.

It may be easily verified that we have, in this problem, three singular points:

$$O: \quad x_1 = x_2 = 0$$

$$A: \quad x_1 = (-\omega_0^2/\mu)^{1/2} \qquad x_2 = 0$$

$$B: \quad x_1 = -(-\omega_0^2/\mu)^{1/2} \qquad x_2 = 0$$

These trajectories are simply related to the curves of Fig. 7, since

$$E_k = \tfrac{1}{2}m(\dot{x})^2 = \tfrac{1}{2}mx_2^2$$

We must keep in mind the fact that these results do not apply to the simple pendulum because (5) was obtained by assuming that x_1 remains small.

As a matter of fact, the exact equations of the simple pendulum,

$$\dot{x}_1 = x_2 \qquad \dot{x}_2 = -\omega_0^2 \sin x_1$$

also exhibit singular points, defined by

$$x_1 = 2n\pi \qquad x_2 = 0$$

$$x_1 = (2n + 1)\pi \qquad x_2 = 0$$

(n is an integer). A family of trajectories is shown in Fig. 10.

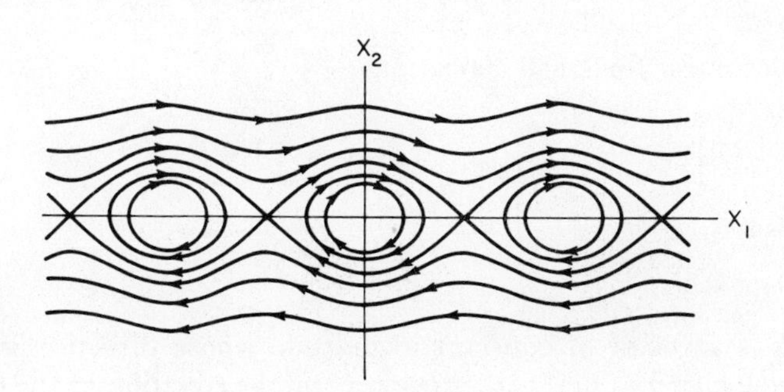

FIG. 10. Trajectories of a simple pendulum.

6. NONLINEAR OSCILLATOR WITH DAMPING

In the preceding sections we have neglected forces of friction. We shall consider them now, to derive a more realistic picture of the oscillatory motion. As a matter of fact, the physical nature of energy losses is so difficult to ascertain that most mathematical approaches are based on empirical observation.

In this section we shall restrict ourselves to simple assumptions by considering two cases:

Case 1. Viscous friction, i.e., friction whose magnitude is proportional to velocity.

Case 2. Solid friction or Coulomb friction, i.e., friction whose magnitude is constant.

Assume that the friction is applied to the bob of the pendulum whose velocity is $\mathbf{v}$ at any time; then the force of friction $\mathbf{f}$ is

In Case 1:

$$\mathbf{f} = -\eta \mathbf{v}$$

η being the viscosity coefficient of the fluid in which the pendulum is swinging. Consequently, we have to introduce in the right side of the equation of the pendulum the damping term $f = -\eta l \dot{x}$ and the equation is rewritten

$$ml\ddot{x} + \eta l \dot{x} + mg \sin x = 0 \tag{51}$$

Again we will put $\omega_0 = (g/l)^{1/2}$ and transform (51) into

$$\ddot{x} + \lambda \dot{x} + \omega_0^2 \sin x = 0 \qquad \text{with} \quad \lambda = \eta/m \tag{52}$$

which becomes, for small deviations,

$$\ddot{x} + \lambda \dot{x} + \omega_0^2 \left(x - \frac{x^3}{6}\right) = 0 \tag{53}$$

In Case 2:

$$\mathbf{f} = \mathbf{k}$$

where $\mathbf{k}$ is a vector of constant magnitude whose direction is opposite that of velocity $\mathbf{v}$. Then $f = -k \operatorname{sgn} \dot{x}$ and the equation of the pendulum is

$$ml\ddot{x} + k \operatorname{sgn} \dot{x} + mg \sin x = 0 \tag{54}$$

Note that, if the starting conditions are $x = \pm a_0 \, (a_0 > 0)$ and $\dot{x} = 0$, the pendulum will move only if the magnitude of the restoring force at the initial time, $mg \sin a_0$, is greater than k, say

$$mg \sin a_0 > k \tag{55}$$

Accordingly, when we make use of the approximation

$$\sin x \simeq x - \frac{x^3}{6} \tag{56}$$

for small values of x, we shall always assume that

(a) $\max x = a_0$ is sufficiently small, so that $\sin a_0 \simeq a_0 - a_0^3/6$
(b) a_0 verifies (55).

Obviously both conditions can be fulfilled only if k/m is sufficiently small.

When these assumptions are taken into consideration, (54) is rewritten

$$\ddot{x} + \mu \operatorname{sgn} \dot{x} + \omega_0^2 \left(x - \frac{x^3}{6}\right) = 0 \tag{57}$$

where $\mu = k/ml$.

Now the work of the different forces which enter the equation of the pendulum, between any two times t_1 and t_2, is easily computed. As a matter of fact, these different terms are

$$-\int_{t_1}^{t_2} mgl\dot{x} \sin x \, dt = mgl \cos x \Big|_{t_1}^{t_2} \tag{58}$$

$$-\int_{t_1}^{t_2} ml^2 \ddot{x}\dot{x} \, dt = -\tfrac{1}{2} ml^2 (\dot{x})^2 \Big|_{t_1}^{t_2} \tag{59}$$

and

In Case 1:

$$\int_{t_1}^{t_2} fl\dot{x} \, dt = -\int_{t_1}^{t_2} \eta l^2 (\dot{x})^2 \, dt \tag{60}$$

In Case 2:

$$\int_{t_1}^{t_2} fl\dot{x} \, dt = -\int_{t_1}^{t_2} kl| \, \dot{x} \, | \, dt \tag{61}$$

Then the equation

$$\int_{t_1}^{t_2} ml^2 \ddot{x}\dot{x} \, dt + \int_{t_1}^{t_2} mgl \sin x \dot{x} \, dt = \int_{t_1}^{t_2} fl\dot{x} \, dt$$

deduced from (52) or (54), expresses the fact that the change in energy of the pendulum is due to the work of forces of friction. Therefore energy losses are given by (60) and (61) in Cases 1 and 2.

When λ and μ, as well as deviation x, are sufficiently small, so that we can make use of the first approximation,

$$x = a_0 \sin \omega t \qquad \text{with} \quad \omega = \omega_0 \left(1 - \frac{a_0^2}{16}\right)$$

We find, by computing (60) and (61) over a period T,

$$-2 \int_0^{T/2} \eta l^2 (\dot{x})^2 \, dt = -\pi \eta l^2 a_0^2 \omega \tag{62}$$

$$-2 \int_0^{T/2} kl |\, \dot{x} \,| \, dt = -4kla_0 \tag{63}$$

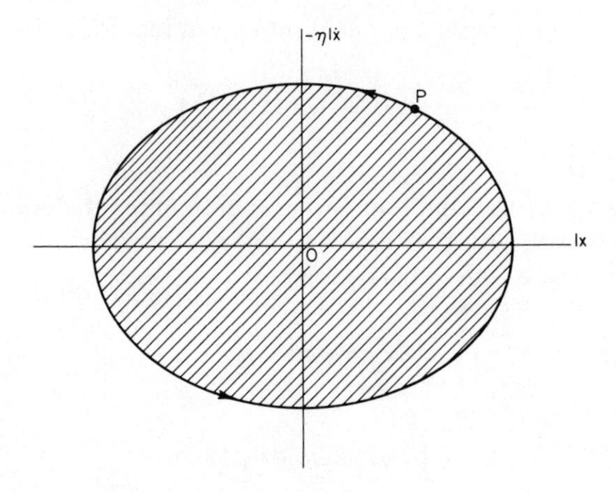

FIG. 11. Geometric representation of losses, viscous friction.

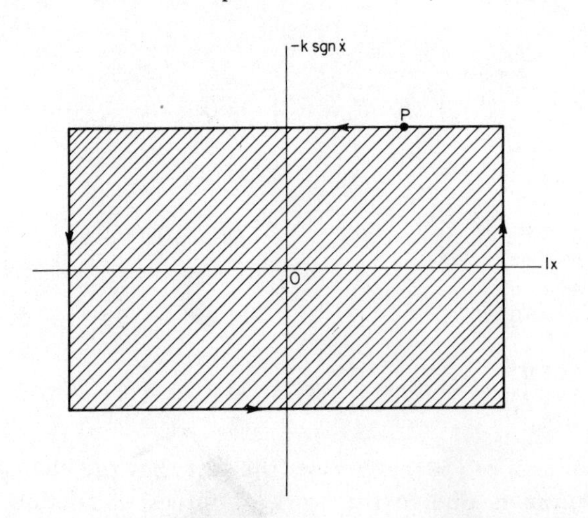

FIG. 12. Geometric representation of losses, Coulomb friction.

A simple geometric representation of these losses is shown in Figs. 11 and 12, in which a moving point P is considered whose coordinates are lx and either $f = -\eta l\dot{x}$ or $f = -k \operatorname{sgn} \dot{x}$, according to the case under consideration.

It may easily be verified that, given amplitude a_0, losses due to friction, over a whole cycle, are measured by the shaded area bounded by the corresponding closed trajectory of P. In Case 1, trajectories of P are ellipses similar to phase trajectories. However, the rotation of P is counterclockwise, whereas it was clockwise on phase trajectories. In Case 2 trajectories are rectangular cycles.

On the other hand, with our assumptions, the total energy of the pendulum corresponding to amplitude a_0 is

$$E = \tfrac{1}{2} ml^2\omega^2 a_0{}^2$$

Accordingly, the change in energy per cycle is

$$\Delta E = ml^2\omega^2 a_0\, \Delta a_0 \,.$$

where Δa_0 is the corresponding change in amplitude.

Another expression of this energy loss is (62) in Case 1, and (63) in Case 2. Therefore:

In Case 1:

$$ml^2\omega^2 a_0\, \Delta a_0 = -\pi\eta l^2 a_0{}^2\omega$$

from which follows

$$\frac{\Delta a_0}{a_0} = -\frac{\eta}{2m} T = -\frac{\lambda}{2} T \tag{64}$$

In Case 2:

$$ml^2\omega^2 a_0\, \Delta a_0 = -4kla_0$$

from which we deduce

$$\Delta a_0 = -\frac{2k}{\pi ml\omega} T = -\frac{2\mu}{\pi\omega} T \tag{65}$$

Variations of a_0 with time are shown in Figs. 13 and 14 for each case.

Note that in Case 1 the damping term in (53) is linear, and amplitude a_0 is exponentially decreasing with time; in Case 2 the damping term in (57) is nonlinear, and the decreasing of amplitude a_0 is linear.

Because λ and μ are assumed to be small, a_0 is a slowly varying function

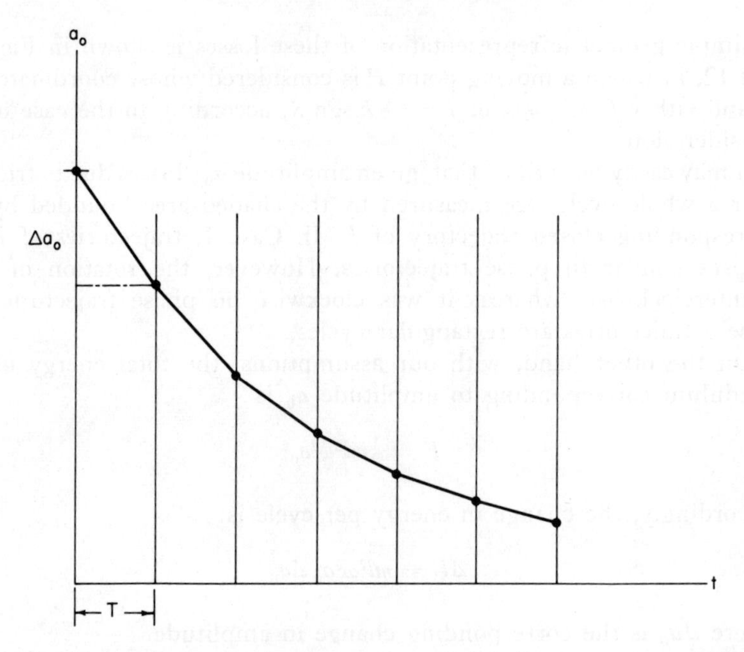

FIG. 13. Variation of a_0 with time, viscous friction. The equation of motion is linear and the damping is exponential.

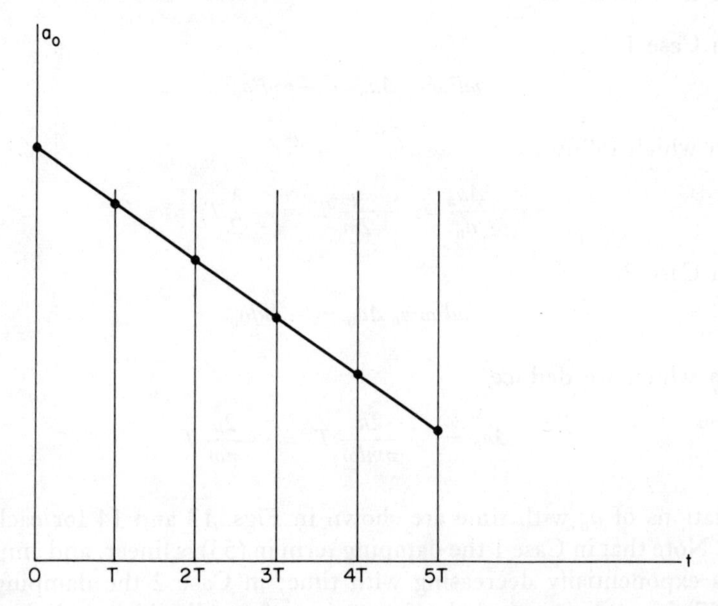

FIG. 14. Variation of a_0 with time, Coulomb friction. The equation of motion is nonlinear and the damping is linear.

of time, and it may be convenient to smooth the curves of Figs. 13 and 14 by replacing (64) and (65) by the differential equations

$$\frac{da_0}{dt} = -\frac{\lambda}{2} a_0 \tag{66}$$

$$\frac{da_0}{dt} = -\frac{2\mu}{\pi\omega} \tag{67}$$

This transformation is obtained by considering Δa_0 and T as small intervals which can be conveniently (though nonrigorously) replaced by da_0 and dt. Indeed, we must keep in mind the fact that this method is only an approximate one. However, it can be justified by using asymptotic methods.

As was pointed out above, there exist other kinds of damping forces. For example, it may be that a solid friction has a coefficient k which depends on the position of the pendulum; i.e., k is a function of x. Assume the proportionality

$$k = Kx$$

Then Fig. 12 is replaced by Fig. 15, and the energy loss per cycle is

$$-2Kla_0^2$$

Accordingly, we have

$$ml^2\omega^2 a_0\, \Delta a_0 = -2Kla_0^2$$

say

$$\frac{\Delta a_0}{a_0} = -\frac{K}{\pi m l \omega} T \tag{68}$$

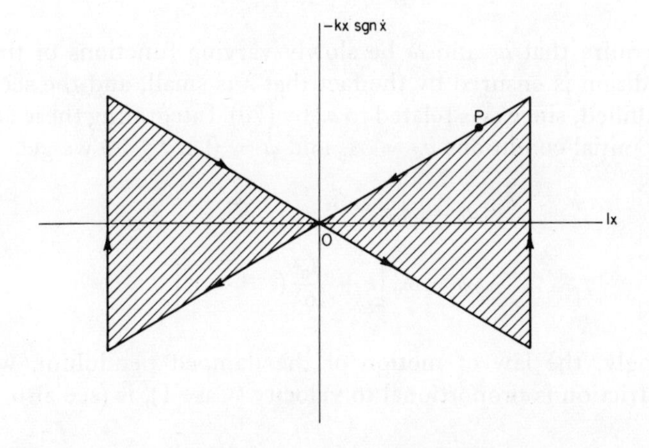

FIG. 15. Geometric representation of losses, solid friction coefficient proportional to x.

which again will be replaced by

$$\frac{da_0}{dt} = -\frac{K}{\pi m l \omega} a_0 \tag{69}$$

We obtain in this case an exponential damping which exhibits similarities to viscous damping. However ω occurs in the time constant $\pi m l \omega / K$, whereas the time constant of a viscous damping, $2/\lambda = 2m/\eta$, is independent of ω.

6.1. Approximate Solution of the Pendulum Problem with Viscous Damping

Let us return to (53) with the above assumptions; say λ and a_0 are small, and again use the first approximation,

$$x = a_0 \sin \psi$$

with

$$\psi = \omega t \qquad \text{and} \qquad \omega = \omega_0 \left(1 - \frac{a_0^2}{16}\right) \tag{70}$$

Now let us write

$$\frac{da_0}{dt} = -\frac{\lambda}{2} a_0 \qquad \frac{d\psi}{dt} = \omega_0 \left(1 - \frac{a_0^2}{16}\right)$$

which require that a_0 and ω be slowly varying functions of time. The first condition is ensured by the fact that λ is small, and the second one is also fulfilled, since ω is related to a_0 by (70). Integrating these equations with the initial conditions $a_0 = A_0$ and $\psi = 0$ at $t = 0$ we get

$$a_0 = A_0 e^{-(\lambda/2)t}$$

$$\psi = \omega_0 \left[t + \frac{A_0^2}{16\lambda} (e^{-\lambda t} - 1)\right]$$

Accordingly, the law of motion of the damped pendulum, when the force of friction is proportional to velocity (Case 1), is (see also Ref. [7])

$$x = A_0 e^{-(\lambda/2)t} \sin \left\{\omega_0 \left[t + \frac{A_0^2}{16\lambda} (e^{-\lambda t} - 1)\right]\right\}$$

6.2. Approximate Solution of the Pendulum Problem with Coulomb Damping

On the other hand, when the force of friction has constant magnitude (Case 2) we have

$$\frac{da_0}{dt} = -\frac{2\mu}{\pi\omega} = -\frac{2\mu}{\pi\omega_0[1-(a_0^2/16)]} \simeq -\frac{2\mu[1+(a_0^2/16)]}{\pi\omega_0} \tag{71}$$

$$\frac{d\psi}{dt} = \omega_0\left(1-\frac{a_0^2}{16}\right) \tag{72}$$

Since μ and a_0 are small, we can neglect μa_0^2 in (71), which can be replaced by

$$\frac{da_0}{dt} = -\frac{2\mu}{\pi\omega_0} \tag{73}$$

Then we obtain by integration

$$a_0 = -\frac{2\mu}{\pi\omega_0}t + A_0$$

with initial condition $a_0 = A_0$ at $t = 0$.

Integrating (72) with $\psi = 0$ at $t = 0$ we get

$$\psi = \omega_0\left[\left(1-\frac{A_0^2}{16}\right)t + \frac{\mu A_0}{8\pi\omega_0}t^2\right]$$

where μ^2 has been neglected. In this case, the law of motion of the pendulum is

$$x = \left(-\frac{2\mu}{\pi\omega_0}t + A_0\right)\sin\left\{\omega_0\left[\left(1-\frac{A_0^2}{16}\right)t + \frac{\mu A_0}{8\pi\omega_0}t^2\right]\right\}$$

Let us also note that, if we rewrite (57) in the form

$$\ddot{x} + \omega_0^2\left(x - \frac{x^3}{6}\right) = -\mu \operatorname{sgn} \dot{x} \tag{74}$$

and again use the first approximation,

$$x = a_0 \sin \omega t \qquad \omega = \omega_0\left(1-\frac{a_0^2}{16}\right)$$

the driving force on the right side of (74) is a square wave whose phase lag with respect to velocity $\dot{x}$ is π.

The first harmonic of this wave is

$$\frac{4\mu}{\pi} \cos \omega t$$

The other harmonics do not play an important role, because of the high selectivity of the undamped oscillator on which the driving force is acting (74). Furthermore, the Fourier series of the square wave contains only odd harmonics whose amplitudes decrease according to a $1/n^2$ law, ($n = 3, 5, 7, \ldots$ order of the harmonics). Then (74) is replaced by

$$\ddot{x} + \omega_0^2 \left(x - \frac{x^3}{6}\right) = -\frac{4\mu}{\pi} \cos \omega t \tag{75}$$

which can also be written

$$\ddot{x} + \omega_0^2 \left(x - \frac{x^3}{6}\right) = -\frac{4\mu}{\pi a_0 \omega} \dot{x} \tag{76}$$

or

$$\ddot{x} + \frac{4\mu}{\pi a_0 \omega} \dot{x} + \omega_0^2 \left(x - \frac{x^3}{6}\right) = 0 \tag{77}$$

This equation is similar to the one which governs a pendulum with viscous damping but in which the viscosity coefficient would depend on a_0. Putting $\lambda = 4\mu/\pi a_0 \omega$, we have

$$\frac{da_0}{dt} = -\frac{\lambda}{2} a_0 = -\frac{2\mu}{\pi\omega} \simeq -\frac{2\mu}{\pi\omega_0}$$

and we again find

$$a_0 = -\frac{2\mu}{\pi\omega_0} t + A_0$$

$$\psi = \int_0^t \omega \, dt = \omega_0 \left[\left(1 - \frac{A_0^2}{16}\right) t + \frac{\mu A_0}{8\pi\omega_0} t^2\right]$$

7. SIMPLE PENDULUM WITH FORCING FUNCTION. RESONANCE

7.1. Pendulum without Damping, Duffing's Method†

Let us return to the equation

$$\ddot{x} + \omega_0^2 x + \mu x^3 = 0 \qquad \mu = -\frac{\omega_0^2}{6}$$

† See [3, 6].

which describes the free oscillation of a pendulum, for small amplitudes, when the forces of friction are neglected.

Now suppose that an external force, which we shall assume to be a sine function of time, say $\mathscr{F} \sin \omega t$, is acting on the pendulum. The law of motion becomes

$$\ddot{x} + \omega_0^2 x + \mu x^3 = F \sin \omega t \tag{78}$$

where $F = \mathscr{F}/ml$ in the case of a simple pendulum (m is the mass, and l is the length).

If we start with the first approximation,

$$x^* = a_0 \sin \omega t \qquad a_0 \gtrless 0$$

where ω is now the angular frequency of the external force, we obtain, by substituting in the nonlinear term μx^3,

$$\mu(x^*)^3 = \mu a_0^3 \sin^3 \omega t = \mu a_0^3(\tfrac{3}{4} \sin \omega t - \tfrac{1}{4} \sin 3\omega t)$$

Then, substituting in (78) we get

$$\ddot{x} = (F - \omega_0^2 a_0 - \tfrac{3}{4}\mu a_0^3) \sin \omega t + \frac{\mu a_0^3}{4} \sin 3\omega t$$

This equation is readily integrated, and, assuming that $x(t)$ is a periodic function, it gives

$$x^{**} = \frac{1}{\omega^2}\left[\omega_0^2 + \tfrac{3}{4}\mu a_0^2 - \frac{F}{a_0}\right] a_0 \sin \omega t - \frac{1}{36}\frac{\mu}{\omega^2} a_0^2 \sin 3\omega t \tag{79}$$

Indeed, this expression is a better approximation of the solution of (78), which again might be improved by iterating the process. However, it would then be necessary to pay attention to the convergence of the expansion thus obtained. As a matter of fact, it may easily be seen that this convergence can be ensured only if μ, F, and a_0 are sufficiently small; accordingly, iteration is of no use in the neighborhood of resonance.

So, instead of computing other terms, we shall follow Duffing's procedure [3, 6], by letting the amplitude a_0 of the first approximation x^* be equal to the amplitude of the first harmonic of the second approximation x^{**}, say

$$a_0 = \frac{1}{\omega^2}\left(\omega_0^2 + \tfrac{3}{4}\mu a_0^2 - \frac{F}{a_0}\right) a_0$$

or

$$\omega^2 = \omega_0^2 + \tfrac{3}{4}\mu a_0^2 - \frac{F}{a_0} \tag{80}$$

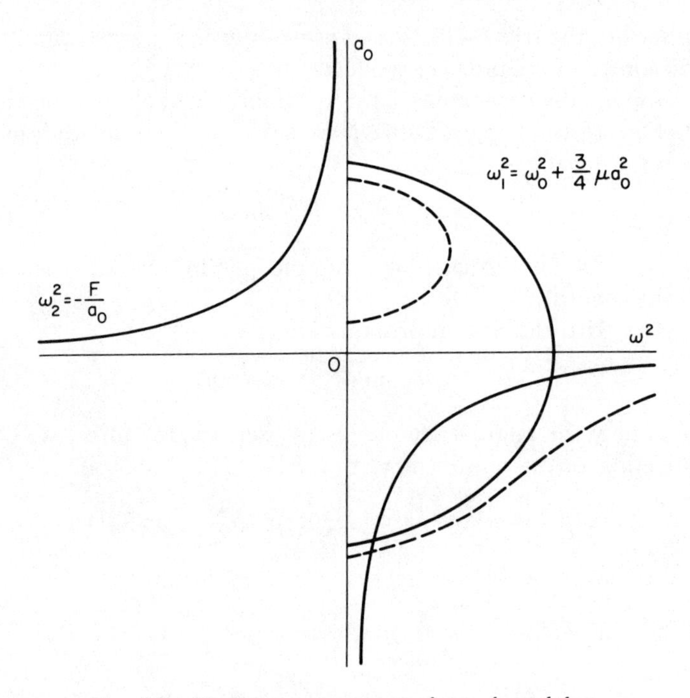

FIG. 16. Nonlinear resonance, undamped pendulum.

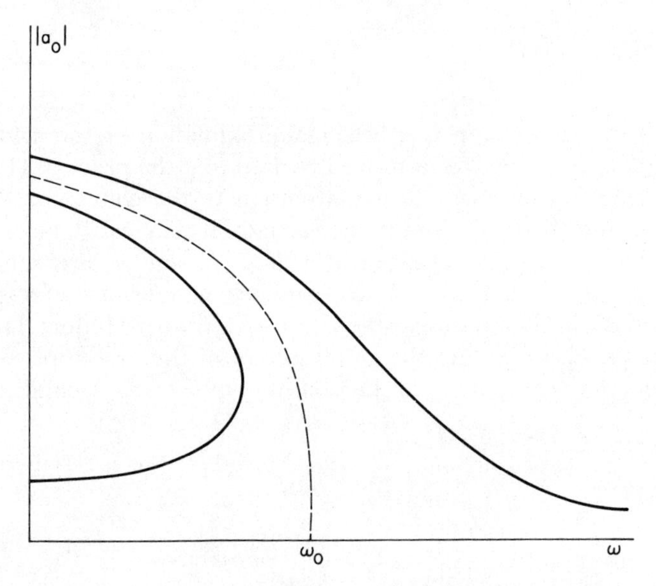

FIG. 17. Nonlinear resonance, $\mu < 0$; undamped pendulum.

When $F = 0$, we get the relation between the frequency ω and amplitude a_0 for a freely swinging pendulum, formula (23). More generally (80) is the relation between ω and a_0 for a given F.

In Figs. 16 and 17 we have plotted a_0 against ω^2 and $|a_0|$ against ω,

$$a_0 = h(\omega^2) \qquad \text{and} \qquad |a_0| = g(w)$$

The curve of Fig. 16 is obtained by separately plotting

$$\omega_1{}^2 = \omega_0{}^2 + \tfrac{3}{4}\mu a_0{}^2 \qquad \text{and} \qquad \omega_2{}^2 = -\frac{F}{a_0}$$

Obviously the above arguments hold whatever the sign of μ. Accordingly, in Fig. 18 we have also plotted $|a_0|$ against ω, for the case $\mu > 0$.

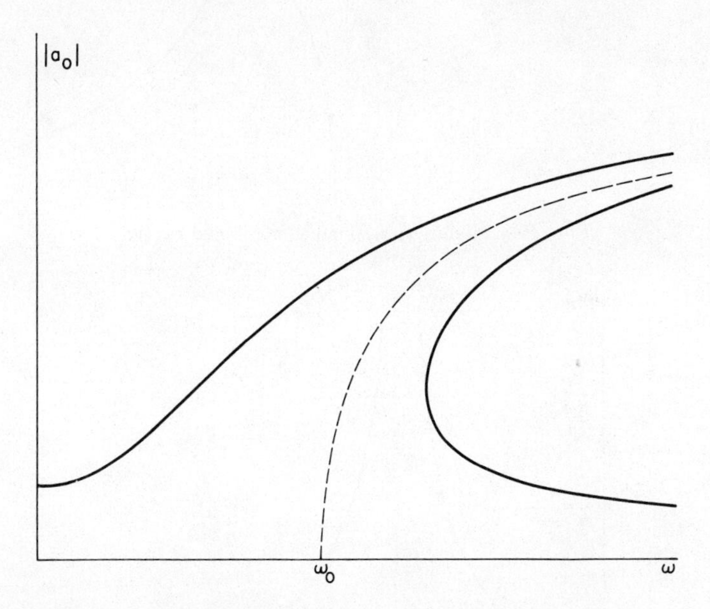

FIG. 18. Nonlinear resonance, $\mu > 0$; undamped pendulum.

Note that the cases $\mu > 0$ and $\mu < 0$ are not strictly similar, because the resonance curve which is bent to the left ($\mu < 0$) has three intersection points with axis $\omega = 0$, whereas the one which is bent to the right ($\mu > 0$) has only one intersection point (Fig. 19). $\omega \geqslant 0$ in both cases.

It is interesting to compare the resonance curves of Fig. 17 and 18 with the one obtained in the linear case where $\mu = 0$, Fig. 20.

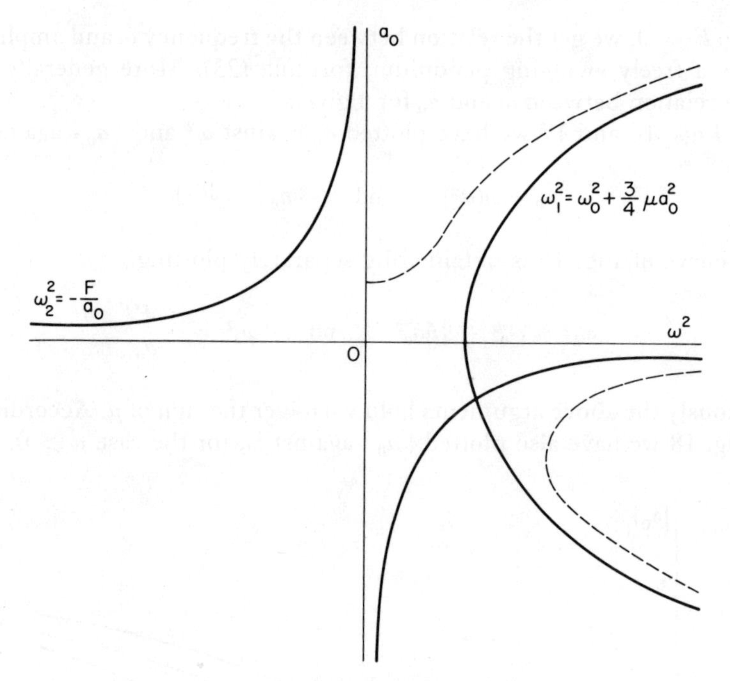

FIG. 19. Nonlinear resonance, combined curves.

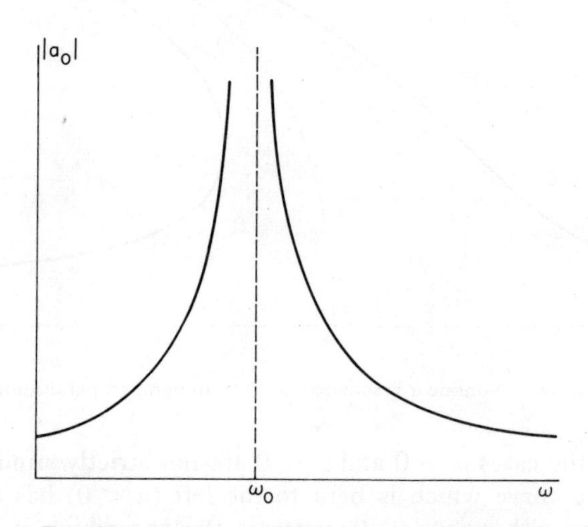

FIG. 20. Linear resonance; undamped pendulum.

As a matter of fact, in the nonlinear case, we may get three values of $|a_0|$ for a given ω. However, not all these solutions correspond to stable oscillation. The jump phenomenon, which we shall describe later, is a consequence of this situation, but does not occur in linear oscillators.

Finally, note that by substituting (80) in (79) for the second approximation we obtain

$$x^{**} = a_0 \sin \omega t - \frac{1}{36} \frac{\mu a_0^2}{\omega_0^2 + \frac{3}{4}\mu a_0^2 - (F/a_0)} \sin 3\omega t$$

7.2. Pendulum with Damping

When forces of friction are neglected, the phase lag between the forcing function and the oscillation of the pendulum is either 0 or π. Indeed, while the forcing function is $F \sin \omega t$, the first harmonic of the oscillation is $a_0 \sin \omega t$, where $a_0 \gtrless 0$.

Now let us write the equation of the damped pendulum driven by the forcing function:

$$\ddot{x} + \lambda \dot{x} + \omega_0^2 x + \mu x^3 = F_1 \sin \omega t + F_2 \cos \omega t \tag{81}$$

and, again, assume that

$$x^* = a_0 \sin \omega t \tag{82}$$

is a first approximation while the ratio F_2/F_1 is not yet defined. However, the amplitude of the forcing function $F = (F_1^2 + F_2^2)^{1/2}$ is given.

Substituting (82) in (81) we get by identification

$$(\omega_0^2 - \omega^2)a_0 + \tfrac{3}{4}\mu a_0^3 = F_1 \qquad \lambda \omega a_0 = F_2$$

from which follow

$$[(\omega_0^2 - \omega^2)a_0 + \tfrac{3}{4}\mu a_0^3]^2 + \lambda^2 \omega^2 a_0^2 = F_1^2 + F_2^2 = F^2 \tag{83}$$

and

$$\tan \varphi = \frac{F_2}{F_1} = \frac{\lambda \omega a_0}{(\omega_0^2 - \omega^2)a_0 + \frac{3}{4}\mu a_0^3} \tag{84}$$

Equation (83) is the relation between angular frequency ω and amplitude a_0, and the phase lag between the forcing function and the oscillation of the pendulum is given by (84).

In Figs. 21, 22, and 23 we have plotted $|a_0|$ against ω, when λ is small, in the three cases corresponding, respectively, to $\mu > 0$, $\mu < 0$, and $\mu = 0$. These curves should be compared with the ones of Figs. 17 and 18, sketched for $\lambda = 0$.

In Figs. 21 and 22, points where the tangent has infinite slope are indicated, as well as the locus of all such points when $F = (F_1^2 + F_2^2)^{1/2}$ is varied. It can be shown that this locus is the boundary of a domain—

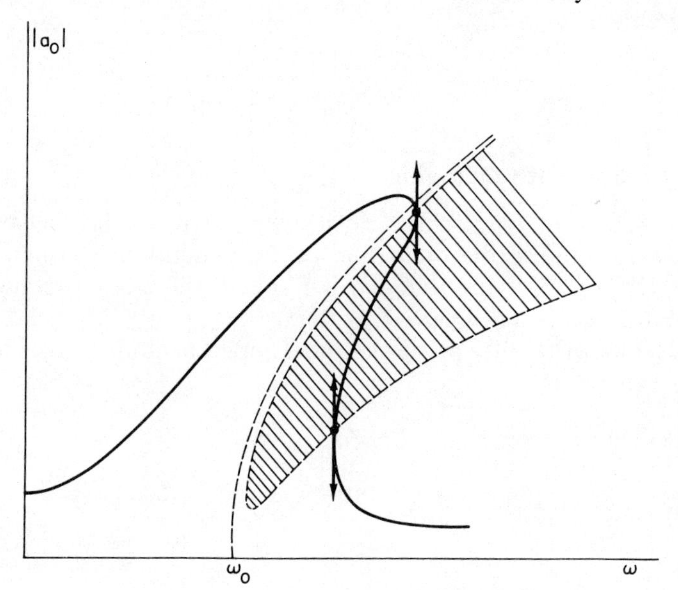

FIG. 21. Nonlinear resonance, $\mu > 0$; damped pendulum.

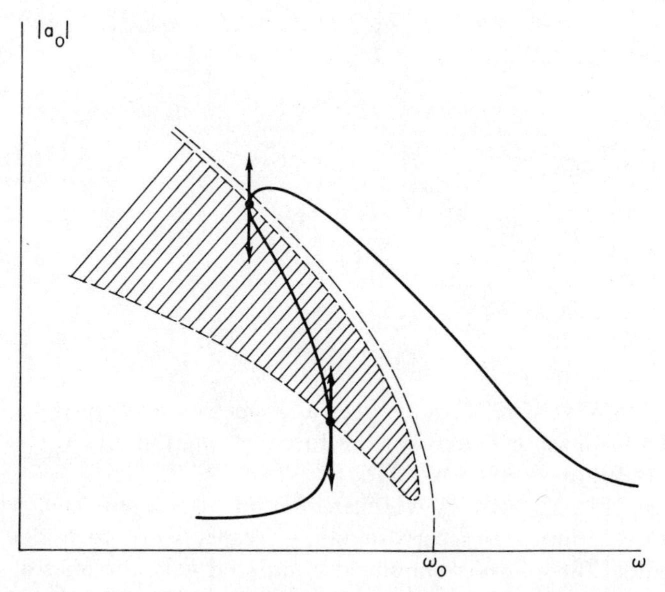

FIG. 22. Nonlinear resonance, $\mu < 0$; damped pendulum.

the shaded area of Figs. 21 and 22—whose interior points correspond to unstable solutions. Accordingly, to each given value of ω is associated either one or two stable solutions—the third solution, when it exists inside the shaded domain, being an instable one.

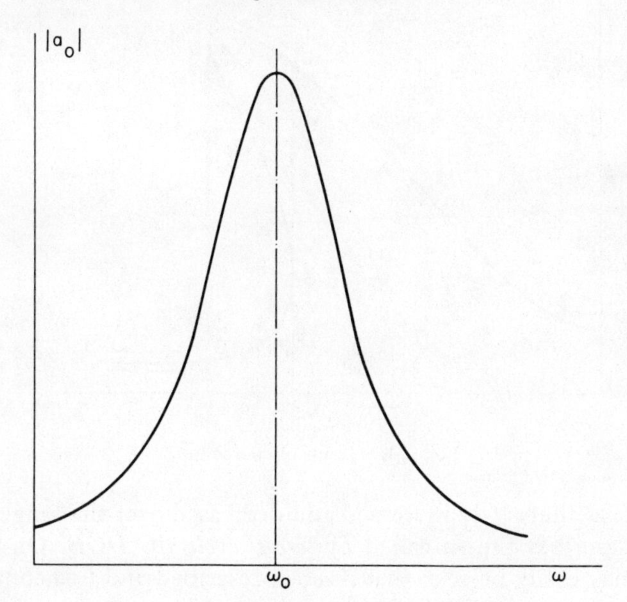

FIG. 23. Linear resonance; damped pendulum.

7.3. The Jump Phenomenon

Now assume that, for a specified F, ω is given continuously increasing values, from $\omega = 0$ to infinity, and let ω' and ω'' be angular frequencies which correspond to points B and D where the tangent to the resonance curve has infinite slope.

If we consider, for example, Fig. 24, where $\mu > 0$, we see that point P, whose coordinates are $| a_0 |$ and ω, moves continuously along the resonance curve from starting point E to point D. When ω goes on increasing beyond ω'', the only possible solution lies on the piece of curve GH. Accordingly, $| a_0 |$ undergoes a jump which brings moving point P from D to G along the vertical dashed line DG.

Then assume that ω is continuously decreased from any value $\omega > \omega''$ to $\omega = 0$. First P comes back to G, where it does not experience any discontinuity, and goes beyond until B is reached. At point B, $| a_0 |$ undergoes another jump, which brings P from B to G', after which point P follows the arrow to the left along the continuous piece of curve $G'E$.

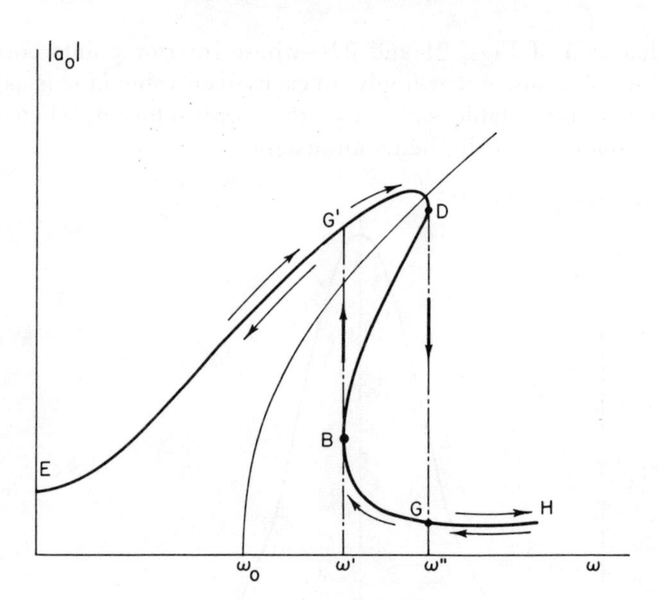

FIG. 24. Jump phenomenon.

When ω is alternately increased and decreased over the range $[\omega', \omega'']$, point P describes the so-called *hysteresis cycle BG'DGB*. On the other hand, it may easily be seen that if ω is prescribed and F is continuously increased, the jump phenomenon again occurs, as a consequence of the continuous deformation of the resonance curve.

7.4. Rauscher's Method†

Rauscher's method may be applied to systems whose equations have the form

$$\ddot{x} + f(x) = F \sin \omega t \tag{85}$$

where $f(x)$ is a continuous function of x. When F is small, a first approximation of the forced motion $x(t)$ is obtained by solving the equation which governs the free motion of the system, say

$$\ddot{x} + f(x) = 0 \tag{86}$$

As pointed out earlier, the system may be thought of as a mass $m = 1$, which hangs from a nonlinear spring. Multiplying (86) by $\dot{x}$, we get

$$\dot{x}\ddot{x} + \dot{x}f(x) = 0 \tag{87}$$

† See [4].

Then integrating (87) we get

$$\tfrac{1}{2}(\dot{x})^2 - \tfrac{1}{2}(\dot{x}_0)^2 = -\int_{x_0}^{x} f(x)\,dx$$

$$= -[E_p(x) - E_p(x_0)] \tag{88}$$

where x_0 and $\dot{x}_0$ are initial conditions, $E_p(x)$ the potential energy of the system, and $\frac{1}{2}(\dot{x})^2$ its kinetic energy. It follows that

$$\dot{x} = \pm\{(\dot{x}_0)^2 + 2[E_p(x_0) - E_p(x)]\}^{1/2}$$

which gives by integration

$$t = t_0 + \int_{x_0}^{x} \frac{dx}{\pm\{(\dot{x}_0)^2 + 2[E_p(x_0) - E_p(x)]\}^{1/2}} \tag{89}$$

Next, by assuming that $x = x_0$ and $\dot{x} = \dot{x}_0 = 0$ at time $t_0 = 0$, we reduce (89) to the simpler form

$$t = \int_{x_0}^{x} \frac{dx}{\pm\{2[E_p(x_0) - E_p(x)]\}^{1/2}} \tag{90}$$

A second approximation will be obtained by substituting (90) in the right side of (85). ω is readily computed from (90) by noting that the transfer time between state $x = x_0$, which corresponds to zero velocity, and state $x = 0$, which corresponds to maximum velocity, is equal to $T/2$, i.e.,

$$\frac{T}{4} = \frac{\pi}{2\omega} = -\int_{x_0}^{0} \frac{dx}{\{2[E_p(x_0) - E_p(x)]\}^{1/2}} \tag{91}$$

Accordingly, ωt is a function of x which can be determined by (90) and (91), say

$$\omega t = g_1(x)$$

Then (85) is rewritten

$$\ddot{x} + f(x) - F\sin[g_1(x)] = 0 \tag{92}$$

and again we find a similar problem by putting

$$f_1(x) = f(x) - F\sin[g_1(x)]$$

The solution is improved by iterating the process.

7.5. Optimal Linearization Method†

The method which we introduced in Section 3.5 may be conveniently applied to the study of forced motion. Let us start, for example, with the equation

$$\ddot{x} + \omega_0^2 \sin x = F \sin \omega t$$

and use the optimal linear equation with constant coefficient which we have associated with it, namely

$$\ddot{x} + \lambda x = F \sin \omega t \qquad \text{with} \quad \lambda = \omega_0^2 \frac{\overline{x \sin x}}{\overline{x^2}} \tag{93}$$

If we assume that the forced oscillation is approximately

$$x = a_0 \sin \omega t$$

where ω is the angular frequency of the forcing function and a_0 the unknown amplitude, we get, as previously,

$$\lambda = \omega_0^2 \frac{2J_1(a_0)}{a_0}$$

and it follows from (93) that

$$a_0 = \frac{F}{\lambda - \omega^2} = \frac{F}{\omega_0^2[2J_1(a_0)/a_0] - \omega^2}$$

Finally, we obtain the relation between amplitude a_0 of the first approximation and ω:

$$\omega^2 = \omega_0^2 \frac{2J_1(a_0)}{a_0} - \frac{F}{a_0}$$

If we keep the first two terms of the series expansion of $J_1(a_0)$, when assuming that a_0 is small,

$$J_1(a_0) = \frac{a_0}{2}\left(1 - \frac{a_0^2}{2^2 \cdot 2} + \frac{a_0^4}{2 \cdot 2^4 \cdot 2 \cdot 3} - \cdots\right)$$

we get

$$\omega^2 = \omega_0^2\left(1 - \frac{a_0^2}{8}\right) - \frac{F}{a_0}$$

which is relation (80) with $\mu = -\omega_0^2/6$.

† See [12].

7.6. Subharmonic Oscillations

Another property of nonlinear oscillators is the generation of subharmonics, i.e., of oscillations whose frequencies are related to that of the driving function by

$$\omega_n = \frac{\omega}{n} \tag{94}$$

where ω is the frequency of the driving function, ω_n the frequency of a subharmonic oscillation, and n an integer: 2, 3, 4, ...
This phenomenon, which is also called *frequency demultiplication*, was discovered by Helmholtz and described in his theory of physiological acoustics [2].

Helmholtz pointed out that sometimes the ear is influenced by sounds whose frequencies are not contained in the incoming acoustic wave but are related to its frequency by relation (94), and he explained this phenomenon by nonlinearity of the tympanic membrane.

We shall illustrate this property by studying the generation of a subharmonic of order $\frac{1}{3}$, in the oscillation of a pendulum governed by the equation (see also [8])

$$\ddot{x} + \omega_0^2 x + \mu x^3 = F \sin 3\omega t \tag{95}$$

As a matter of fact, we have to show that (95) has a solution which contains a component whose frequency is ω. Accordingly, we shall try to find an approximate solution of the form

$$x = a_0 \sin \omega t + a_3 \sin 3\omega t \qquad a_0 \gtrless 0 \quad a_3 \gtrless 0 \tag{96}$$

Note that this phenomenon exhibits some similarity with the superposition of free and forced vibrations in a linear oscillator. Indeed in a linear oscillator which is driven by a forcing function, the law of motion is obtained by superposing:

(a) A free vibration, with the characteristic frequencies of the oscillator.

(b) A forced vibration whose frequency is that of the forcing function.

As we shall see, in the nonlinear oscillator we shall consider, the frequency of the subharmonic oscillation is close to the characteristic frequency of the oscillator, and this oscillation is sustained by the forcing function whose frequency is a multiple of the characteristic one. However, a strong difference lies in the fact that, in a linear system there is not necessarily a relation between frequencies of free and forced vibrations.

To study subharmonic oscillation it will be convenient to use the perturbation method. Let us assume that μ is small, and write

$$x = x_0(t) + \mu x_1(t) + o(\mu)$$
$$\omega^2 = \omega_0^2 + \mu\xi_1(a_0, a_3) + o(\mu) \tag{97}$$

Substituting in (95) we find

$$\ddot{x}_0 + \omega^2 x_0 + \mu(\ddot{x}_1 + \omega^2 x_1 - \xi_1 x_0 + x_0^3) + o(\mu) = F \sin 3\omega t$$

Then the *generating solution* is given by

$$\ddot{x}_0 + \omega^2 x_0 = F \sin 3\omega t$$

It is

$$x_0 = a_0 \sin \omega t + a_3 \sin 3\omega t \tag{98}$$

with

$$a_3 = -\frac{F}{8\omega^2} \tag{99}$$

On the other hand, the correcting function x_1 is deduced from

$$\ddot{x}_1 + \omega^2 x_1 = \xi_1 x_0 - x_0^3 \tag{100}$$

or, by substituting the generating function (98) into the right side of (100), from

$$\begin{aligned}\ddot{x}_1 + \omega^2 x_1 = {} & (\xi_1 - \tfrac{3}{4}a_0^2 + \tfrac{3}{4}a_0 a_3 - \tfrac{3}{2}a_3^2)a_0 \sin \omega t \\ & + (\xi_1 a_3 + \tfrac{1}{4}a_0^3 - \tfrac{3}{2}a_0^2 a_3 - \tfrac{3}{4}a_3^3) \sin 3\omega t \\ & + (\tfrac{3}{4}a_0 a_3 - \tfrac{3}{4}a_3^2)\, a_0 \sin 5\omega t \\ & + \tfrac{1}{4}a_3^3 \sin 6\omega t + \tfrac{3}{4}a_0 a_3^3 \sin 7\omega t\end{aligned}$$

Finally, the secular term is ruled out by putting

$$\xi_1 = \tfrac{3}{4}(a_0^2 - a_0 a_3 + 2a_3^2)$$

from which follows

$$\omega^2 = \omega_0^2 + \frac{3\mu}{4}(a_0^2 - a_0 a_3 + 2a_3^2) \tag{101}$$

By substituting (99) in (101) we find

$$\omega^6 - \omega^4\omega_0^2 = \frac{3\mu F^2}{128}\left(1 + 4\,\frac{a_0\omega^2}{F} + 32\,\frac{a_0^2\omega^4}{F^2}\right)$$

This relation between amplitude and frequency of the subharmonic oscillation is represented in Fig. 25, where a_0 has been plotted as a function of ω.

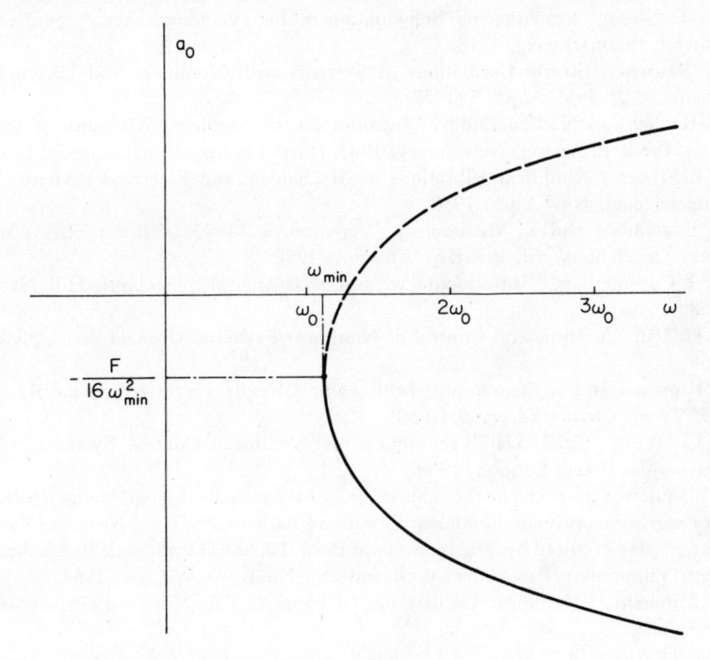

FIG. 25. Relation between amplitude and frequency of the subharmonic oscillation of order 3, for Duffing's pendulum.

It is possible to prove that points at which the tangent has a positive slope correspond to unstable solutions. The minimum value of ω is easily computed; it is

$$\omega_{\min} \simeq \omega_0 + \frac{21\mu F^2}{2048\omega_0{}^5}$$

It is related to the minimum value of a_0 by

$$(a_0)_{\min} = \frac{-F}{16\omega^2_{\min}}$$

By similar arguments it would be possible to study subharmonics of order $\frac{1}{2}$. In general they are obtained less easily than subharmonics of order $\frac{1}{3}$ in systems governed by Duffing's equation. In any case the starting conditions play a very important role.

BIBLIOGRAPHY

1. A. Lindstedt, *Mem. Acad. Imp. St. Petersburg* **31** (1883).
2. H. Helmholtz, "Sensation of Tone." Longmans, Green, London, 1895.
3. G. Duffing, "Erzwungene Schwingungen bei veränderlicher Eigenfrequenz." Vieweg, Braunschweig, 1918.
4. M. Rauscher, Steady Oscillations of Systems with Nonlinear and Unsymmetrical Elasticity. *J. Appl. Mech.* **5** (1938).
5. N. Krylov and N. Bogoliubov, "Introduction to Nonlinear Mechanics." Princeton Univ. Press, Princeton, New Jersey, 1943. (First published in Russian in 1937.)
6. J. J. Stoker, "Nonlinear Vibrations in Mechanical and Electrical Systems." Wiley (Interscience), New York, 1950.
7. N. Bogoliubov and Y. Mitropolsky, "Asymptotic Methods in the Theory of Nonlinear Oscillations" (in Russian). Moscow, 1958.
8. W. J. Cunningham, "Introduction to Nonlinear Analysis." McGraw-Hill, New York, 1958.
9. Y. H. Ku, "Analysis and Control of Nonlinear Systems." Ronald Press, New York, 1958.
10. S. Kumagai and S. Kawamoto, Multistable Circuits Using Nonlinear Reactances. *IRE Trans. Circuit Theory* **7** (1960).
11. J. C. West, "Analytical Techniques for Nonlinear Control Systems." English Universities Press, London, 1960.
12. A. Blaquière, Une nouvelle méthode de linéarisation locale des opérateurs nonlinéaires; approximation optimale. Nonlinear Vibration Problems. *2nd Conf. Nonlinear Vibrations, Warsaw, 1962* (edited by The Institute of Basic Technical Problems, Polish Acad. Sci., Dept. Vibrations), Państwowe Wydawnictwo Naukowe, Warsaw, 1964.
13. N. Minorsky, "Nonlinear Oscillations," Chapt. 2. Van Nostrand, Princeton, New Jersey, 1962.

CHAPTER II

Self-Oscillatory Systems

Before discussing the theory of self-oscillatory systems, let us return to the law of motion of a pendulum with Coulomb damping [Chapter I, Eq. (57)], in the simplified form

$$\ddot{x} + \mu \operatorname{sgn} \dot{x} + \omega_0^2 x = 0 \qquad \mu > 0 \tag{1}$$

Equation (1) can be separated into two equations,

$$\ddot{x} + \mu + \omega_0^2 x = 0 \qquad \dot{x} > 0$$
$$\ddot{x} - \mu + \omega_0^2 x = 0 \qquad \dot{x} < 0$$

which suit a phase-plane representation very well. Indeed, rewriting these equations

$$\begin{aligned} \ddot{x} + \omega_0^2(x + k) = 0 \qquad \dot{x} > 0 \\ \ddot{x} + \omega_0^2(x - k) = 0 \qquad \dot{x} < 0 \end{aligned} \qquad \text{with} \quad k = \frac{\mu}{\omega_0^2}$$

and multiplying by $2\dot{x}$, a straightforward integration gives

$$(\dot{x})^2 + \omega_0^2(x + k)^2 = R_{i-1}^2 \qquad \dot{x} > 0$$
$$(\dot{x})^2 + \omega_0^2(x - k)^2 = R_i^2 \qquad \dot{x} < 0$$

where R_{i-1}^2 and R_i^2 are constants of integration which are to be computed at each switching of *sgn* $\dot{x}$.

Accordingly, the trajectory of point P whose coordinates are

$$x_1 = \omega_0 x \qquad x_2 = \dot{x}$$

is made of half-circles, whose centers are alternately located at $x_1 = -k\omega_0$ and $x_1 = k\omega_0$, and which join one to the other at points $P_1, P_2, \ldots, P_n$ on the horizontal axis (Fig. 1).

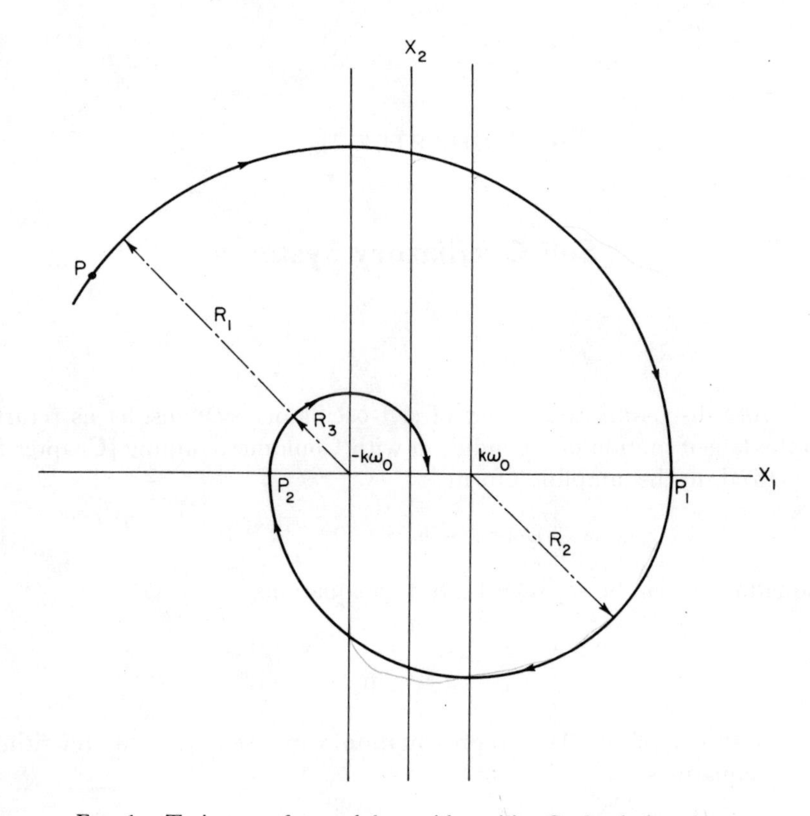

FIG. 1. Trajectory of a pendulum with positive Coulomb damping.

The radii of these circles, R_1, R_2, ..., R_n, decrease according to the formula

$$R_i = R_{i-1} - 2k\omega_0 \qquad (i = 1, 2, ..., n) \tag{2}$$

and the system will stop when

$$R_n \leqslant 2k\omega_0$$

Later we shall discuss an oscillator whose law of motion is similar to (1), but with $\mu < 0$.

Then, by using the same representation, it may easily be seen that the radii of the consecutive circles continue to increase, and that (2) is replaced by

$$R_i = R_{i-1} + 2k\omega_0 \qquad (i = 1, 2, ..., n) \tag{3}$$

Accordingly the amplitude of the oscillation increases (Fig. 2).

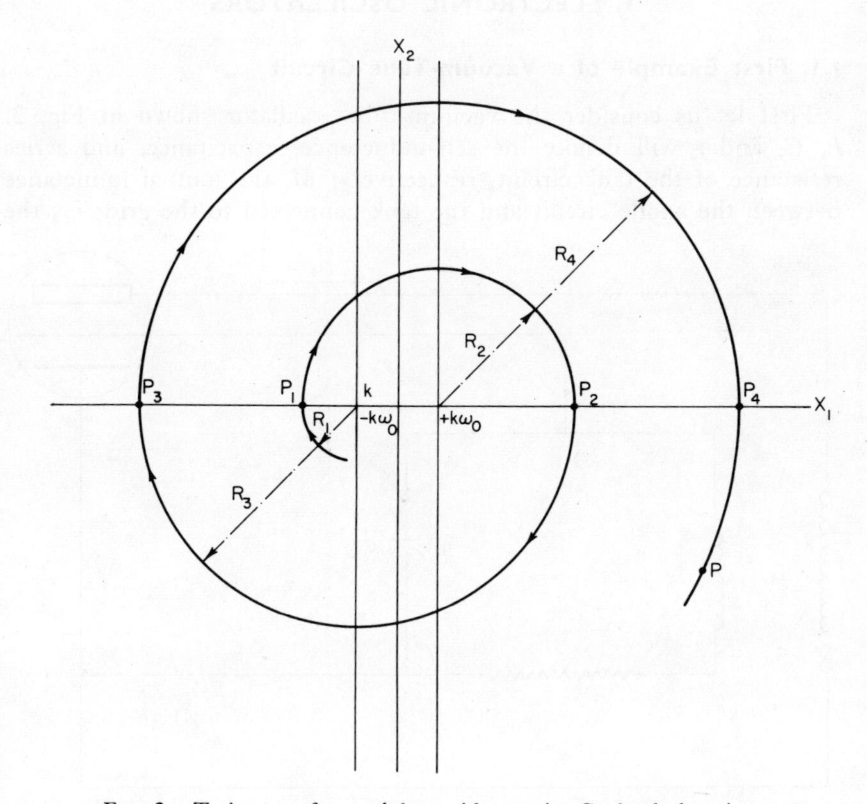

FIG. 2. Trajectory of a pendulum with negative Coulomb damping.

The term μ sgn $\dot{x}$, with $\mu < 0$, which occurs in the equation of this oscillator and which is responsible for the increase in amplitude, is called the *negative damping term*.

First of all, let us note that this kind of oscillator needs a source of energy. However, as will be seen later, this source has no definite periodicity and the physical theory of the oscillatory motion is strongly different from the one which we considered in Chapter 1 when discussing the operation of a periodic forcing function.

On the other hand, in practical devices, amplitude does not increase indefinitely, as will be explained by taking into account the nonlinear characteristics of components such as electronic tubes. These oscillators reach a steady behavior when the gain in energy, which will appear to be a function of amplitude, compensates the losses, over each cycle, in the mean. Such oscillators are called *self-oscillators*.

1. ELECTRONIC OSCILLATORS

1.1. First Example of a Vacuum-Tube Circuit

First let us consider the vacuum-tube oscillator shown in Fig. 3. L, C, and r will denote the self-inductance, capacitance, and series resistance of the tank circuit, respectively; M, the mutual inductance between the anode circuit and the tank connected to the grid; i_a, the

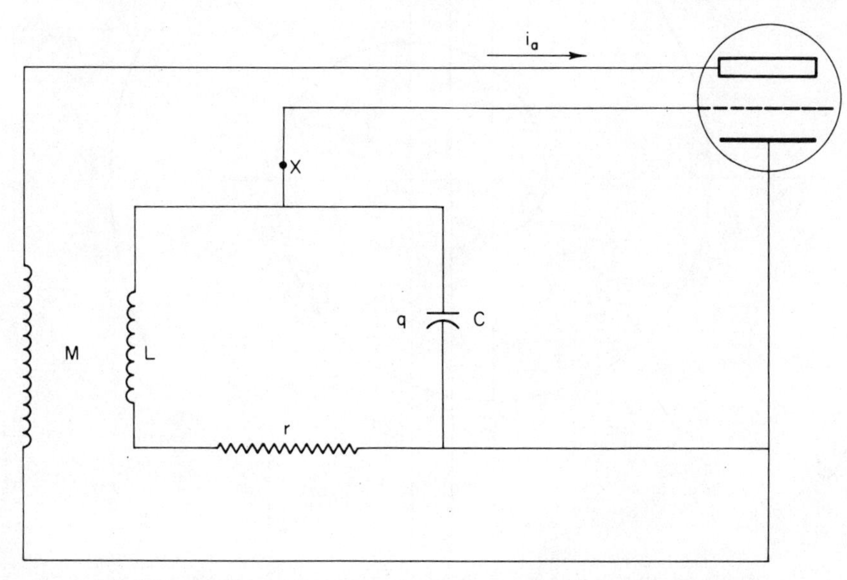

FIG. 3. Vacuum-tube oscillator, tank circuit connected to the grid.

small variations of anode current; x, the grid voltage (i.e., the voltage between the end points of the capacitor); and q, the charge, of the capacitor.

We get, from Kirchhoff's laws,

$$L\ddot{q} + r\dot{q} + \frac{1}{C}q = M\frac{di_a}{dt}$$

or, replacing q by Cx,

$$LC\ddot{x} + rC\dot{x} + x = M\frac{di_a}{dt} \tag{4}$$

The small variations of plate current i_a and grid voltage x around the quiescent point are linked by the relation

$$i_a = \varphi(x) \tag{5}$$

which is the equation of the characteristic curve of the tube. We suppose here for the sake of simplicity that the tube has a very high internal resistance, which enables us to neglect the anode reaction.

Then putting

$$S(x) = \frac{d\varphi(x)}{dx}$$

and substituting $di_a/dt = S(x)\dot{x}$ in (4) we arrive at the equation which describes the oscillations of the circuit:

$$LC\ddot{x} + [rC - MS(x)]\dot{x} + x = 0 \tag{6}$$

In the linear approximation, when assuming that $i_a = S_0 x$, where S is the slope of the tube, (6) reduces to

$$LC\ddot{x} + (rC - MS_0)\dot{x} + x = 0 \tag{7}$$

Now if $\varphi(x)$ can be approximated by an expansion of the form

$$i_a = S_0 x + S_1 x^2 + S_2 x^3 \tag{8}$$

we are led to

$$LC\ddot{x} + (rC - MS_0 - 2MS_1 x - 3MS_2 x^2)\dot{x} + x = 0 \tag{9}$$

As a matter of fact, the calculations will be significantly simplified by dropping the term $S_1 x^2$ in (8) and using

$$i_a = S_0 x + S_2 x^3 \tag{10}$$

Then (9) is rewritten

$$LC\ddot{x} + (rC - MS_0 - 3MS_2 x^2)\dot{x} + x = 0 \tag{11}$$

or

$$\ddot{x} + A(1 - Bx^2)\dot{x} + \omega_0^2 x = 0 \tag{12}$$

with

$$\omega_0^2 = 1/LC$$
$$A = (rC - MS_0)\omega_0^2$$
$$B = 3MS_2/(rC - MS_0)$$

Many interesting results can be obtained from (12), which is Van der Pol's equation [4, 5, 8].

In practical applications [33], expansions (8) and (10) may sometimes be quite misleading. Indeed, closer inspection of the characteristic curve of a

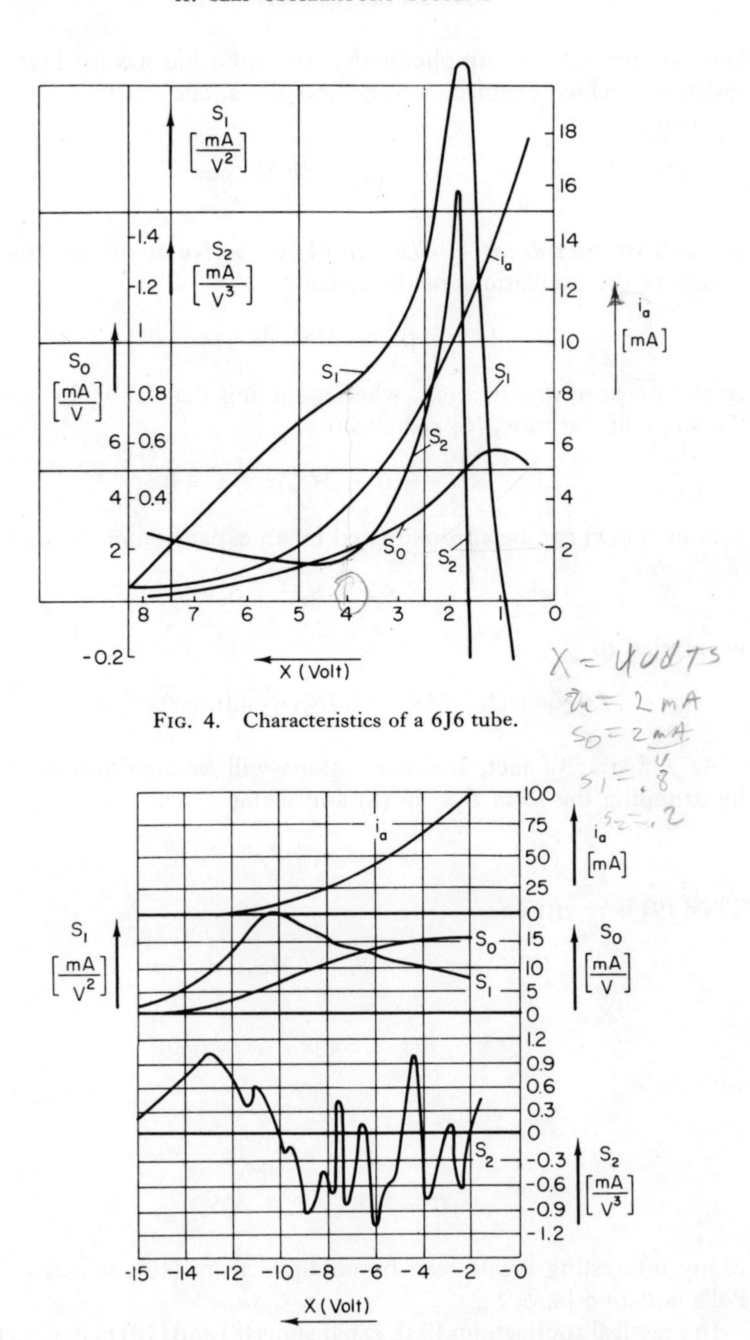

FIG. 4. Characteristics of a 6J6 tube.

FIG. 5. Characteristics of a pentode tube.

vacuum tube shows that coefficients S_1, and S_2 cannot always be considered constant during normal operation of the tube. This appears clearly in Fig. 4, which shows the behavior of S_0, S_1, and S_2 for the 6J6 triode. We conclude that the working point must be chosen carefully. This choice would be much less easy for a pentode tube, because S_2 shows a number of oscillations caused by an electron optical effect in the region between the screen grid and suppressor (Fig. 5).

1.2. Feedback Loop of a Self-Oscillator

To simplify the analysis of self-oscillatory circuits, it is convenient, from a theoretical point of view, to separate active and passive elements

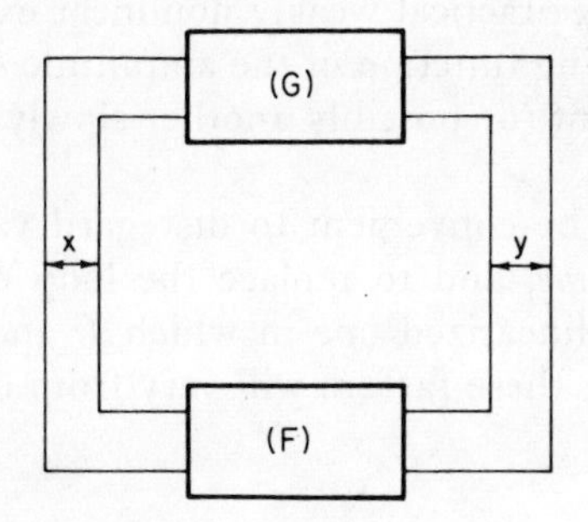

FIG. 6. Feedback loop.

into two components (Fig. 6):

(a) An active network (G), which will be characterized by its *gain without feedback* Γ.

(b) A passive network (F), whose *feedback coefficient* will be called Φ.

The product $\Gamma\Phi$ is the *loop transmission*. It will play a very important role in the theory of the generalized Nyquist diagram.

Now if x and y are the input and the output voltages of (G), respectively, we have

$$\frac{y}{x} = \Gamma \tag{13}$$

and likewise, across (F),

$$\frac{x}{y} = \Phi \tag{14}$$

Then from (13) and (14) we deduce the equation of the loop,

$$(1 - \Gamma\Phi)x = 0 \tag{15}$$

If (G) and (F) are linear networks, as will be the case in the following example, Γ and Φ can be considered either as constant coefficients (complex in general) or as differential operators, in which case (15) is a linear differential equation with respect to the variable x. Note that when Γ and Φ are considered as algebraic factors, they generaly depend on some parameter ω which is introduced by assuming at the outset that $x(t)$ is a sine function with angular frequency ω, or that $x(t)$ is a periodic function whose harmonics can be considered independent of each other in view of the superposition property of linear systems. Accordingly, representation of Γ and Φ by differential operators is more general and, in any case, it is the only one which can be applied to typical nonlinear systems, in which the superposition property does not hold. In this case (15) is a nonlinear differential equation.

Later we shall see that, in many practical weakly nonlinear examples, Γ can be considered a slowly varying function of the amplitude a_0 of the input voltage, while Φ is a constant (or possibly another slowly varying function of a_0).

In the present situation it will be convenient to disregard variations of a_0, *during small intervals of time*, and to replace the loop equation over each of these intervals by a linearized one in which Γ and Φ will again be algebraic factors. Indeed, these factors will vary from one time

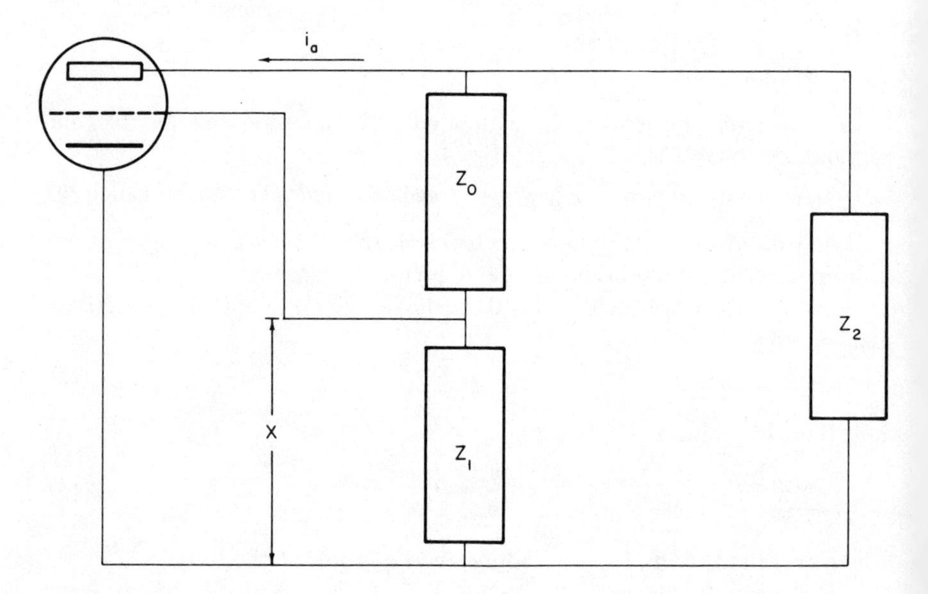

FIG. 7. Electronic oscillator.

interval to the next. When the problem under consideration fits such linearization techniques, we say that it is *quasi-linear*.

Figure 7 shows another example of an electronic oscillator, in which the two components mentioned above can be easily identified [30]. The active network is the tube T, while the feedback component (say the passive network the elements of which have impedances Z_0, Z_1, and Z_2) is a potentiometer whose coefficient Φ is

$$\Phi = \frac{Z_1}{Z_0 + Z_1}$$

and whose impedance between the plate and the cathode is

$$Z = \frac{Z_2(Z_0 + Z_1)}{Z_0 + Z_1 + Z_2}$$

On the other hand, if, as a first approximation, we linearize the relation between the anode current and the grid voltage,

$$i_a = S_0 x$$

we have

$$y = -Zi_a = -\frac{Z_2(Z_0 + Z_1)}{Z_0 + Z_1 + Z_2} S_0 x$$

and

$$\Gamma = \frac{y}{x} = -\frac{Z_2(Z_0 + Z_1)}{Z_0 + Z_1 + Z_2} S_0$$

Finally, substituting in (15), we get

$$\left(1 + \frac{S_0 Z_1 Z_2}{Z_0 + Z_1 + Z_2}\right) x = 0$$

or

$$(Z_0 + Z_1 + Z_2 + S_0 Z_1 Z_2)x = 0 \tag{16}$$

which is the equation of the loop in the linear approximation.

When impedances Z_0, Z_1, and Z_2 are written in complex form, (16) is an algebraic equation, whereas it is a differential equation when impedances are considered as differential operators.

A staightforward application of (16) to the Colpitt's oscillator (Fig. 8) leads to

$$-LC\omega^2 + j\omega\left(rC - LS_0 \frac{C_1}{C_2}\right) + 1 = 0 \qquad \text{with} \quad C = \frac{C_1 C_2}{C_1 + C_2} \tag{17}$$

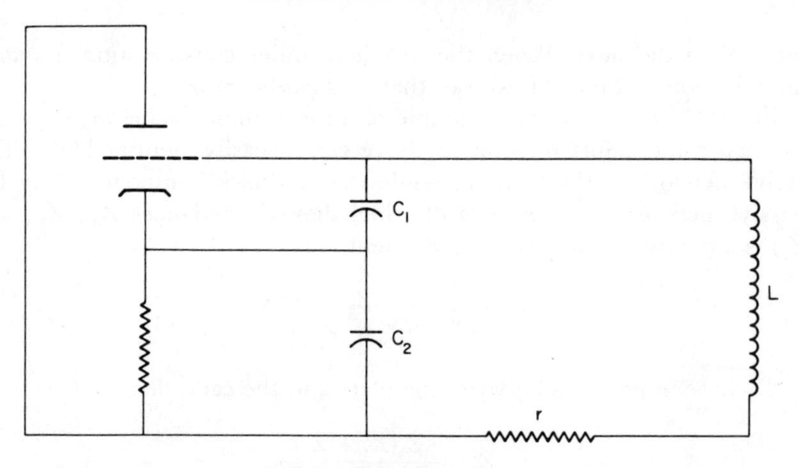

FIG. 8. Colpitt's oscillator.

when impedances Z_0, Z_1, and Z_2 are identified with complex coefficients

$$Z_0 = r + j\omega L \qquad Z_1 = \frac{1}{j\omega C_1} \qquad Z_2 = \frac{1}{j\omega C_2}$$

If they are considered differential operators,

$$Z_0 = r + L\frac{d}{dt} \qquad Z_1 = \frac{1}{C_1}\int_0^t (\quad)\,dt \qquad Z_2 = \frac{1}{C_2}\int_0^t (\quad)\,dt$$

(16) leads to the linear differential equation

$$LC\ddot{x} + \left(rC - LS_0\frac{C_1}{C_2}\right)\dot{x} + x = 0 \tag{18}$$

1.3. Negative Resistance. Threshold Condition

Again we find a negative damping term in (18), $-LS_0(C_1/C_2)\dot{x}$, which will be responsible for the growth of oscillations. It expresses the fact that the oscillator is receiving energy from an outside source, whereas positive damping is the consequence of energy losses. Indeed in (18) energy losses are also taken into account, in connection with the term $rC\dot{x}$. It is clear that steady motion can exist only if energy losses are exactly balanced by energy gains, namely when the resulting damping term is canceled.

We arrive at the threshold condition

$$rC = LS_0\frac{C_1}{C_2}$$

Note that if this condition is fulfilled, the amplitude can neither increase nor decrease and, accordingly, there will exist a steady motion in the system only if heretofore it has been steered to any given nonzero level. Then the amplitude will remain constant, whatever this initial level.

On the other hand:

(a) If $rC > LS_0(C_1/C_2)$, the amplitude of the oscillation will gradually decrease to zero, from any initial nonzero value.

(b) If $rC < LS_0(C_1/C_2)$, the amplitude will increase indefinitely, provided an initial perturbation has started the oscillator.

Next we shall take account of nonlinear terms in the equation of the loop, and show how the amplitude can reach a steady bounded value in practical devices. From a physical point of view, it will be convenient to describe the gain in energy in terms of a negative resistance whose role will be similar to the one which is played by positive resistance in the process of energy loss.

In the above example, the negative resistance is

$$\rho = -\frac{1}{C} LS_0 \frac{C_1}{C_2}$$

Accordingly, the threshold condition is rewritten

$$r + \rho = 0$$

In some diagrams the shunt resistance R of the tank circuit is considered rather than the series resistance r. It is related to r by the well-known formula

$$R = Q^2 r = \frac{L}{rC}$$

Q is the quality factor of the tank circuit,

$$Q = \frac{L\omega_0}{r} \qquad \omega_0 = (LC)^{-1/2}$$

Putting

$$G = \frac{1}{R} \qquad \text{(shunt conductance of the tank)}$$

$$m = \frac{C_1}{C_2}$$

we arrive at the new threshold condition

$$L(G - mS_0) = 0 \qquad \text{or} \qquad G - mS_0 = 0$$

On the other hand, (18) is rewritten

$$LC\ddot{x} + L(G - mS_0)\dot{x} + x = 0 \tag{19}$$

When the threshold condition is not fulfilled, we find it advisable to introduce another parameter, the actual *degree of coupling n*, which is defined by

$$n = \frac{m - m_0}{m_0}$$

where m_0 is the threshold value for barely sustained oscillations.

Indeed:

(a) When $-1 < n < 0$, the damping is positive.

(b) When $0 < n$, the oscillator is self-sustained.

1.4. Quasi Linearization

Equation (19) applies only to a linear loop. Let us take account of the nonlinearity of the tube by assuming that the anode current i_a is related to the grid voltage x by

$$i_a = S_0 x + S_1 x^2 + S_2 x^3$$

Suppose that the tube has been insulated from the tank circuit, and that we apply to the grid a sine voltage

$$x = a_0 \sin \omega t$$

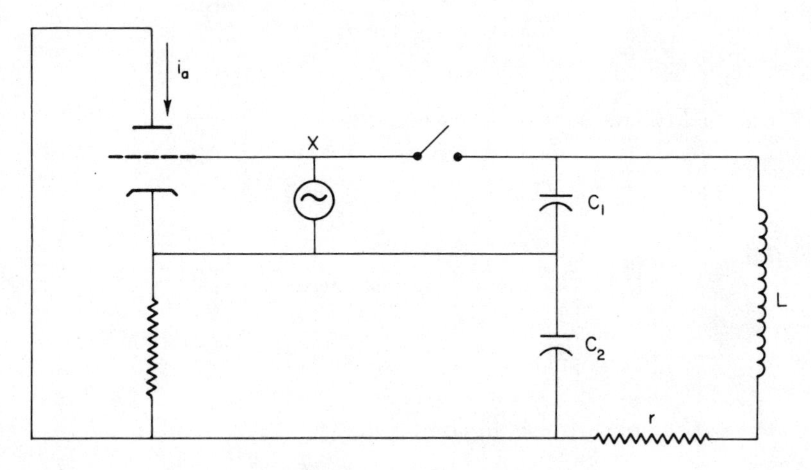

FIG. 9. Anode current in insulated tube.

Then the anode current (Fig. 9) will be

$$i_a = \left(S_0 + \frac{3S_2}{4} a_0^2\right) a_0 \sin \omega t - \frac{S_1}{2} a_0^2 \cos 2\omega t - \frac{S_2}{4} a_0^3 \sin 3\omega t$$

If we disregard the second and third harmonics—which will be canceled out, in practice, when the selective tank circuit (which will play the same role as the present sinusoidal source), is connected to the grid—we get

$$i_a = \left(S_0 + \frac{3S_2}{4} a_0^2\right) x$$

Finally, the nonlinearity of the tube is expressed by the fact that the slope is now a function of amplitude a_0,

$$S(a_0^2) = S_0 + \frac{3S_2}{4} a_0^2 \tag{20}$$

and (19) is replaced by

$$LC\ddot{x} + L\left[G - m\left(S_0 + \frac{3S_2}{4} a_0^2\right)\right] \dot{x} + x = 0 \tag{21}$$

which is a quasi-linear equation.

Later we shall obtain the same result from the Van der Pol equation, by taking into consideration only the first harmonic, by arguments similar to those above.

1.5. Robinson's Oscillator†

In this section we shall examine a radioelectric oscillator whose equation is similar to the one of a pendulum with Coulomb damping, namely (1), but in which the coefficient μ of the Coulomb damping term, $\mu \operatorname{sgn} \dot{x}$, is negative.

As was shown at the begining of the chapter, the motion of this self-oscillator can be easily studied by using the phase-plane representation. Now we shall derive an approximate solution by the same arguments as those used Chapter I, Section 6.2; i.e., we shall transform the damping term due to the first-harmonic technique, which fits pretty well with this kind of problem, when μ is small, because of the high selectivity of the tank circuit.

† See [26, 30, 33].

This scheme was introduced by Robinson in the domain of nuclear resonance spectrographs, and the first aim was to assure a stable oscillation with a very low amplitude of the tank-circuit voltage (Fig. 10).

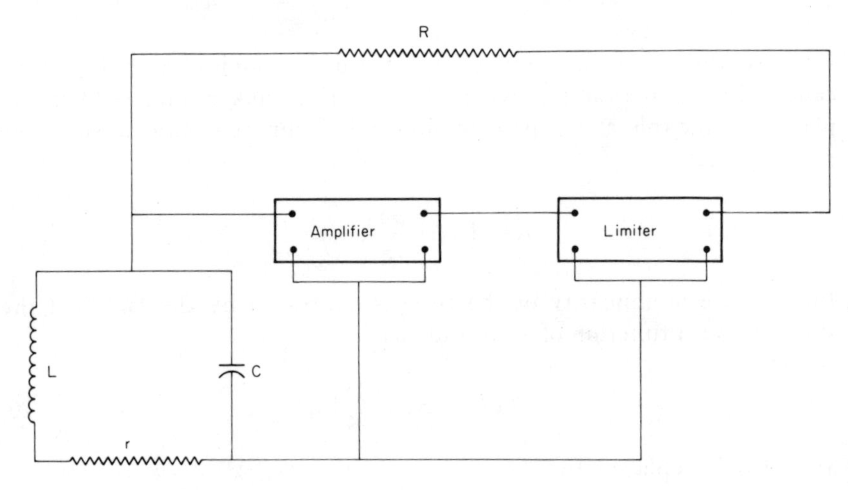

FIG. 10. Robinson's oscillator.

The tank is connected to the grid of the first tube of an amplifier, and the amplitude is so small that the third term in (8), S_2x^3, is negligible. This first "linear" tube is followed by a second stage of linear amplification, which gives a linear over-all characteristic. Then an amplitude-defining mechanism is introduced in the form of a symmetrical clipper, connected at the output of the linear amplifier, which controls the feedback loop. The result is that the tank is feedback with a square wave, instead of a sinusoidal wave.

As a matter of fact, this situation is not at all unusual in the domain of clocks. Indeed, in the best mechanical clocks, the pendulum is kept in motion by a regular series of short pulses, which is mathematically near the derivative of a square wave (Fig. 11). It may easily be shown that the following arguments are also relevant to this operation.

As was explained in the preceding section the first harmonic technique is equivalent to replacing the nonlinear valve by a fictitious valve, endowed with a linear *dynamic* characteristic; in fact we have only to introduce a variable mean slope $S(a_0)$, which can be easily determined.

The tank circuit, because of its narrow bandwidth, nearly selects the first harmonic of the square wave at the output of the clipper, and the amplitude of this harmonic, as well as the amplitude of the square wave, is independent of the input voltage $x = a_0 \sin \omega t$ on the grid.

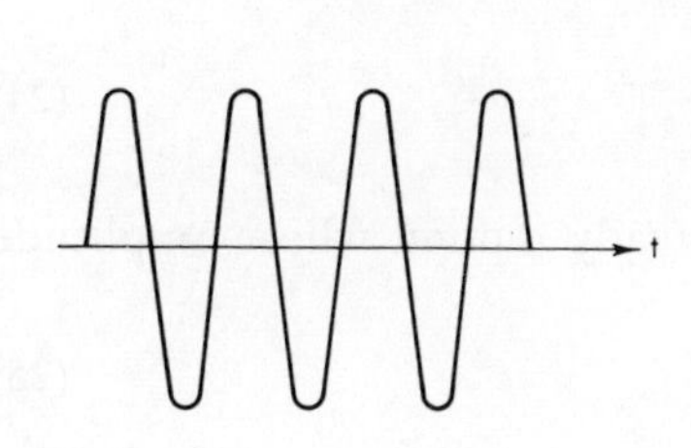

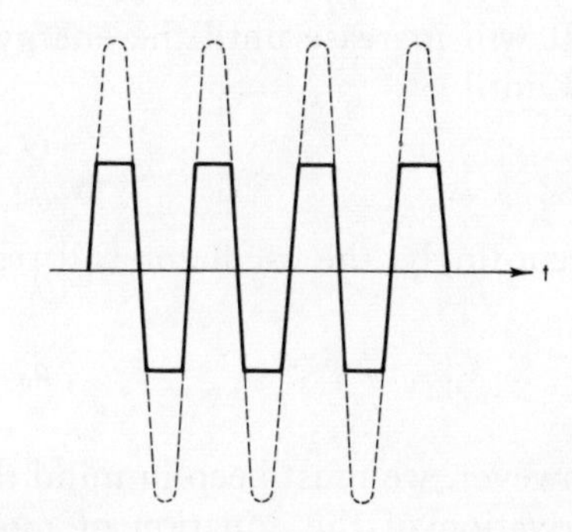

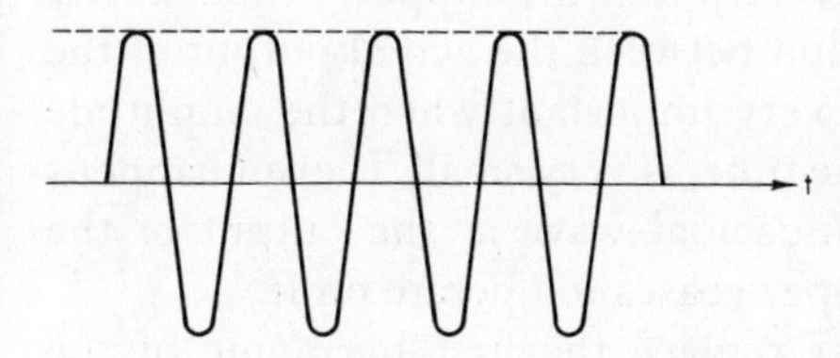

FIG. 11. Outputs of the linear amplifier, and amplitude-defining mechanism. (When the amplitude of the sinusoidal wave is sufficiently small, the clipper ceases to operate on it.)

Accordingly, the first harmonic is

$$i = K_0 \sin \omega t = \frac{K_0}{a_0} a_0 \sin \omega t$$

(where K_0 is a constant depending only on the clipper). This equation shows that the over-all characteristic of the plate circuit obeys the law

$$S(a_0) = \frac{K_0}{a_0} \tag{22}$$

Then, substituting in (19), we get the equation of the loop†:

$$LC\ddot{x} + L\left(G - \frac{mK_0}{a_0}\right)\dot{x} + x = 0 \tag{23}$$

1.6. Threshold Condition of Robinson's Oscillator

At first sight this oscillator seems to have no threshold condition. Indeed, when a_0 is very small, the negative coefficient in the damping term, $-mK_0/a_0$, has a very large absolute value. This means that whatever the energy loss, i.e., whatever $G > 0$, the amplitude of the oscillation will start increasing.

† It may be easily verified that, when no simplification is introduced, the equation of motion is

$$LCx + L\left(G\dot{x} - \frac{\pi}{4} mK_0 \operatorname{sgn} \dot{x}\right) + x = 0 \qquad \text{with} \quad mK_0 > 0 \tag{23*}$$

It will increase until the energy gain is balanced by the energy loss, say until

$$G = \frac{mK_0}{a_0} \tag{24}$$

Accordingly, the oscillator will reach a steady motion whose amplitude is

$$a_0 = \frac{mK_0}{G} \tag{25}$$

However, we must keep in mind the fact that we have introduced in the derivation of the equation of motion a simplification, which we shall now discuss.

As a matter of fact, the wave, at the output of the clipper, is not exactly a square wave (Fig. 12). The deviation between the actual output of the clipper and a square wave becomes very important when the amplitude of the input voltage, on the grid of the tube, is very small. It even happens that, when the amplitude of the sinusoidal wave at the output of the linear amplifier is too small, the clipper ceases to operate on it.

To improve the theory, we must replace the first harmonic of the above square wave by the first harmonic of the actual wave, as shown in Fig. 12, i.e., by

$$i = \left(\frac{2\Gamma a_0 \tau}{\pi} - \frac{2\Gamma a_0}{\pi} \sin\tau \cos\tau + \frac{4K_1}{\pi} \cos\tau\right) \sin\omega t \tag{26}$$

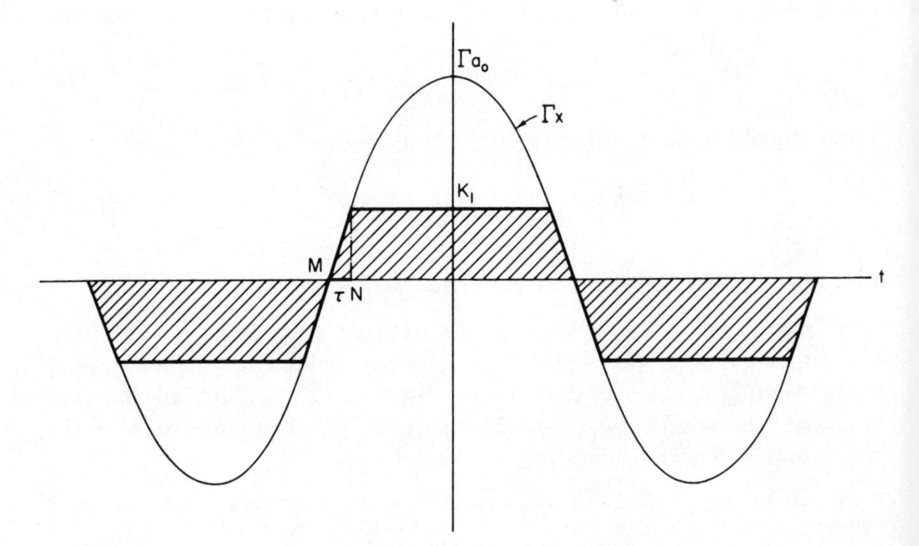

FIG. 12. Wave at the output of a clipper.

The meaning of τ and K_1 is pointed out in Fig. 12: $\tau = MN$. The symbol K_1 represents the amplitude limitation prescribed by the clipper. It is related to K_0 by

$$K_1 = \frac{\pi}{4} K_0$$

The symbol Γ is used for the amplification of the active component of the loop, namely the tube followed by the linear amplifier. Here Γ is assumed to be a real number.

From (26) we deduce a more general expression for the slope of the over-all characteristic:

$$S(a_0) = \frac{2\Gamma}{\pi} (\tau - \sin \tau \cos \tau) + \frac{4K_1}{\pi a_0} \cos \tau \tag{27}$$

from which, by substituting in (19) we get the new equation of motion,

$$LC\ddot{x} + L\left[G - \frac{2m\Gamma}{\pi} (\tau - \sin \tau \cos \tau) - \frac{4mK_1}{\pi a_0} \cos \tau\right] \dot{x} + x = 0 \tag{28}$$

Note that τ decreases from $\pi/2$ to 0, when Γa_0 increases from K_1 to infinity. Accordingly:

(a) When $\Gamma a_0 = K_1$,

$$S(a_0) = \Gamma$$

(b) When $\Gamma a_0 < K_1$, the clipper is ineffective, and the amplificator being assumed to be linear, we again have

$$S(a_0) = \Gamma$$

The above remarks enable us to plot the curve $S(a_0)$ (Fig. 13), from

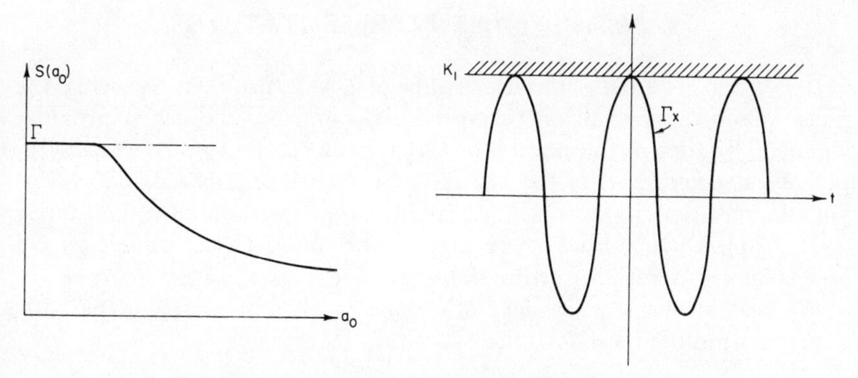

FIG. 13. Threshold condition of Robinson's oscillator.

which we conclude that the negative conductance $-mS(a_0)$ does not tend to infinity when a_0 tends to zero, as was previously indicated by the square-wave approximation. Obviously this approximation led to an unfair extrapolation of the actual law $S(a_0)$ in the domain of very small amplitudes.

Now we obtain the threshold condition

$$\Gamma = \frac{G}{m}$$

where the clipper just begins to be effective (Fig. 14).

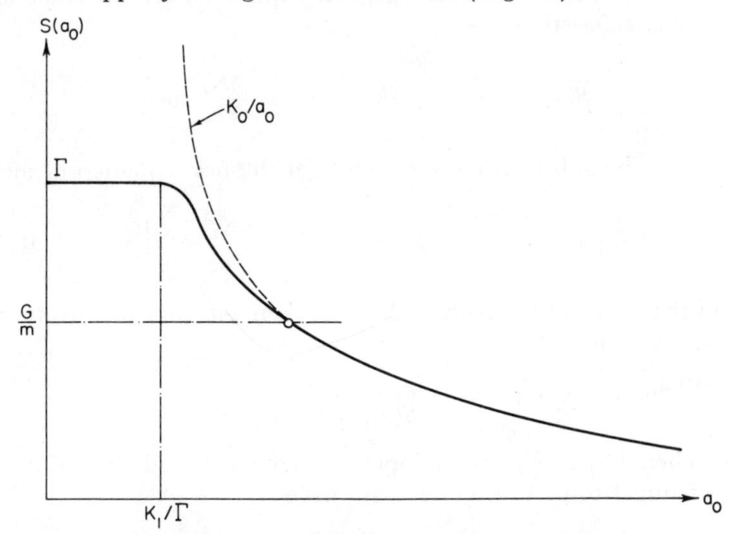

FIG. 14. Operating point of Robinson's oscillator.

2. PHASE-PLANE REPRESENTATION

It is often useful to start the study of a new problem by using the phase-plane representation (Chapter I, Section 5), because it provides a general picture of the motion of the dynamic system under consideration. In a second part of the analysis, qualitative methods based on this flexible representation will have to be complemented by quantitative ones, from which numerical results can be obtained but which do not give such a general idea of the different aspects of the motion.

We shall start with Van der Pol's equation (12), or more conveniently with its simplified version,

$$\ddot{x} - \mu(1 - x^2)\dot{x} + x = 0 \qquad \mu = -(rC - MS_0)\omega_0 \tag{29}$$

which is obtained from (12) by changing t and x into dimensionless variables,

$$t^* = \omega_0 t \qquad \text{and} \qquad x^* = x\sqrt{B}$$

(We shall drop the star in what follows.)

We shall assume that $A < 0$, $S_2 < 0$, and consequently $B > 0$. This also implies $\mu > 0$.

2.1. Isoclynes' Method

From (29) we deduce the normal set

$$\dot{x}_1 = x_2 \qquad \dot{x}_2 = -x_1 + \mu(1 - x_1{}^2)x_2 \tag{30}$$

Then, as pointed out in Chapter I, Section 5, at each nonsingular point P_ν of the phase plane the slope of the tangent to the integral curve which passes through that point is

$$\frac{dx_2}{dx_1} = \frac{\dot{x}_2}{\dot{x}_1} = \frac{-x_1 + \mu(1 - x_1{}^2)x_2}{x_2} \tag{31}$$

where x_1 and x_2 are the coordinates of P_ν.

Accordingly, if we replace a small piece of this integral curve, in the close neighborhood of P_ν, by a small piece of its tangent, we can proceed along this tangent to a neighboring point $P_{\nu+1}$, then iterate the process [5, 28]. This is quite easy, because although the integral curve is unknown, the tangent at point P_ν is perfectly determined by (31). Obviously this iterative construction of the integral curve is an approximate one, whose accuracy depends on the smallness of the basic intervals.

This construction can be put in a more convenient form by first drawing the locus of all points which correspond to the same value of the slope m, namely an *isoclyne curve*, then by giving m different numerical values, which generates a family. From (31) the equation of the family of isoclyne curves is

$$x_2 = \frac{x_1}{\mu(1 - x_1{}^2) - m} \tag{32}$$

Such a family is portrayed in Fig. 15 for the case $\mu = 1$, together with three integral curves. Solutions for (29) by the isoclynes' method were first obtained by Van der Pol.

The different aspects of the families of isoclyne and integral curves when the parameter μ is given different values will not be discussed here, since this discussion is classical and can be found in many reference books.

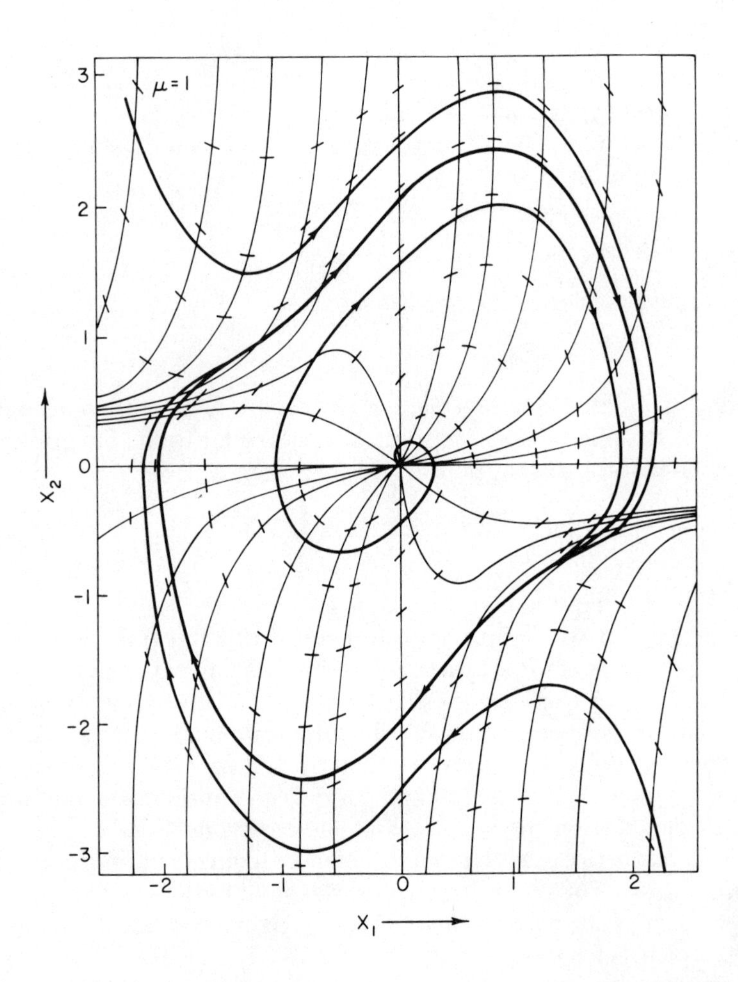

FIG. 15. Isoclynes' method for a Van der Pol oscillator, trajectories and limit cycle.

The isoclynes' method, as outlined above, can be easily extended to any other autonomous system.

2.2. Lienard's Graphical Construction†

More generally, the equation which Lienard considers is

$$\ddot{x} + f(x)\dot{x} + g(x) = 0 \tag{33}$$

† See [6, 25].

and he introduces a new normal set, i.e., a new phase plane, by putting

$$x_1 = x \qquad x_2 = \dot{x} + \int_0^{x_1} f(x)\, dx \tag{34}$$

from which it follows that

$$\dot{x}_1 = x_2 - \int_0^{x_1} f(x)\, dx \qquad \dot{x}_2 = -g(x_1) \tag{35}$$

Accordingly, the slope of the tangent to the integral curve, at each point of Lienard's phase plane, is

$$\frac{dx_2}{dx_1} = -\frac{g(x_1)}{x_2 - \int_0^{x_1} f(x)\, dx} \tag{36}$$

Then, from (36), he deduces a method of graphical integration as follows (Fig. 16).

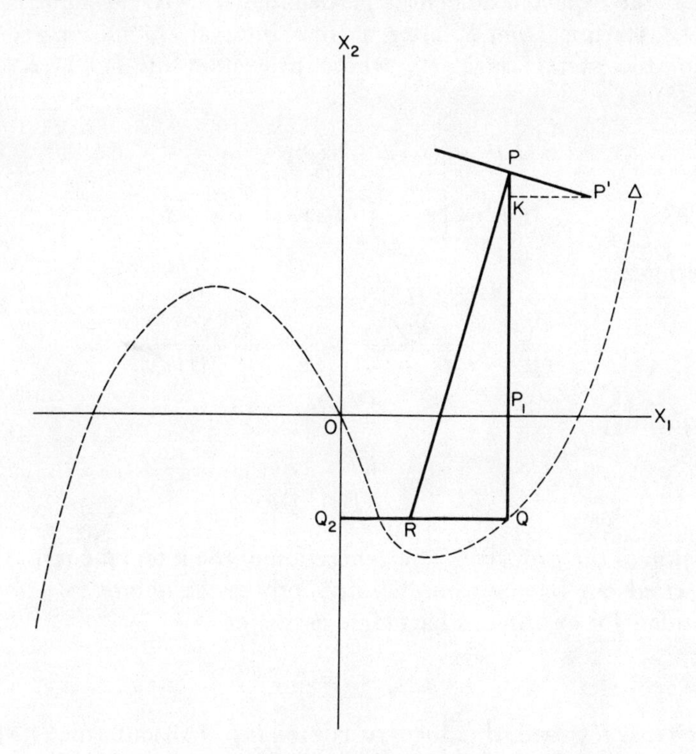

FIG. 16. Lienard's graphical construction.

Consider the curve Δ whose equation is

$$x_2 = \int_0^{x_1} f(x)\, dx$$

and a point P of the phase plane, where we wish to determine the tangent to the integral curve. For example, in the case of Van der Pol's equation, with $\mu = 1$, we have

$$f(x) = x^2 - 1 \qquad x_2 = \frac{x_1^{\,3}}{3} - x_1$$

From P draw a perpendicular PP_1 on axis $0x_1$ and let Q be its intersection with Δ. Also draw a perpendicular QQ_2 to axis $0x_2$, and let R be a point on this perpendicular such that

$$\overline{RQ} = g(x_1)$$

Then it may easily be shown that RP is normal to the integral curve at P; i.e., the expected tangent is perpendicular to RP at point P.

Indeed, starting from P, after a time interval dt, the representative point of the system is at P', whose projection on PQ is K. Then from (35),

$$\overline{KP} = g(x_1)\, dt + o(dt)$$

$$\overline{KP'} = \left[x_2 - \int_0^{x_1} f(x)\, dx\right] dt + o(dt)$$

Furthermore,

$$\frac{\overline{RQ}}{\overline{QP}} = \frac{\overline{RQ}}{\overline{P_1P} + \overline{QP_1}} = \frac{g(x_1)}{x_2 - \int_0^{x_1} f(x)\, dx}$$

Consequently,

$$\lim_{dt \to 0} \frac{\overline{KP}}{\overline{KP'}} = \frac{\overline{RQ}}{\overline{QP}}$$

which proves the property. The generation of the integral curve through P is carried out by the same iterative process as before.

Consider, for example, a harmonic oscillator

$$\ddot{x} + x = 0$$

in which case $f(x) \equiv 0$. Δ reduces to axis $0x_1$ ($x_2 = 0$) and, since $g(x) = x$, point R coincides with the origin 0 of the phase plane, whatever the locus of P (Fig. 17). Accordingly, integral curves are circles whose center is the origin.

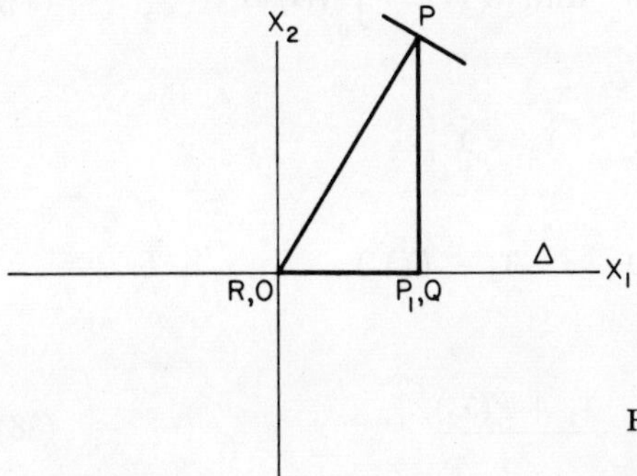

FIG. 17. Lienard's graphical construction for a harmonic oscillator.

As a matter of fact, the construction is much simplified if we assume $g(x) \equiv x$, since then R and Q_2 are in coincidence (Fig. 18). Many practical systems fit this assumption, for example, Van der Pol's oscillator, and in this situation it may sometimes be convenient to change variable x into

$$\mathfrak{x} = \int_0^t x\, dt$$

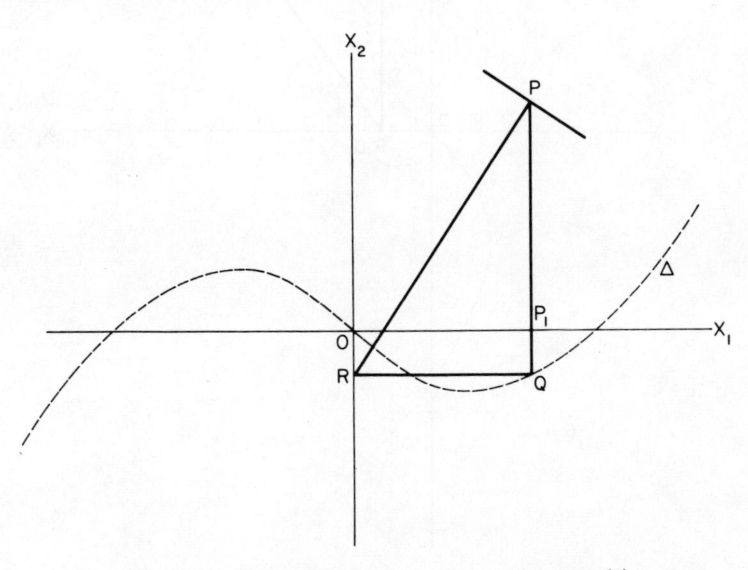

FIG. 18. Lienard's graphical construction in the case $g(x) \equiv x$.

according to which (33) is rewritten

$$\ddot{\mathfrak{x}} + F(\dot{\mathfrak{x}}) + \mathfrak{x} = 0 \qquad \text{with} \quad F(x) = \int_0^x f(s)\, ds \tag{37}$$

Putting

$$\mathfrak{x}_1 = \mathfrak{x} \qquad \mathfrak{x}_2 = \dot{\mathfrak{x}}$$

we get

$$\dot{\mathfrak{x}}_1 = \mathfrak{x}_2 \qquad \dot{\mathfrak{x}}_2 = -\mathfrak{x}_1 - F(\mathfrak{x}_2)$$

from which follows

$$\frac{d\mathfrak{x}_2}{d\mathfrak{x}_1} = -\frac{\mathfrak{x}_1 + F(\mathfrak{x}_2)}{\mathfrak{x}_2} \tag{38}$$

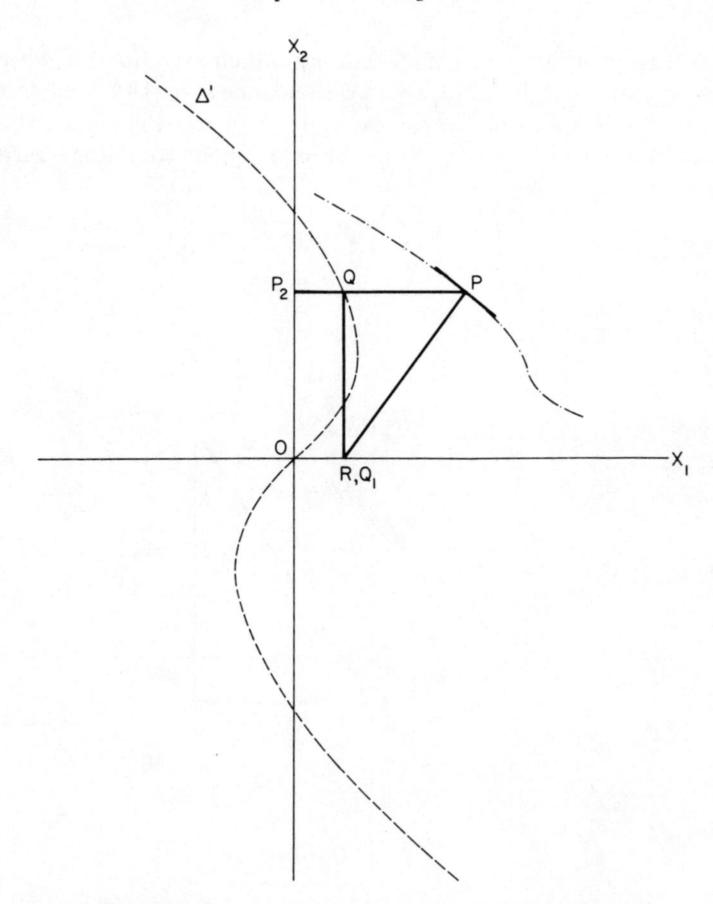

FIG. 19. Modified version of Lienard's graphical construction.

A geometric construction, similar to the one above, can be deduced from (38) by using the curve Δ' instead of Δ:

$$\mathfrak{x}_1 + F(\mathfrak{x}_2) = 0$$

The new construction is explained in Fig. 19.

By using the new variable $\mathfrak{x}$ in Van der Pol's equation we obtain

$$\ddot{\mathfrak{x}} - \mu \left(1 - \frac{(\dot{\mathfrak{x}})^2}{3}\right) \dot{\mathfrak{x}} + \mathfrak{x} = 0$$

and Δ' is defined by

$$\mathfrak{x}_1 = \mu \left(1 - \frac{\mathfrak{x}_2{}^2}{3}\right) \mathfrak{x}_2$$

As another example, consider Robinson's oscillator, whose equation (23*) can be reduced to

$$\ddot{x} + (A\dot{x} - B \operatorname{sgn} \dot{x}) + x = 0 \qquad \text{with} \quad A, B > 0 \tag{39}$$

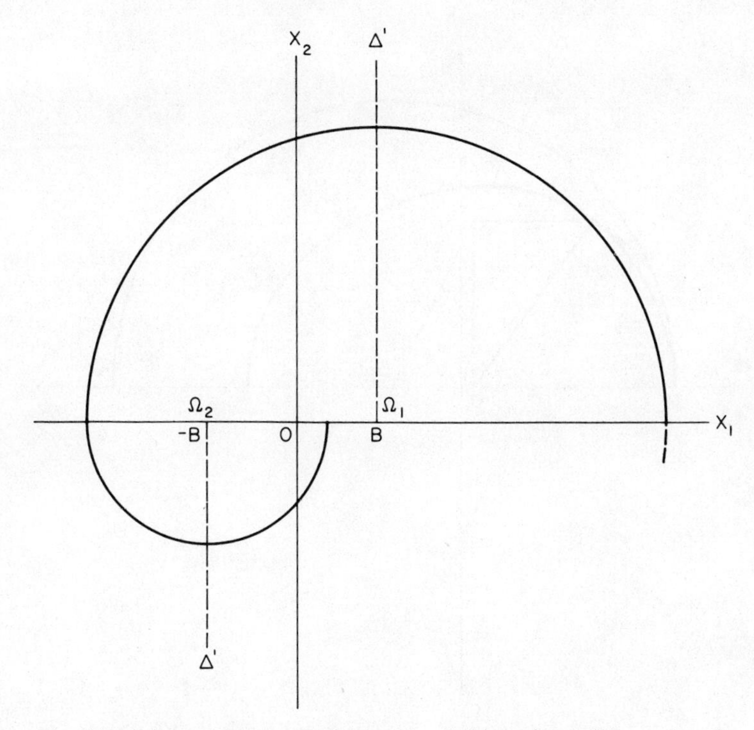

FIG. 20. Lienard's graphical construction for Robinson's oscillator: $A = 0$.

by again using the dimensionless variable $t^* = \omega_0 t$. Equation (39) is similar to (37):

$$F(\dot{x}) = A\dot{x} - B \operatorname{sgn} \dot{x}$$

and Δ' is defined by

$$x_1 = B - Ax_2 \qquad \text{when} \quad x_2 > 0$$
$$x_1 = -B - Ax_2 \qquad \text{when} \quad x_2 < 0$$

(Fig. 20). Note that when $A = 0$, i.e., in the case of pure negative Coulomb damping, Δ' is defined by

$$x_1 = B \qquad \text{when} \quad x_2 > 0$$
$$x_1 = -B \qquad \text{when} \quad x_2 < 0$$

Then Lienard's construction shows clearly that the trajectory of P, i.e., the integral curve, is an expanding spiral made of half-circles with centers at $x_1 = B$ and $x_1 = -B$.

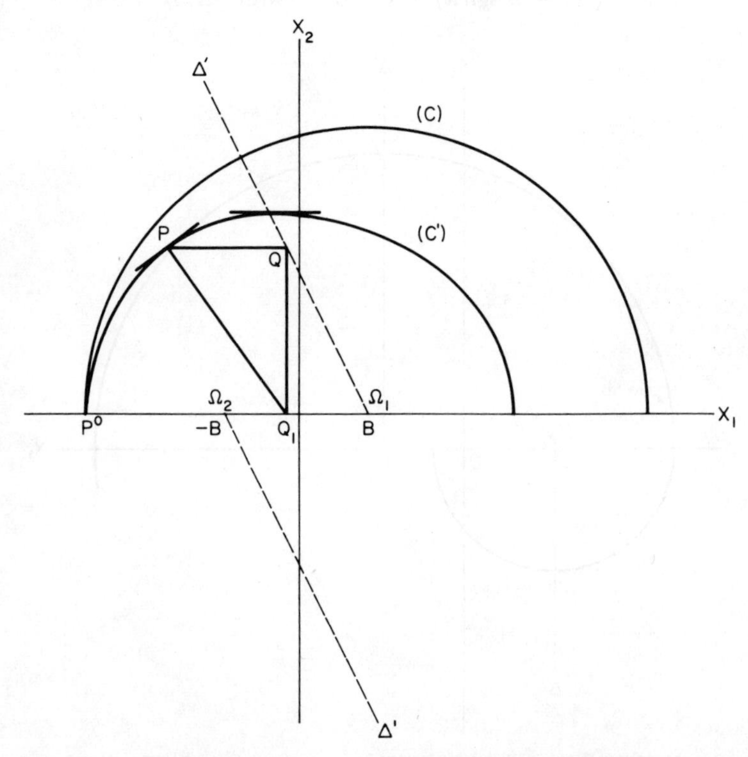

FIG. 21. Lienard's graphical construction for Robinson's oscillator: $A \neq 0$.

Now if we consider a trajectory whose starting point is P^0 on axis $0x_1$ (Fig. 21), we see that

(a) If $A = 0$, a first piece of trajectory will be the half-circle (C) whose center is the point $x_1 = B$.

(b) If $A \neq 0$, the center will slip toward the left, for example from Ω_1 to Q_1, and at the same time the radius will decrease, from $\Omega_1 P^0$ to $Q_1 P$, in such a way that (C) will be replaced by the piece of spiral (C').

In case (b) it may be shown that the spiral thus generated will remain in a bounded domain, which is in conformity with an earlier discussion (Section 1.6)—the oscillator will tend to a steady motion whose amplitude was computed previously.

There also exists a position of P^0 on axis $0x_1$ such that the trajectory issuing at that point is a "closed loop." Because of the symmetry of the above construction, this property is verified if the intersection of (C′) with $0x_1$ is symmetric from P^0 with respect to 0.

We wish to emphasize two important features of the above examples, to which we shall devote the next few paragraphs:

(a) By each nonsingular point P^0 of the phase plane, there exists one and only one integral curve through this point.

(b) When point P^0 is properly chosen, the integral curve which passes through it is a "closed loop"; i.e., if moving point P starts from P^0 at time t_0, there exists an interval of time T such that

$$P(t_0 + T) = P(t_0)$$

In this situation $x(t)$ is a periodic function of time. This closed loop is called a *limit cycle*, and it may be verified that, in the above examples, trajectories whose starting point is not on the limit cycle are not periodic and tend to it asymptotically as $t \to \infty$.

3. CAUCHY-LIPSCHITZ THEOREM

Thus far we have considered specific examples. More generally assume that the state of the dynamic system under study is defined by n variables $x_1, \ldots, x_n$, which satisfy a set of differential equations

$$\dot{x}_j = f_j(x_1, \ldots, x_n, t) \qquad (j = 1, 2, \ldots, n) \tag{40}$$

Henceforth we shall denote the position of P in the phase space R^n by the vector $\mathbf{x}$ with components $x_1, \ldots, x_n$; we shall write $P = P(\mathbf{x})$.

If point P is a point of a trajectory, its position is a function of time t, given by $\mathbf{x} = \mathbf{x}(t)$; we shall write $P = P(\mathbf{x}(t))$. Vector function $\mathbf{x}(t)$ has components $x_1(t), ..., x_n(t)$, where $x_1(t), ..., x_n(t)$ denote a solution of the dynamic equations of the system.

Now, given P^0 at time t_0, how can we guarantee, *in a neighborhood of* t_0, the *existence* and *uniqueness* of a trajectory in R^n through P^0?

The answer to this important question is given by the Cauchy-Lipschitz theorem [32]:

THEOREM 1. (a) *A function* $\varphi(s_1, ..., s_m)$ *of* m *variables* $s_1, ..., s_m$ *is said to satisfy the Lipschitz condition at point* $\mathscr{S}^0$: $s_1^0, ..., s_m^0$, *if, in a neighborhood* Δ: $| s_i - s_i^0 | < \delta$ $(i = 1, ..., m)$ *of* $\mathscr{S}^0$, *there exists* $\alpha > 0$, *whose value depends only on* $s_1^0, ..., s_m^0$, *and not on* $s_1, ..., s_m$, *such that*

$$| \varphi(s_1, ..., s_m) - \varphi(s_1^0, ..., s_m^0)| \leqslant \alpha \sum_{i=1}^{m} | s_i - s_i^0 | \qquad \forall \mathscr{S} \in \Delta$$

(b) *Given* P^0 *and time* t_0, (40) *has a unique solution* $P(\mathbf{x}(t))$, *defined in a neighborhood of* $t = t_0$, *such that* $P(\mathbf{x}(t_0)) = P_0$, *provided the functions* $f_j(x_1, ..., x_n, t)$ $(j = 1, ..., n)$, *satisfy the Lipschitz condition with respect to the* $n + 1$ *variables* $x_1, ..., x_n, t$.

When this is true (40) is called a *Lipschitzian differential system*. It may easily be proved that if $\partial f_j/\partial t$ and $\partial f_j/\partial x_\alpha$ $(\alpha = 1, ..., n)$, are defined and continuous on a domain (D), (40) is Lipschitzian on (D).

Indeed, in the case of autonomous systems, t does not occur in the equations. For example, consider equations (9), Chapter I:

$$\begin{aligned} \dot{x}_1 &= x_2 \\ \dot{x}_2 &= -\omega_0^2 x_1 - \mu x_1^3 \qquad \mu < 0 \end{aligned}$$

and a family of trajectories as shown in Fig. 9 of Chapter I. We have

$$\frac{\partial f_1}{\partial x_1} = 0 \qquad \frac{\partial f_1}{\partial x_2} = 1$$

$$\frac{\partial f_2}{\partial x_1} = -\omega_0^2 - 3\mu x_1^2 \qquad \frac{\partial f_2}{\partial x_2} = 0$$

These partial derivatives are defined and continuous throughout the whole phase plane. Therefore the system is Lipschitzian everywhere.

However at each of the singular points, A and B, it appears that *two branches* are intersecting (Fig. 22), and at the same time the Cauchy-Lipschitz theorem applies. We can see that there is no paradox here, since the theorem guarantees existence and *uniqueness* of a *trajectory* in a *neighborhood of* t_0. For example, if $P^0 = B$ at time t_0, then $P(\mathbf{x}(t)) = B, \forall t \geqslant t_0$, since B is an equilibrium point.†

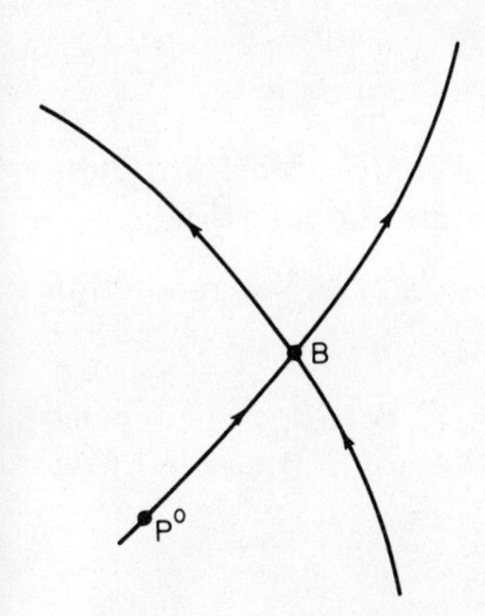

FIG. 22. Intersecting branches in the phase plane.

On the other hand if we start, at time t_0, at a point P^0 which belongs to any one of the two branches which are converging toward B, the theorem again applies, since the branching point B cannot be reached:

$$P \to B \quad \text{as} \quad t \to \infty$$

4. GEOMETRIC STUDY OF PERIODIC SOLUTIONS

It is important to determine under which conditions (33) has a limit cycle. Here we shall discuss this question from a geometric viewpoint, following the arguments of Bogoliubov and Mitropolsky [25], whose starting point is the work of Lienard [6, 25, 34].

Let us return to equations (35) or

$$\begin{aligned} \dot{x}_1 &= x_2 - F(x_1) \qquad \text{with} \quad F(x_1) = \int_0^{x_1} f(x)\, dx \\ \dot{x}_2 &= -g(x_1) \end{aligned} \tag{41}$$

and assume that:

Assumption 1. $F(x)$ and $g(x)$ are odd single-valued continuous functions of x, $-\infty < x < +\infty$, such that

$$xg(x) > 0 \qquad \forall\, x \neq 0$$
$$F(x) \to \pm\infty \qquad \text{as} \quad x \to \pm\infty$$

† Here the "trajectory" reduces to *one* point.

Assumption 2. $F(x) = 0$ has one and only one positive root $x = a$. Furthermore, for $x \geqslant a$, $F(x)$ increases monotonically with x, and

$$xF(x) < 0 \qquad \forall\, x; \quad x \neq 0; \quad -a < x < +a$$

Assumption 3. $F(x)$ and $g(x)$ satisfy the Lipschitz condition.

A lot of interesting geometric properties can be deduced from these simple and realistic assumptions, among which are the following:

Property 1. If $\mathbf{x}(t)$ is a solution of (41), then $-\mathbf{x}(t)$ is also a solution of this system.

Accordingly, to every piece of trajectory there corresponds a piece of trajectory symmetric to it with respect to the origin 0 of the phase plane.

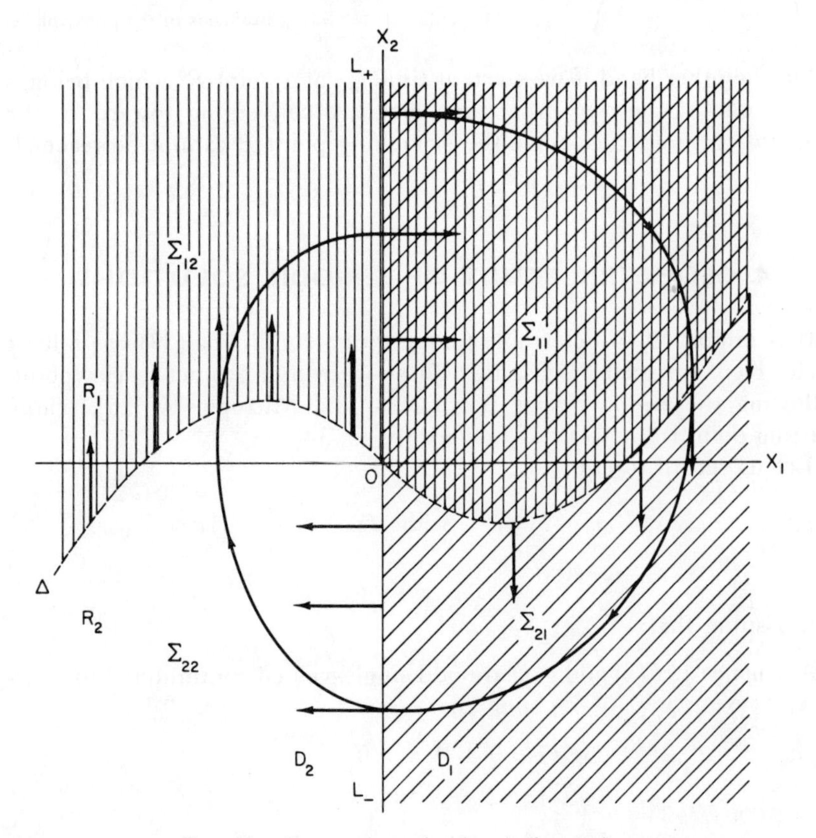

FIG. 23. Geometric study of periodic solutions.

Property 2. Curve Δ: $x_2 = F(x_1)$ separates the phase plane into two connected open regions, R_1 and R_2:

$$\text{In region } R_1\text{: } x_2 - F(x_1) > 0 \qquad \text{and thus} \quad \dot{x}_1 > 0$$

$$\text{In region } R_2\text{: } x_2 - F(x_1) < 0 \qquad \text{whence} \quad \dot{x}_1 < 0$$

(Fig. 23). Also the x_2 axis is separated into two open half-rays, L_+ and L_-: L_+ corresponds to positive values and L_- to negative values of x_2. Since $x_1 = 0 \Rightarrow F(x_1) = 0$,

$$L_+ \subset R_1 \qquad \text{and} \qquad L_- \subset R_2$$

Property 3. The x_2 axis separates the phase plane into two connected open regions D_1 and D_2:

$$\text{In region } D_1\text{: } x_1 > 0 \qquad \text{which implies} \quad g(x_1) > 0 \text{ and } \dot{x}_2 < 0$$

$$\text{In region } D_2\text{: } x_1 < 0 \qquad \text{whence} \quad g(x_1) < 0 \text{ and } \dot{x}_2 > 0$$

Property 4. According to Properties 2 and 3, curve Δ, together with the x_2 axis, separate the phase plane into four connected regions:

$$\sum_{11} = R_1 \cap D_1 \qquad \sum_{21} = R_2 \cap D_1$$

$$\sum_{12} = R_1 \cap D_2 \qquad \sum_{22} = R_2 \cap D_2$$

The signs of $\dot{x}_1$ and $\dot{x}_2$ in these different regions are given in the following table:

	Σ_{11}	Σ_{21}	Σ_{22}	Σ_{12}
$\dot{x}_1$	+	−	−	+
x_1	↗	↘	↘	↗
x_2	↘	↘	↗	↗
$\dot{x}_2$	−	−	+	+

An arrow indicates if functions $x_1(t)$ and $x_2(t)$ in that region are increasing, or decreasing, functions of time.

Furthermore, at every point of Δ,

$$\dot{x}_1 = 0$$

i.e., velocity $\dot{\mathbf{x}}$ is parallel to the x_2 axis, and at every point of the x_2 axis

$$\dot{x}_2 = 0$$

i.e., velocity $\dot{\mathbf{x}}$ is parallel to the x_1 axis. These results are summarized in Fig. 23.

Then, according to these conclusions and to Assumption 2, we can see that a trajectory whose starting point is P_+^0, on L_+, necessarily has the form shown in Fig. 24—it is a loop which surrounds the origin 0

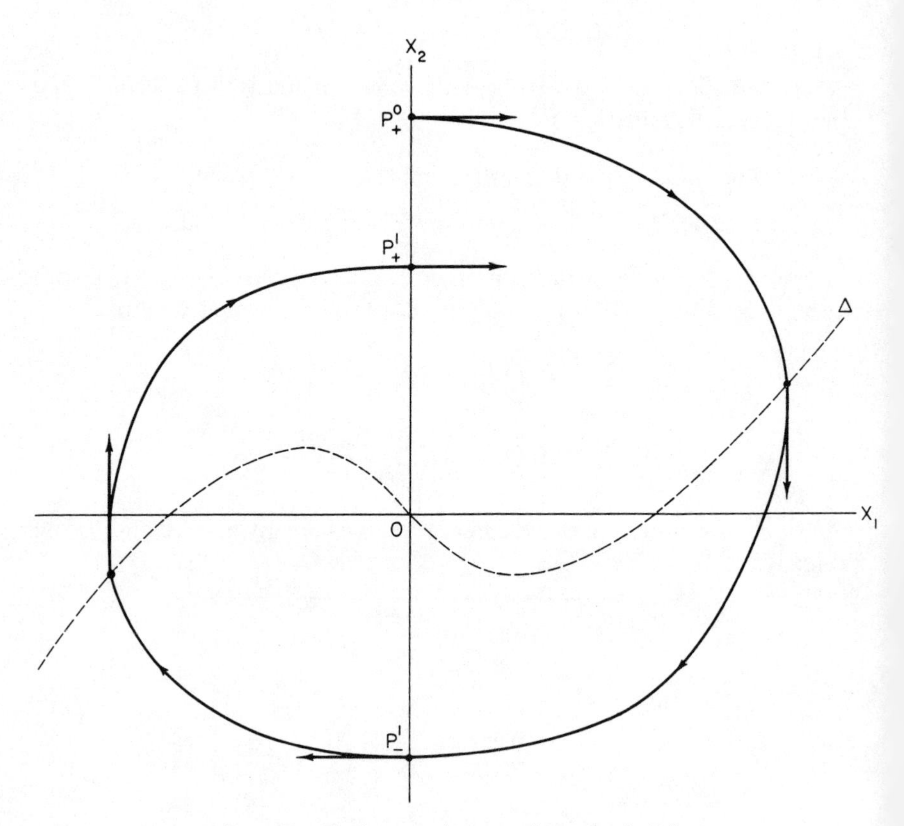

FIG. 24. Trajectory where starting point is P_+^0.

and crosses L_- for the first time at point P_-^1, then again L_+ at point P_+^1. Now, if we extend this trajectory by considering P_+^1 as a new starting point, we get the next crossing points, P_-^2, P_+^2, and so on.

From Assumption 3 we deduce the fact that no piece of trajectory can *cross* any other piece of the same or another trajectory, hence:

Property 5. The above trajectory will spiral around the origin and cross L_+ at points

$$P_+^0, P_+^1, P_+^2, \ldots, P_+^k, \ldots$$

If $OP_+^1 < OP_+^0$ then

$$OP_+^k < \cdots < OP_+^2 < OP_+^1 < OP_+^0$$

If $OP_+^1 > OP_+^0$ then

$$OP_+^k > \cdots > OP_+^2 > OP_+^1 > OP_+^0$$

We arrive at the following lemmas:

LEMMA 1. *For the trajectory to be a* closed loop, *i.e., a* limit cycle, *it is necessary and sufficient that*

$$\overline{OP_+^1} = \overline{OP_+^0}$$

This lemma follows directly from Property 5.

LEMMA 2. *In order that the trajectory be a closed loop, it is necessary and sufficient that*

$$\overline{OP_-^1} = -\overline{OP_+^0}$$

Lemma 2 can be proved by the following arguments.

Consider a closed loop which intersects L_+ and L_- at points P_+^0 and P_-^1, respectively, and assume

$$\overline{OP_-^1} \neq -\overline{OP_+^0}$$

In this situation, according to Property 1, we have another solution which is also a closed loop, the loop symmetric to the above one with respect to 0 (Fig. 25). It intersects L_+ and L_- at points $\mathscr{P}_+^0$ and $\mathscr{P}_-^1$ such that

$$\overline{O\mathscr{P}_+^0} = -\overline{OP_-^1} \qquad \overline{O\mathscr{P}_-^1} = -\overline{OP_+^0}$$

But this implies that the two closed loops cross one another, which is impossible, since system (41) is Lipschitzian. Accordingly,

$$\overline{OP_-^1} = -\overline{OP_+^0}$$

which proves the first part of Lemma 2.

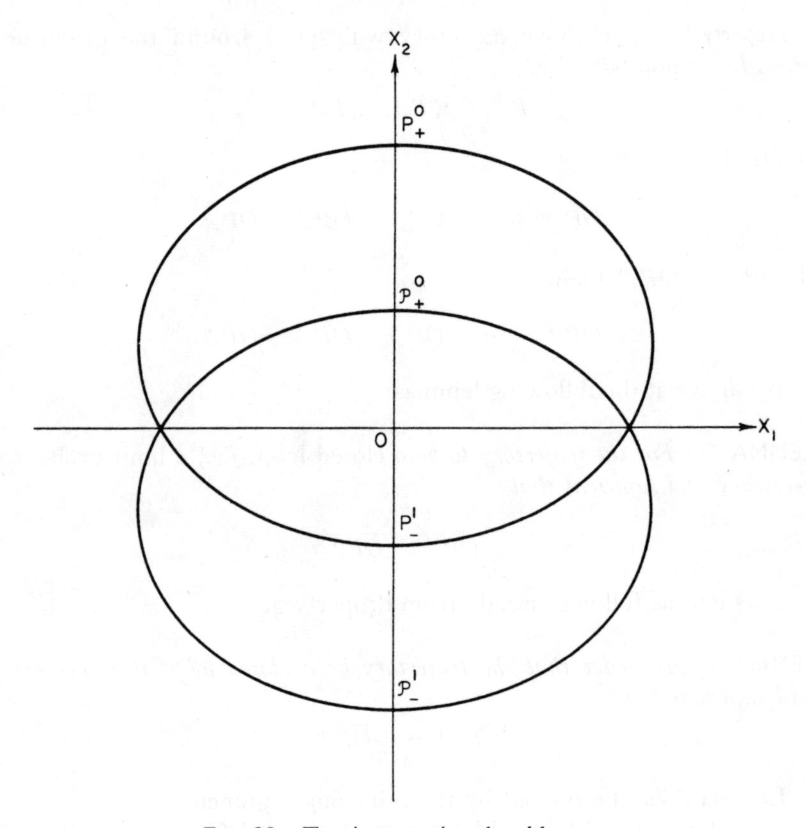

FIG. 25. Two intersecting closed loops.

Now assume $\overline{OP_-{}^1} = -\overline{OP_+{}^0}$, and consider the piece of trajectory which connects $P_+{}^0$ with $P_-{}^1$ $(x_1 \geqslant 0)$ (Fig. 26). We know that there corresponds to this one another piece of trajectory, symmetric to it with respect to the origin, which will cross L_+ at point $P_+{}^0$ and thus Lemma 2 is established.

Next let us prove the following theorem:

THEOREM 2. *Under Assumptions* 1–3, (33) *has one and only one limit cycle.*

First of all, let us note that the total energy of the oscillator when the damping term is canceled, say for $f(x) \equiv 0$, is

$$E = E_p + E_k = \frac{x_2{}^2}{2} + \int_0^{x_1} g(x)\, dx \tag{42}$$

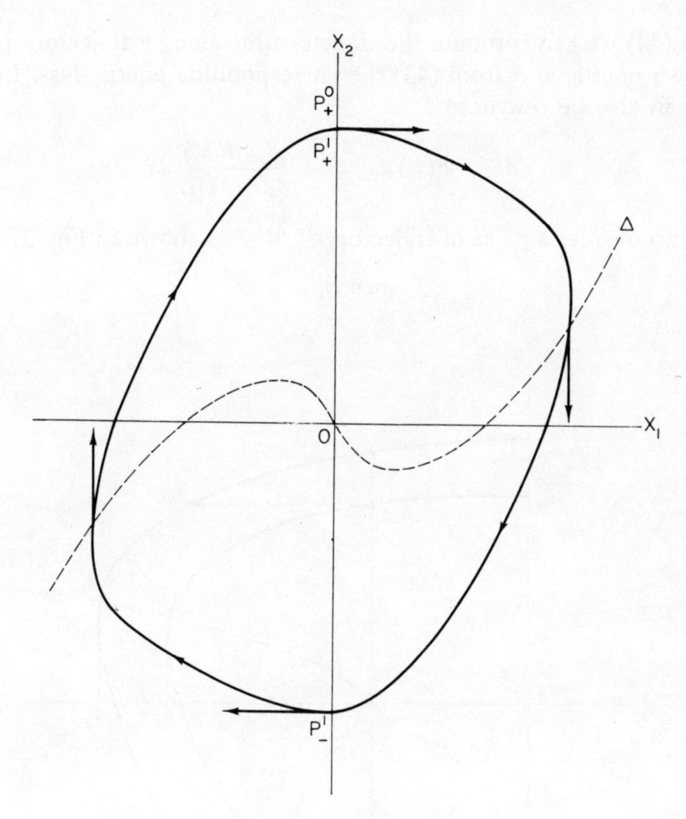

FIG. 26. Limit cycle.

This energy is the sum of potential energy E_p and kinetic energy E_k. It is a constant of the motion, since there is no dissipation in this case.

When $f(x) \not\equiv 0$, forces of friction result in energy dissipation, in such a way that during any time interval dt the change in total energy of the oscillator is equal to the work of forces of friction.

From (42) we deduce, along a trajectory,

$$\frac{dE}{dt} = \frac{\partial E}{\partial x_1}\dot{x}_1 + \frac{\partial E}{\partial x_2}\dot{x}_2 = g(x_1)\dot{x}_1 + x_2\dot{x}_2$$

and from (41),

$$\frac{dE}{dt} = -g(x_1)F(x_1) \tag{43}$$

$$dt = \frac{dx_1}{x_2 - F(x_1)} = \frac{dx_2}{-g(x_1)} \tag{44}$$

From (44) we can compute the transfer time along a trajectory between any two points, and from (43) the corresponding energy loss. Equation (43) can also be rewritten

$$dE = F(x_1)\,dx_2 = -\frac{g(x_1)F(x_1)}{x_2 - F(x_1)}\,dx_1 \tag{45}$$

Now consider a piece of trajectory $P_+^0P_-^1$ as shown in Fig. 27, and let

$$\max x_1 = \alpha$$

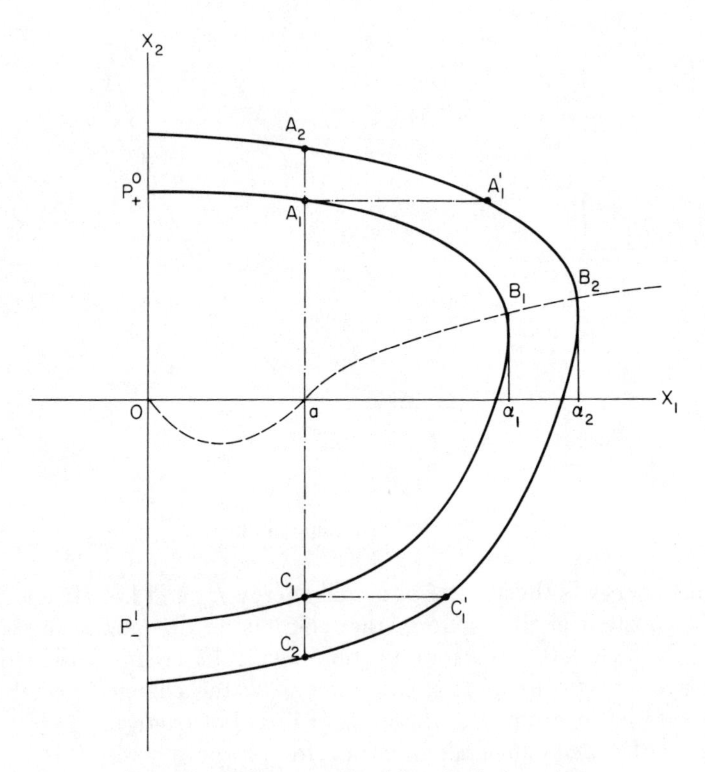

FIG. 27. Existence and unicity of a limit cycle.

From (43) and Assumptions 1–3 we have

$$0 < x_1 < a \quad \Rightarrow \quad \frac{dE}{dt} > 0$$

That is, if $\alpha \leqslant a$, the energy of the oscillator will be an increasing function of time. From a physical viewpoint this means that, when the amplitude is small, the oscillator does not dissipate as much energy as it receives, so its amplitude will increase with time. Then, when the amplitude is sufficiently large, as we shall see, energy losses overtake energy gains and the system reaches steady motion.

To prove the above theorem, we shall now assume that

$$\alpha = \alpha_1 \qquad \alpha_1 > a$$

and we shall consider the change in energy ΔE along $P_+^0 P_-^1$.

As a matter of fact, ΔE is a function of α. We shall use the notation $\Delta_\alpha E$ and will have to determine a value of α, say α_0, such that $\Delta_{\alpha_0} E = 0$. Indeed, for this value, the energy E will be the same at P_+^0 and P_-^1; accordingly, from (42) with $x_1 = 0$, x_2^2 will be the same at P_+^0 and P_-^1,

$$(OP_+^0)^2 = (OP_-^1)^2$$

From Lemma 2 we shall associate a limit cycle to α_0. Furthermore, we shall prove that this limit cycle is unique.

Let us write

$$\Delta E = \Delta^{(1)} E + \Delta^{(2)} E + \Delta^{(3)} E$$

where (Fig. 27)

$\Delta^{(1)} E$ is obtained by integration along $P_+^0 A_1$

$\Delta^{(2)} E$ is obtained by integration along $A_1 B_1 C_1$

$\Delta^{(3)} E$ is obtained by integration along $C_1 P_-^1$

(B_1: $x_1 = \alpha_1$; $\quad A_1$, C_1: $x_1 = a$). According to (43) we have

$$\frac{dE}{dt} > 0 \qquad \text{along} \quad P_+^0 A_1 \quad \text{and} \quad C_1 P_-^1, \qquad 0 < x_1 < a$$

$$\frac{dE}{dt} = < 0 \qquad \text{along} \quad A_1 B_1 C_1, \qquad x_1 \neq a$$

and consequently,

$$\Delta^{(1)} E > 0 \qquad \Delta^{(2)} E < 0 \qquad \Delta^{(3)} E > 0$$

We will see that $\Delta_\alpha E$ is a decreasing function of α by giving α two different values:

$$\alpha = \alpha_1 \quad \alpha = \alpha_2 \qquad \alpha_2 > \alpha_1$$

The corresponding trajectories are shown in Fig. 27.

As concerns $\Delta^{(1)}E$, the limits of integration along the x_1 axis are 0, a for $\alpha = \alpha_1$ and $\alpha = \alpha_2$, and for each given x_1, $0 \leqslant x_1 \leqslant a$, we have

$$(x_2)_{\alpha=\alpha_2} > (x_2)_{\alpha=\alpha_1} \quad \text{and} \quad \left[\frac{1}{x_2 - F(x_1)}\right]_{x_1, \alpha=\alpha_2} < \left[\frac{1}{x_2 - F(x_1)}\right]_{x_1, \alpha=\alpha_1}$$

Moreover, since $-g(x_1)\, F(x_1) > 0$ for $0 < x_1 < a$, we get from (45)

$$\alpha_2 > \alpha_1 \Rightarrow \Delta^{(1)}_{\alpha_2}E < \Delta^{(1)}_{\alpha_1}E$$

By similar arguments along C_1P_{-1} one gets

$$\alpha_2 > \alpha_1 \Rightarrow \Delta^{(3)}_{\alpha_2}E < \Delta^{(3)}_{\alpha_1}E$$

i.e., $\Delta^{(1)}E$ and $\Delta^{(3)}E$ are monotonically decreasing functions of α.

Now, again using (45) we have

$$\Delta^{(2)}_{\alpha_1}E = \int_{A_1B_1C_1} F(x_1)\, dx_2$$

and since $\Delta^{(2)}E$ is negative,

$$\Delta^{(2)}_{\alpha_2}E = \int_{A_2B_2C_2} F(x_1)\, dx_2 < \int_{A_1'B_2C_1'} F(x_1)\, dx_2$$

Moreover

$$\int_{A_1'B_2C_1'} F(x_1)\, dx_2 < \int_{A_1B_1C_1} F(x_1)\, dx_2 < 0$$

since the limits of integration along the x_2 axis are the same for both integrals, and the piece of trajectory which corresponds to α_2 lies entirely on the right side of the one which corresponds to α_1; i.e., for each given x_2,

$$F[x_1(x_2)]_{\alpha_2} > F[x_1(x_2)]_{\alpha_1}$$

Finally,

$$\alpha_2 > \alpha_1 \Rightarrow \Delta^{(2)}_{\alpha_2}E < \Delta^{(2)}_{\alpha_1}E$$

It follows that *ΔE is a monotonically decreasing function of* α, and we have seen at the begining that

$$\Delta E > 0 \quad \text{when} \quad 0 < \alpha \leqslant a$$

On the other hand, it may easily be shown that

$$\Delta E \to -\infty \qquad \text{when} \quad \alpha \to +\infty$$

as a straightforward consequence of the assumption

$$F(x_1) \to +\infty \qquad \text{as} \quad x_1 \to +\infty$$

Hence there exists one and only one value $\alpha = \alpha_0$ for which

$$\Delta_{\alpha_0} E = 0$$

and in view of Lemma 2 there exists one and only one limit cycle which corresponds to $\alpha = \alpha_0$.

For more on the above, see the works of Ivanov, Levinson, Smith, and Dragilev. We give below the theorem of Dragilev and the theorem of Levinson and Smith [10, 11, 14, 15, 19, 22]:

THEOREM OF DRAGILEV [14]. *The differential equation* (33) *has at least one limit cycle if:*

(a) *$g(x)$ satisfies the Lipschitz condition, and*

$$xg(x) > 0 \qquad \forall\, x \neq 0$$

$$G(x) = \int_0^x g(s)\, ds \to \infty \qquad \textit{as} \quad x \to \infty$$

(b) *$F(x)$ is a single-valued functions of x, $-\infty < x < +\infty$, and satisfies the Lipschitz condition in any finite interval, and $xF(x) < 0$ for sufficiently small values of x.*

(c) *There exist numbers M, k, and k', $k' < k$, such that*

$$x > M \Rightarrow F(x) \geqslant k$$

$$x < -M \Rightarrow F(x) \leqslant k'$$

THEOREM OF LEVINSON AND SMITH [11]. *The differential equation* (33) *has one and only one limit cycle if:*

(a) *$g(x)$ is an odd function, and*

$$g(x) > 0 \qquad \textit{for} \quad x > 0$$

(b) *$F(x)$ is an odd function, and there exists a value $x = a$ such that*

$$0 < x < a \Rightarrow F(x) < 0$$
$$a \leqslant x \Rightarrow F(x) \geqslant 0$$

For $x \geqslant a$, $F(x)$ is a monotonically increasing function of x.

(c)
$$F(x) \to \infty \quad \text{as} \quad x \to \infty$$
$$G(x) = \int_0^x g(s)\, ds \to \infty \quad \text{as} \quad x \to \infty$$

(d) *$f(x)$ and $g(x)$ satisfy the Lipschitz condition in any finite interval.*

5. ANALYTIC APPROACHES TO PERIODIC PHENOMENA

5.1. Perturbation Method†

The names Lindstedt and Poincaré are associated with the perturbation method, which they chiefly developed because of its applications in the field of celestial mechanics.

The work of Poincaré was probably the first to emphasize the importance of the notion of periodicity, and it thus opened a new period, during which the theory has been generalized and made more rigorous. It applies to equations of motion which depend on a small parameter μ.

Next we shall consider only autonomous systems, of the general form

$$\ddot{x} + x = \mu f(x, \dot{x}) \tag{46}$$

The starting point of the theory is the *problem of Poincaré.*

If $\mu = 0$, (46) has periodic solutions whatever the initial conditions. On the other hand, if $\mu \neq 0$, very simple examples show that this situation no longer occurs in general; however, it may be that there exist periodic solutions for some functions $f(x, \dot{x})$, and for proper initial conditions. The crux of the problem is to acertain under which conditions (46) will exhibit periodic solutions.

Since μ is small, if a periodic solution exists, it is reasonable to expect that it will be similar to a solution of the harmonic equation

$$\ddot{x} + x = 0 \tag{47}$$

† See [1, 2, 12].

Accordingly, it will be convenient to introduce at the outset a solution of (47), the so-called *generating solution*, with respect to which the expected periodic solution of (46) will be shaped. That is, the difference between the two solutions will be expressed in the form of a series expansion with respect to μ, whose coefficients will be computed. However, since the angular frequency of the periodic solution of (46) is different from the basic frequency $\omega_0 = 1$ of the harmonic oscillator (47), and since, in any case, it is also an unknown function of μ:

(a) We shall change variable t to angular variable

$$\psi = \omega t$$

where ω is the unknown angular frequency of the periodic solution of (46). Then (46) is rewritten

$$\omega^2 \frac{d^2x}{d\psi^2} + x = \mu f\left(x, \omega \frac{dx}{d\psi}\right) \tag{48}$$

(b) We shall expand ω with respect to μ, say

$$\omega = 1 + \sum_{n=1}^{\infty} \mu^n \omega^{(n)} \tag{49}$$

Also it will be convenient to write

$$\omega^2 = 1 + \sum_{n=1}^{\infty} \mu^n \Omega^{(n)} \tag{50}$$

where the expansion (50) can be readily deduced from (49).

Now we shall denote the periodic solution of (46) under study by $x(\psi, \mu)$, since it depends on the parameter μ, and, as mentionned above, we shall introduce its series expansion:

$$x(\psi, \mu) = x^{(0)}(\psi) + \sum_{n=1}^{\infty} \mu^n x^{(n)}(\psi) \tag{51}$$

where $x^{(n)}(\psi)$ have the period 2π in ψ and $x^{(0)}(\psi)$ is the generating solution.

Indeed this procedure requires that μ be sufficiently small, to guarantee the convergence of the series (49)–(51). As a matter of fact, it is very difficult to make the assumption more precise, because the range of μ over which this requirement is fulfilled can hardly be determined.

The method is carried out by determining the series expansion of $dx/d\psi$ from (51), and by substituting it, together with (49), into the right side of (48). We get

$$f\left(x, \omega \frac{dx}{d\psi}\right) = f^{(0)}(\psi) + \sum_{n=1}^{\infty} \mu^n f^{(n)}(\psi) \tag{52}$$

where again $f^{(n)}(\psi)$ have the period 2π in ψ and

$$f^{(0)}(\psi) = f\left[x^{(0)}(\psi), \frac{dx^{(0)}(\psi)}{d\psi}\right] \tag{53}$$

Finally, substituting (50) and (51) in the left side of (48) and taking account of (52) in the right side, we get a set of recursive equations:

$$\frac{d^2x^{(0)}}{d\psi^2} + x^{(0)} = 0 \tag{54a}$$

$$\frac{d^2x^{(1)}}{d\psi^2} + x^{(1)} = -\Omega^{(1)} \frac{d^2x^{(0)}}{d\psi^2} + f^{(0)}(\psi) \tag{54b}$$

$$\frac{d^2x^{(2)}}{d\psi^2} + x^{(2)} = -\Omega^{(2)} \frac{d^2x^{(0)}}{d\psi^2} - \Omega^{(1)} \frac{d^2x^{(1)}}{d\psi^2} + f^{(1)}(\psi) \tag{54c}$$

$$\vdots$$

$$\frac{d^2x^{(n)}}{d\psi^2} + x^{(n)} = F^{(n)}(\psi) \tag{54d}$$

where $F^{(n)}(\psi)$ is a function of $\Omega^{(1)}, \ldots, \Omega^{(n)}$; $x^{(0)}, \ldots, x^{(n-1)}$ and their derivatives.

Before discussing these equations, let us investigate the starting conditions, i.e., the values of $x^{(0)}, x^{(1)}, \ldots, x^{(n)}$ and $x(\psi, \mu)$, at initial time $t = 0$, say at $\psi = 0$. As pointed out earlier, a cyclic trajectory can be obtained *only if* the starting point P^0 in the phase plane is properly chosen.

Indeed, for the sake of convenience we can assume that P^0 is chosen on the x_1 axis, which corresponds to zero initial velocity. Then if the trajectory issuing at P^0 is a limit cycle we get a periodic solution, whereas other trajectories starting at neighboring points, $P_1{}^0$ and $P_2{}^0$, for instance (Fig. 28), are in general noncyclic trajectories; i.e., they do not represent periodic solutions. That is the primary difference between the nonlinear systems as described by (46) (*even when μ is small*) and the linear harmonic oscillator.

Conversely, we can expect to locate the limit cycle by writing the periodicity conditions:

$$x^{(n)}(2\pi) - x^{(n)}(0) = 0 \qquad (n = 0, 1, 2, ...) \tag{55}$$

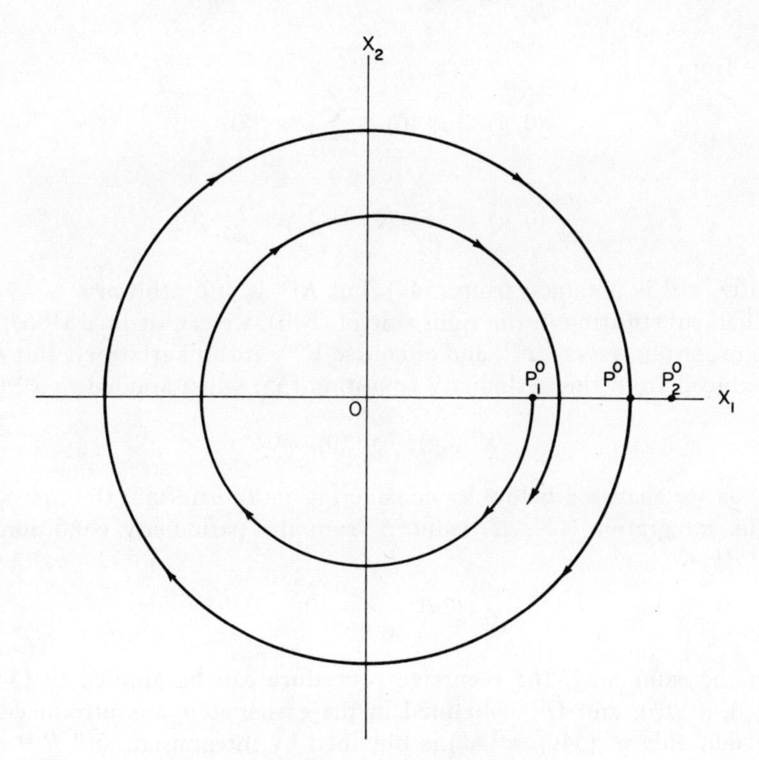

FIG. 28. Cyclic and noncyclic trajectories

together with

$$x(0, \mu) = R(\mu) \qquad \frac{dx}{d\psi}(0, \mu) = 0 \tag{56}$$

which define the expected position of P^0 on the x_1 axis, with zero initial velocity.

Then if we consider the series expansion of $R(\mu)$, say

$$R(\mu) = \sum_{n=1}^{\infty} \mu^n R^{(n)} \tag{57}$$

we see that the starting conditions get in (54) in the form

$$x^{(n)}(0) = R^{(n)} \qquad \frac{dx^{(n)}}{d\psi}(0) = 0 \qquad (n = 0, 1, 2, \ldots) \tag{58}$$

since from (51),

$$x(0, \mu) = x^{(0)}(0) + \sum_{n=1}^{\infty} \mu^n x^{(n)}(0)$$

$$\frac{dx}{d\psi}(0, \mu) = \frac{dx^{(0)}}{d\psi}(0) + \sum_{n=1}^{\infty} \mu^n \frac{dx^{(n)}}{d\psi}(0)$$

Finally, $x^{(0)}$ is obtained from (54*a*), but $R^{(0)}$ is still arbitrary.

Then substituting in the right side of (54b), we can deduce $x^{(1)}(\psi)$ by integration; however, $\Omega^{(1)}$, and of course $R^{(0)}$, are still arbitrary. But $R^{(0)}$ is deduced from the periodicity condition (55) when applied to $x^{(1)}(\psi)$:

$$x^{(1)}(2\pi) - x^{(1)}(0) = 0$$

and, as we shall see below by considering more carefully the first step of the integration, $\Omega^{(1)}$ is deduced from the periodicity condition of $dx^{(1)}/d\psi$:

$$\frac{dx^{(1)}}{d\psi}(2\pi) - \frac{dx^{(1)}}{d\psi}(0) = 0$$

In the same way, the recursive procedure can be applied to (54c): $x^{(0)}(\psi)$, $x^{(1)}(\psi)$, and $\Omega^{(1)}$, obtained in the earlier step, are introduced in the right side of (54c). $x^{(2)}(\psi)$ is obtained by integration, and $R^{(2)}$ and $\Omega^{(2)}$ are again given by the periodicity condition of $x^{(2)}(\psi)$ and $dx^{(2)}/d\psi$, and so on.

For example, let us carry out the recursive process to the first order. Equation (54a) gives

$$x^{(0)} = a \cos \psi + b \sin \psi$$

From (58) we get

$$a = R^{(0)} \qquad b = 0 \qquad x^{(0)} = R^{(0)} \cos \psi$$

which is the generating solution.

We have also

$$\frac{dx^{(0)}}{d\psi} = -R^{(0)} \sin \psi$$

Then by substituting in (54b) we get

$$\frac{d^2x^{(1)}}{d\psi^2} + x^{(1)} = \Omega^{(1)}R^{(0)} \cos \psi + f\,[R^{(0)} \cos \psi, -R^{(0)} \sin \psi]$$

Next, $x^{(1)}(\psi)$ is obtained by usual integration techniques. As a matter of fact, it will be convenient to deduce the solution of this equation by means of Duhamel's formula; if, for example, we consider the more general equation [cf. (54d)]

$$\frac{d^2x^{(n)}}{d\psi^2} + x^{(n)} = F^{(n)}(\psi)$$

the solution is

$$x^{(n)} = X^{(n)}(\psi) + \int_0^{\psi} \sin(\psi - s)F^{(n)}(s)\, ds$$

and

$$\frac{dx^{(n)}}{d\psi} = \frac{dX^{(n)}(\psi)}{d\psi} + \int_0^{\psi} \cos(\psi - s)F^{(n)}(s)\, ds$$

$X^{(n)}(\psi)$ is the general solution of the homogeneous equation.

Now since $X^{(n)}(\psi)$ and $dX^{(n)}(\psi) \mid d\psi$ have period 2π in ψ, the periodicity condition is rewritten

$$x^{(n)}(2\pi) - x^{(n)}(0) = -\int_0^{2\pi} \sin s\, F^{(n)}(s)\, ds = 0$$

$$\frac{dx^{(n)}}{d\psi}(2\pi) - \frac{dx^{(n)}}{d\psi}(0) = \int_0^{2\pi} \cos s\, F^{(n)}(s)\, ds = 0$$

Accordingly, we obtain, at the first step of the integration,

$$F^{(n)}(s) = \Omega^{(1)}R^{(0)} \cos s + f\,[R^{(0)} \cos s, -R^{(0)} \sin s]$$

$$x^{(1)}(2\pi) - x^{(1)}(0) = -\int_0^{2\pi} \sin s\, f\,[R^{(0)} \cos s, -R^{(0)} \sin s]\, ds$$

$$\triangleq -\Phi[R^{(0)}] = 0 \tag{59}$$

$$\frac{dx^{(1)}}{d\psi}(2\pi) - \frac{dx^{(1)}}{d\psi}(0) = \pi\Omega^{(1)}R^{(0)} + \int_0^{2\pi} \cos s\, f\,[R^{(0)} \cos s, -R^{(0)} \sin s]\, ds = 0 \tag{60}$$

A positive solution of (59) is called a *generating amplitude*, and obviously (60) provides $\Omega^{(1)}$. The process can be carried out to obtain higher-order approximations; however, the calculations become rather complicated.

When these results are applied to Van der Pol's equation (29), we find

$$f(x, \dot{x}) \equiv (1 - x^2)\dot{x}$$

$$-f[R^{(0)} \cos s, -R^{(0)} \sin s] \equiv (1 - [R^{(0)}]^2 \cos^2 s)R^{(0)} \sin s$$

$$\Phi[R^{(0)}] \equiv -\pi R^{(0)} \left(1 - \frac{[R^{(0)}]^2}{4}\right) = 0$$

$$\Omega^{(1)} = 0$$

This gives

$$R^{(0)} = 2$$

A second-order approximation can be readily obtained in this case. It leads to

$$R^{(1)} = 0$$

$$\Omega^{(2)} = -\tfrac{1}{8}$$

from which follows

$$\omega^2 = 1 - \frac{\mu^2}{8} + o(\mu^2)$$

5.2. The Van der Pol and Krylov-Bogoliubov Methods†

Independent of this approach there appeared other analytic methods, also devoted to the existence and stability of periodic solutions—one due to Van der Pol and the other to Krylov and Bogoliubov. These two methods are similar in many ways. At the beginning they were mostly confined to a first approximation. Recently the second one was extended and put on a more rigorous basis in a treatise by Bogoliubov and Mitropolsky.

These methods are very convenient in the case of autonomous systems (and rather complicated in the nonautonomous case), so we shall again consider the equation

$$\ddot{x} + x = \mu f(x, \dot{x}) \tag{61}$$

and we shall try to fit the solution

$$x(t) = R(t) \cos \psi(t) \tag{62}$$

$$\dot{x}(t) = -R(t) \sin \psi(t) \tag{63}$$

† See [8, 9, 25].

where $\psi(t) = t + \varphi(t)$. Since

$$\dot{x}(t) = -R(t)\sin\psi(t) + \dot{R}(t)\cos\psi(t) - R\dot{\varphi}(t)\sin\psi(t)$$

this introduces the additional requirement

$$\dot{R}\cos\psi - R\dot{\varphi}\sin\psi = 0 \qquad \forall\, t \tag{64}$$

Now from (63) compute $\ddot{x}(t)$ and substitute its expression, together with (62) and (63), in (61). One gets

$$-\dot{R}\sin\psi - R\dot{\varphi}\cos\psi = \mu f(R\cos\psi, -R\sin\psi) \tag{65}$$

from which, from (64) and (65),

$$\begin{aligned}\dot{R} &= -\mu\sin\psi\, f(R\cos\psi, -R\sin\psi)\\ \dot{\varphi} &= -\mu\,\frac{\cos\psi}{R}\, f(R\cos\psi, -R\sin\psi)\end{aligned} \tag{66}$$

As a matter of fact, (66) describes the motion of the oscillator by means of polar coordinates, and the periodicity condition will be expressed by

$$R(2\pi) - R(0) = 0$$

say

$$-\mu\int_0^{2\pi} \sin s\, f(R\cos s, -R\sin s)\, ds = 0 \tag{67}$$

This is just equation (59), as obtained by the method of Poincaré; that is, the amplitudes $R^{(0)}$ of the steady states are the solutions of

$$\Phi(R^{(0)}) = 0$$

Another interesting feature of the present method is that it also gives valuable information about transient behavior. Indeed, if R and φ are slowly varying functions of time, we can approximate

$$\frac{R(2\pi) - R(0)}{2\pi} \quad \text{by} \quad \dot{R}$$

$$\frac{\varphi(2\pi) - \varphi(0)}{2\pi} \quad \text{by} \quad \dot{\varphi}$$

and rewrite the equations of motion

$$\dot{R} = -\frac{\mu}{2\pi}\,\Phi(R) \tag{68a}$$

$$\dot{\psi} = 1 + \frac{\mu}{2\pi}\,\Lambda(R) \tag{68b}$$

with

$$\Phi(R) \triangleq \int_0^{2\pi} \sin s\, f(R \cos s, -R \sin s)\, ds$$

$$\Lambda(R) \triangleq -\int_0^{2\pi} \frac{\cos s}{R} f(R \cos s, -R \sin s)\, ds$$

This averaging procedure is very important for engineers, since (68) expresses the law which governs changes in amplitude R and the relation between frequency ω and amplitude. Indeed, since ω is a slowly varying function,

$$\dot{\psi} \simeq \omega \qquad \omega = 1 + \frac{\mu}{2\pi} \Lambda(R)$$

Problems of stability of motion can be approached by means of (68), to which we shall come back later, when we shall wish to analyze the effects of random noise on the behavior of self-sustained oscillators, in the near neighborhood of the steady state. In such problems, the amplitude undergoes small deviations from a steady value $R^{(0)}$, and the law of recovering will play a very important role. Indeed we shall assume that the system returns to the steady state after each small disturbance. In this case the steady state is said to be *orbitally asymptotically stable.*

Problems concerning stability will not be considered in this chapter; however, it will be interesting to deduce from (68a) the condition for orbital stability. Let $R^{(0)}$ be a steady value and z a small deviation. Substituting

$$R = R^{(0)} + z$$

into (68a), we obtain

$$\dot{z} = -\frac{\mu}{2\pi} \left[\Phi(R^{(0)}) + \left(\frac{d\Phi}{dR} \right)_0 z + o(z) \right]$$

But $\Phi(R^{(0)}) = 0$, so that we have

$$\dot{z} = -\frac{\mu}{2\pi} \left(\frac{d\Phi}{dR} \right)_0 z + o(z) \tag{69}$$

Then if $\mu > 0$, and

if $\left(\frac{d\Phi}{dR} \right)_0 > 0$ the steady behavior is asymptotically stable

if $\left(\frac{d\Phi}{dR} \right)_0 < 0$ it is unstable

Furthermore, if τ is the time constant of the law of recovering, we deduce, from (69),

$$\frac{1}{\tau} = \frac{\mu}{2\pi}\left(\frac{d\Phi}{dR}\right)_0$$

As long as one is concerned with the first approximation, the method of Krylov-Bogoliubov is more flexible than the one of Poincaré. However, it is less general, since in the theory of Poincaré, higher-order approximations can be obtained from the recursive equations, which is not the case in the above version of the K.-B. method. As pointed out above, the method has been generalized by Bogoliubov and Mitropolsky.

5.3. Stroboscopic Method†

Minorsky and Schiffer have introduced a convenient modification of the K.-B. method, the so-called *stroboscopic method.*

Suppose that (66) is rewritten, more generally,

$$\dot{R} = \mu F(R, \psi) \qquad \dot{\psi} = 1 + \mu G(R, \psi) \tag{70}$$

and consider a trajectory different from a limit cycle.

Since μ is small, amplitude R and angular frequency ω are slowly varying functions of time, so that the trajectory of moving point $P(R, \psi)$ is very similar to a circle during a few revolutions. However during the time interval T_0, $T_0 = 2\pi$, which is *not exactly* the duration of a full revolution, R and ψ experience very small changes:

$$\begin{aligned} \Delta R &\simeq 2\pi\dot{R} = 2\mu\pi F(R, \psi) \\ \Delta\psi &\simeq 2\pi\dot{\psi} = 2\pi + 2\pi\mu G(R, \psi) \end{aligned} \tag{71}$$

If we consider only the sequence of points which is obtained by illuminating the trajectory in the phase plane by stroboscopic flashes occurring at times

$$t_0 \qquad t_0 + 2\pi \qquad t_0 + 4\pi \qquad \cdots \qquad t_0 + 2n\pi \qquad \cdots$$

(n an integer), we see that this continuous trajectory is replaced by a "dotted" path whose points are governed by the discontinuous law

$$\begin{aligned} \delta R &= \Delta R = 2\pi\mu F(R, \psi) \\ \delta\psi &= \Delta\psi - 2\pi = 2\pi\mu G(R, \psi) \end{aligned} \tag{72}$$

† See [17].

Then, since the distances between the successive points are very small, it is convenient to replace the stroboscopic locus by a continuous one governed by the differential equations

$$\frac{dR}{d\tau} = 2\pi F(R, \psi) \qquad \frac{d\psi}{d\tau} = 2\pi G(R, \psi) \tag{73}$$

where $d\tau = \mu$ or

$$\frac{dR}{d\psi} = \frac{F(R, \psi)}{G(R, \psi)} \tag{74}$$

As a matter of fact Minorsky considers $\rho = R^2 = x^2 + \dot{x}^2$ variable instead of R. It represents the total energy of the system. Many interesting properties have been obtained by Minorsky, starting with the stroboscopic equations. Note that the stroboscopic equations can be easily deduced from (68a) and (68b).

5.4. Optimal Linearization Method†

The optimal linearization method introduced in Chapter I, Section 3.5, can be readily applied to (61):

$$\ddot{x} + x - \mu f(x, \dot{x}) = 0$$

Let us try to approach (61) by the linear equation with constant coefficients λ_1 and λ_2,

$$\ddot{x} + \lambda_1 \dot{x} + \lambda_2 x = 0 \tag{75}$$

With the notation of Chapter I we get

$$\epsilon(\lambda_1, \lambda_2) = \lambda_1 \dot{x} + (\lambda_2 - 1)x + \mu f(x, \dot{x})$$

$$\begin{aligned}\overline{\epsilon^2(\lambda_1, \lambda_2)} &= \lambda_1{}^2\overline{(\dot{x})^2} + (\lambda_2 - 1)^2\overline{x^2} + \mu^2\overline{f^2(x, \dot{x})} \\ &\quad + 2\lambda_1(\lambda_2 - 1)\overline{x\dot{x}} + 2\lambda_1\mu\overline{\dot{x}f(x, \dot{x})} + 2\mu(\lambda_2 - 1)\overline{xf(x, \dot{x})}\end{aligned}$$

Then, according to the optimization technique, as discussed earlier, we shall write the following equations:

$$\frac{\partial\overline{\epsilon^2}}{\partial\lambda_1} = 0 \qquad \frac{\partial\overline{\epsilon^2}}{\partial\lambda_2} = 0$$

† See [31].

say

$$\lambda_1\overline{(\dot{x})^2} + (\lambda_2 - 1)\overline{x\dot{x}} + \mu\overline{\dot{x}f(x, \dot{x})} = 0$$
$$(\lambda_2 - 1)\overline{x^2} + \lambda_1\overline{x\dot{x}} + \mu\overline{xf(x, \dot{x})} = 0 \tag{76}$$

Then λ_1 and λ_2 can be easily computed for *any choice* of function $x(t)$.

Now let us start again with a *steady* first approximation of (61), say

$$x = R^{(0)} \cos \psi \qquad \text{with} \quad \psi = \omega t \tag{77}$$

Then

$$\overline{x\dot{x}} = 0$$

and we get from (76),

$$\lambda_1 = -\mu \frac{\overline{\dot{x}f(x, \dot{x})}}{\overline{(\dot{x})^2}} \tag{78}$$

$$\lambda_2 = 1 - \mu \frac{\overline{xf(x, \dot{x})}}{\overline{x^2}} \tag{79}$$

Furthermore, for (77) to be a solution of (75) we must have

$$\lambda_1 = 0$$

say

$$\overline{\dot{x}f(x, \dot{x})} = -R^{(0)} \int_0^{2\pi} \sin s\, f\,[R^{(0)} \cos s,\ -R^{(0)} \sin s]\, ds = 0 \tag{80}$$

and from (79) we deduce

$$\lambda_2 = 1 - \frac{\mu}{\pi R^{(0)}} \int_0^{2\pi} \cos s\, f\,[R^{(0)} \cos s,\ -R^{(0)} \sin s]\, ds \tag{81}$$

These expressions are identical with (59) and (60), which were deduced from the perturbation method. Indeed, (80) is rewritten

$$\Phi(R^{(0)}) = 0$$

and

$$\lambda_2 = 1 + \mu\Omega^{(1)}$$

from which follows

$$\Omega^{(1)} = -\frac{1}{\pi R^{(0)}} \int_0^{2\pi} \cos s\, f\,[R^{(0)} \cos s,\ -R^{(0)} \sin s]\, ds$$

6. SYNCHRONIZATION OF SELF-OSCILLATORS

In Chapter I, Section 7 we analyzed the motion of a nonlinear pendulum driven by a periodic external force. Here we shall consider a similar problem in the case where the oscillator is self-sustained, and we shall see that under certain conditions the frequency of the free oscillation is canceled out, the latter being replaced by a synchronized oscillation, i.e., by an oscillation whose frequency is that of the forcing function.

The synchronization of a triode oscillator can be easily observed and the theoretical conclusions can be experimentally verified:

(a) For a given amplitude of the forcing function the synchronization effect is observed provided that the forcing frequency is close enough to the frequency ω_0 of the oscillator. On the other hand, the larger the amplitude of the forcing function, the greater is the frequency interval over which the synchronization occurs.

(b) For a given frequency of the forcing function, the oscillator is synchronized provided the forcing amplitude is large enough. The closer the frequency of the forcing function to ω_0, the lower is its threshold amplitude.

When these conditions are not fulfilled, one observes the beats of the frequency ω_0 with the frequency of the forcing function.

Most of the classical publications devoted to the synchronization effect are based on the search for a solution of a nonlinear differential equation. Here we shall first discuss a remarkably simple method which was described by J. Van Slooten in a short and very interesting paper [16].

6.1. Van Slooten's Theory

6.1.1. Synchronization by a Sequence of Pulses

One of the advantages of this method, in comparison with the classical derivations, is that it gives a fair insight into the physical nature of the synchronization effect. It is closely related to another method, which we shall sketch later when analyzing the random fluctuations of a self-sustained oscillator in the presence of noise. The basic idea of this method is the following: Instead of analyzing the effect of a sinusoidal forcing function, let us study synchronization by a periodic sequence of pulses.

The circuit which Van Slooten considered is shown in Fig. 29. Assume that the steady voltage, which one would observe in the absence of pulses, would be

$$x(t) = a_0 \sin(\omega_0 t + \varphi) \tag{82}$$

and denote by $\tilde{x}(t)$ the perturbed voltage when the pulses are acting.

Then, as is well known from the elementary theory of the ballistic galvanometer, each pulse causes a sudden change in dx/dt but no discontinuity in x. This results in a change in amplitude and phase angle which we shall denote by δa and $\delta\varphi$, respectively.

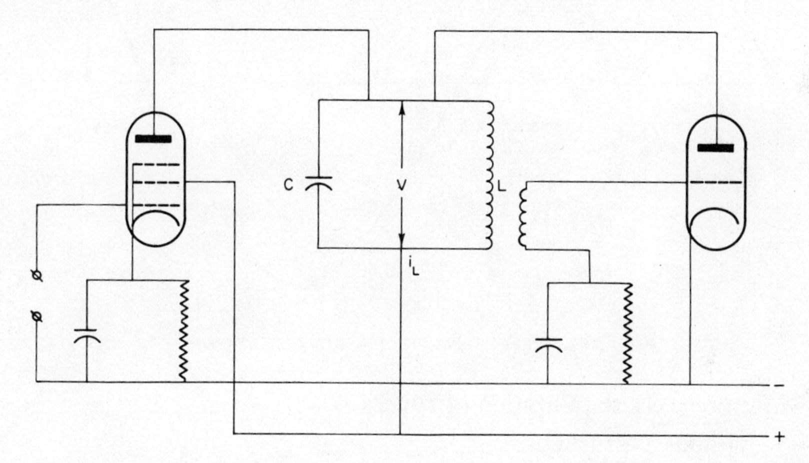

FIG. 29. Circuit for studying the synchronization by a sequence of pulses (from J. Van Slooten).

It may easily be shown that if cycle j begins at time θ_j, and is perturbed by pulse $\mathscr{F}(t - t_j)$ which is assumed to occur at time t_j, then

$$(\delta a)_j = q\omega_0 \cos \omega_0(t_j - \theta_j) \tag{83}$$

$$(\delta\varphi)_j = -q \frac{\omega_0}{a_0} \sin \omega_0(t_j - \theta_j) \tag{84}$$

The symbol q represents the "intensity" of each pulse, say

$$q = \int_{-\infty}^{+\infty} \mathscr{F}(t - t_j)\, dt \qquad q > 0$$

where $\mathscr{F}(t - t_j)$ is the distribution which represents the pulse which occurs at time t_j.

As a matter of fact, the change in amplitude does not play a significant role for the problem we are considering, *provided that it is sufficiently small,* an assumption which will be discussed later.

But, as shown in Fig. 30, the change in phase angle is important, since it can be thought of as a lengthening of the cycle during which the pulse $\mathscr{F}(t - t_j)$ has occurred.

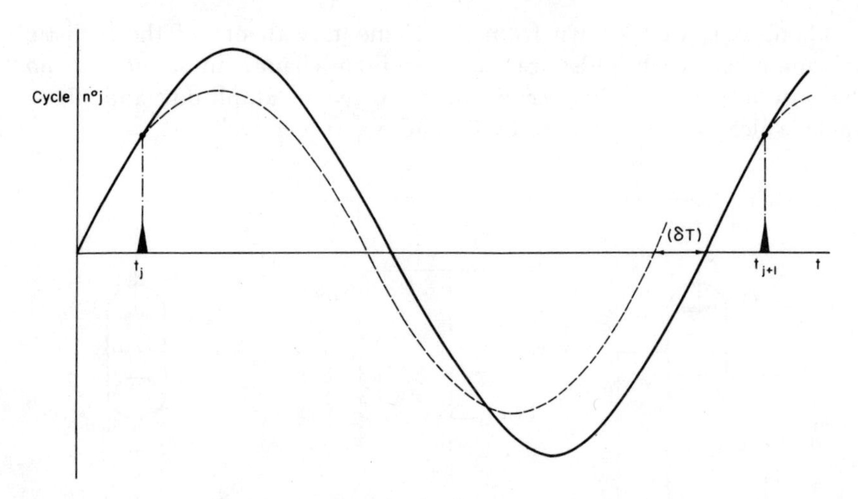

FIG. 30. Synchronization by a sequence of pulses.

More precisely the duration of cycle j is:

(a) Without perturbation: T_0.

(b) When perturbed by a pulse: $T_0 + (\delta T)_j$, where

$$(\delta T)_j = \frac{(\delta\varphi)_j}{\omega_0} \tag{85}$$

Accordingly, the next cycle will begin at time

$$\theta_{j+1} = \theta_j + T_0 + (\delta T)_j$$

If the next pulse, $\mathscr{F}(t - t_{j+1})$, occurs at time

$$t_{j+1} = t_j + T_0 + (\delta T)_j$$

and *if we assume that we can disregard the variation of amplitude* a_0, we have

$$(\delta\varphi)_{j+1} = -q\,\frac{\omega_0}{a_0}\sin\omega_0(t_{j+1} - \theta_{j+1})$$

$$= -q\,\frac{\omega_0}{a_0}\sin\omega_0(t_j - \theta_j)$$

say

$$(\delta\varphi)_{j+1} = (\delta\varphi)_j$$

and

$$(\delta T)_{j+1} = (\delta T)_j$$

The arguments can be iterated, according to which we see that the zero crossings of $\tilde{x}(t)$ from negative to positive values have the same periodicity as the sequence of pulses.

We have shown the existence of a synchronized solution, under certain conditions which we shall now make more precise.

Condition 1. First we will get a relation between the *intensity* q of pulses and the *time lag*

$$\tau_j^* = t_j - \theta_j$$

which is independent of index j. Dropping the index j we deduce from (85) and (84),

$$\delta T = -\frac{q}{a_0} \sin \omega_0 \tau^* \tag{86}$$

Accordingly, if q and δT [or $\delta\omega = -(\omega_0{}^2/2\pi)\,\delta T$] are given, τ^* is determined by (86).

Condition 2. Since

$$|\sin \omega_0 \tau^*| \leqslant 1$$

we have

$$\left|\frac{\delta T}{q}\right| \leqslant \frac{1}{a_0} \quad \text{or} \quad \left|\frac{\delta\omega}{q}\right| \leqslant \frac{\omega_0{}^2}{2\pi a_0}$$

Hence, given the intensity q, the synchronization effect is observed, provided that

$$|\delta\omega| \leqslant \frac{q\omega_0{}^2}{2\pi a_0} \tag{87}$$

and, on the other hand, if the deviation $\delta\omega$ between the frequency of the forcing function and ω_0 is given, the condition is

$$q \geqslant \frac{2\pi a_0}{\omega_0{}^2} |\delta\omega| \tag{88}$$

These are the well-known conclusions.

Note that, at the synchronization threshold we have

$$|\sin \omega_0 \tau^*| = 1$$

say

$$\tau^* = \pm \frac{\pi}{2\omega_0} = \pm \frac{T_0}{4}$$

The meaning of this condition is readily deduced from (84), which shows that the maximum value of $(\delta\varphi)_j$ is obtained for

$$| t_j - \theta_j | = \frac{\pi}{2\omega_0} = \frac{T_0}{4}$$

i.e., when the pulses occur at the maximum deflection of $x(t)$, and thus the maximum change in frequency which can be produced by the impulses is

$$| \delta\omega |_{\max} = -\frac{\omega_0^2}{2\pi} | \delta T |_{\max} \qquad \text{with} \quad | \delta T |_{\max} = \frac{q}{a_0}$$

Accordingly, if the deviation $| \delta T |$ between the period of the sequence of pulses and T_0 is greater than $| \delta T |_{\max}$, there is no longer a possibility of adjustment by operating on the time lag τ^*. This time difference between the pulses and the zero crossings of $\tilde{x}(t)$ will slowly (and periodically) vary.

At last let us discuss the above assumption concerning the changes in amplitude.

Condition 3. A well-known property of regular self-oscillators, which we shall analyze more accurately later is the following: Any deviation from the steady state, provided that it is sufficiently small, tends to decrease with time in such a way that, if no other perturbation occurs, the oscillator will return to the steady state following an exponential law which is characterized by a time constant τ.

Indeed, the exponential law is an approximation in the neighborhood of the steady state, which governs the decrease of small amplitude perturbations.

This is also related to the fact that a self-oscillator, in the neighborhood of the steady state, has the properties of a selective filter whose central angular frequency is ω_0, the frequency of the free oscillation, and whose bandwidth is

$$\Delta\omega \sim \frac{2}{\tau}$$

Accordingly, the small changes in amplitude δa will result in the overall change

$$\Delta a \sim \frac{\delta a}{1 - e^{-T/\tau}}$$

where T is the period of pulses. Then the mechanism discussed above is valid as long as Δa is small with regard to the steady-state amplitude a_0.

6.1.2. Synchronization by a Signal of Arbitrary Wave Form

Since the oscillator, in the neighborhood of steady motion, behaves like a very selective filter, the wave form of the forcing signal is of little consequence. Indeed whatever this form, the oscillator will only admit the frequencies which are sufficiently close to ω_0, the frequency of the free oscillation. If, for instance, the forcing function is a square wave, or a sequence of pulses whose frequency is close to ω_0, the oscillator will strip the harmonics, so that the result will be about the same as if the signal were a sine wave with the same frequency.

Conversely, synchronization by any periodic signal whose frequency is close to ω_0 can be conveniently studied, from a theoretical viewpoint, by replacing the signal by a sequence of pulses with the same periodicity. For example, let the forcing signal be the sine wave

$$E(t) = E_0 \cos \omega t$$

and replace this signal by the sequence of pulses $\mathscr{F}(t - t_j)$, such that q is the intensity of each pulse and the periodicity is the same $T = 2\pi/\omega$. If these pulses occur at times $t_j = jT$ (j an integer), the Fourier series of $\mathscr{F}(t - t_j)$ is

$$\mathscr{F}(t - t_j) = \frac{q}{T} + \frac{2q}{T} \sum_{n=1}^{\infty} \cos n\omega t$$

Accordingly, since the oscillator makes no difference between the fundamental term $(2q/T) \cos \omega t$ and $E_0 \cos \omega t$, we can use the conclusions of the above theory and replace q in it by $(E_0 T/2) \simeq \pi E_0/\omega_0$. Then (87) and (88) are rewritten

$$|\,\delta\omega\,| \leqslant \frac{E_0 \omega_0}{2a_0} \tag{89}$$

$$E_0 \geqslant \frac{2a_0}{\omega_0} |\,\delta\omega\,| \tag{90}$$

Also note that the phase shift between the forcing function $E(t)$ and the synchronized oscillation can be easily deduced from the above considerations. It is

$$\Delta\varphi = \frac{\pi}{2} - \omega_0 \tau^*$$

At the synchronization threshold we have $\omega_0 \tau^* = \pm\pi/2$, and accordingly its value is either

$$\Delta\varphi = 0 \qquad \text{or} \qquad \Delta\varphi = \pi$$

Then the forcing function and the synchronized solution are either in phase or out of phase.

6.2. Van der Pol's Theory[†]

If one introduces a sinusoidal voltage generator $E_0 \sin \omega t$ in the tank circuit of the oscillator shown in Fig. 3, in series with the components L, r, and C, the equation of motion (21) becomes

$$LC\ddot{x} + L\left(G - mS_0 - \frac{3mS_2}{4} a_0^2\right)\dot{x} + x = E_0 \sin \omega t \tag{91}$$

or

$$\ddot{x} - 2\mu\left(1 - \frac{a_0^2}{4V_0^2}\right)\dot{x} + \omega_0^2 x = \omega_0^2 E_0 \sin \omega t \tag{92}$$

with

$$\omega_0^2 = \frac{1}{LC}$$

$$\mu = -\frac{G - mS_0}{2C} \qquad (\mu \text{ is assumed to be positive and very small})$$

$$V_0^2 = \frac{G - mS_0}{3mS_2}$$

As a first approximation, Van der Pol tries to find a solution of the form

$$x = a_1 \sin \omega t + a_2 \cos \omega t \tag{93}$$

where a_1 and a_2 are assumed to be slowly varying functions of time. Therefore, $\dot{a}_1$ and $\dot{a}_2$ will be considered of first-order smallness, and $\ddot{a}_1$ and $\ddot{a}_2$ will be neglected when substituting (93) in (92); this substitution leads to the following equations:

$$\begin{aligned} \dot{a}_1 + a_2\delta\omega - \mu\left(1 - \frac{a_0^2}{4V_0^2}\right) a_1 &= 0 \\ \dot{a}_2 - a_1\delta\omega - \mu\left(1 - \frac{a_0^2}{4V_0^2}\right) a_2 &= -\frac{\omega_0 E_0}{2} \end{aligned} \tag{94}$$

where $\delta\omega = \omega_0 - \omega$ has also been assumed sufficiently small.

In general, since a_1 and a_2 are slowly varying functions of time, $x(t)$ will be almost periodic. However, it may be that there exists a periodic solution, the synchronized solution, as can be seen by putting $\dot{a}_1 = \dot{a}_2 = 0$ in equations (94).

[†] See [5].

Then we deduce, from (94),

$$a_0^2 \left[(\delta\omega)^2 + \mu^2 \left(1 - \frac{a_0^2}{4V_0^2}\right)^2\right] = \frac{E_0^2\omega_0^2}{4} \tag{95}$$

since

$$a_0^2 = a_1^2 + a_2^2.$$

The existence of the synchronized solution can be discussed by rewriting (94) in the form

$$\dot{a}_1 = f_1(a_1, a_2) \qquad \dot{a}_2 = f_2(a_1, a_2) \tag{96}$$

Then we see that a synchronized solution exists provided that system (96) has a singular point, namely if there exist a_1 and a_2 such that

$$f_1(a_1, a_2) = f_2(a_1, a_2) = 0 \tag{97}$$

But their existence is not the only condition for a periodic solution to be a proper one in our problem; moreover their stability must be investigated.

A conclusion can be obtained by writing the variational equations of (96), namely by giving a_1 and a_2 small variations ξ and η in the neighborhood of a solution of (96) and considering the equations that govern these new variables:

$$\dot{\xi} = \frac{\partial f_1}{\partial a_1}\xi + \frac{\partial f_1}{\partial a_2}\eta$$

$$\dot{\eta} = \frac{\partial f_2}{\partial a_1}\xi + \frac{\partial f_2}{\partial a_2}\eta$$

Here we shall not carry on the whole calculation, but it can be performed easily, and leads to the following second-order equation:

$$\ddot{\xi} - 2\mu\left(1 - \frac{a_0^2}{2V_0^2}\right)\dot{\xi} + \left[\mu^2\left(1 - \frac{a_0^2}{4V_0^2}\right)\left(1 - \frac{3a_0^2}{4V_0^2}\right) + (\delta\omega)^2\right]\xi = 0 \tag{98}$$

where $\xi = \xi_1, \xi_2$.

The discussion of stability follows in a straightforward manner. ξ is a decreasing function of t if and only if

$$a_0^2 > 2V_0^2 \tag{99}$$

and

$$\mu^2\left(1 - \frac{a_0^2}{4V_0^2}\right)\left(1 - \frac{3a_0^2}{4V_0^2}\right) + (\delta\omega)^2 > 0 \tag{100}$$

If these conditions are fulfilled, any deviation from the synchronized state will tend to decrease with time in such a way that, if no other perturbation occurs, the oscillation will return to the synchronized steady state. Accordingly, the solution is a stable one. Otherwise it is unstable.

Now let us assume that the amplitude E_0 of the forcing function is small. In the limiting case where $E_0 = 0$, $\delta\omega = 0$, we get from (95),

$$a_0^2 = 0 \qquad a_0^2 = 4V_0^2$$

The first one is unstable; the second one gives the stabilized amplitude of the free oscillation.

If E_0 and $\delta\omega$ are different from zero but small, under certain conditions we obtain three solutions which can be deduced by continuity from the above ones. The smaller one is unstable, since it does not verify (99), and the other ones are close to $4V_0^2$.

From (95) we get the condition

$$\frac{E_0^2\omega_0^2}{4} \geqslant a_0^2(\delta\omega)^2$$

from which follows

$$|\,\delta\omega\,| \leqslant \frac{E_0\omega_0}{2a_0} \tag{101}$$

That is condition (89) obtained above.

When the amplitude of the forcing function is large, another inequality is deduced from the stability condition,

$$\frac{1}{\sqrt{2}}\left|\frac{\omega_0^2 - \omega^2}{\omega^2}\right| < \frac{E_0}{2V_0}$$

The above discussion is summarized in Fig. 31, in which the coordinates are

$$u = \frac{\delta\omega}{\mu} \qquad v = \frac{a_0^2}{4V_0^2}$$

Equation (95) is rewritten in the reduced form

$$v[u^2 + (1 - v)^2] = F^2 \qquad \text{with} \quad F = \frac{E_0\omega_0}{4V_0\mu} \tag{102}$$

and a family of curves represented by (102) is shown for different values of constant F. When $F = 0$ the curve of the family is degenerated into the line $v = 0$ and the point $u = 0$, $v = 1$. When F increases, the curves

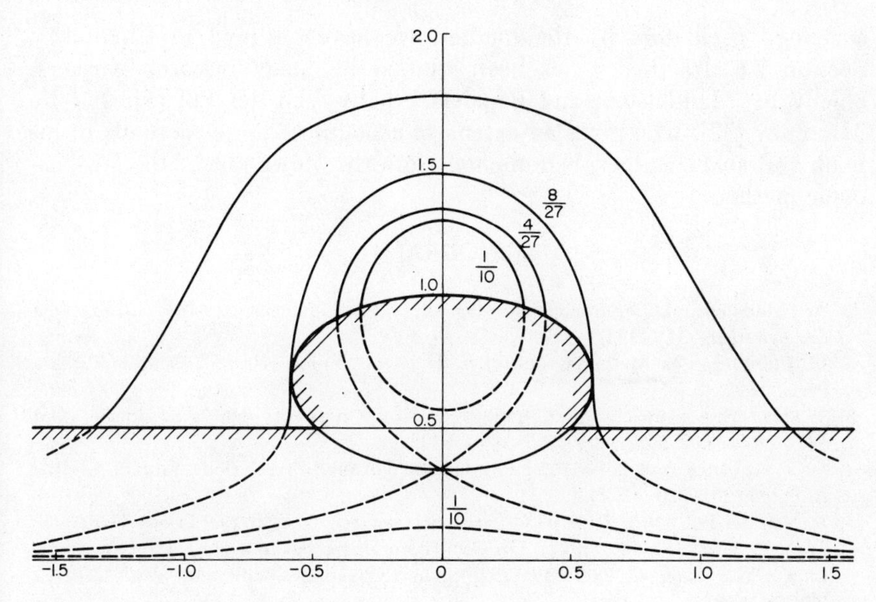

FIG. 31. Existence and stability diagram for synchronized solutions of a Van der Pol oscillator. Coordinates u, v.

first consist of two branches, up to the value for which the two branches join. When F is further increased, the curves consist of a single branch.

On Fig. 31 the line $v = \frac{1}{2}$ is shown together with the ellipse whose equation is

$$(1 - v)(1 - 3v) + u^2 = 0$$

They correspond respectively to the limiting conditions

$$1 - \frac{a_0^2}{2V_0^2} = 0$$

and

$$\mu^2 \left(1 - \frac{a_0^2}{4V_0^2}\right)\left(1 - \frac{3a_0^2}{4V_0^2}\right) + (\delta\omega)^2 = 0$$

The instability region is the shaded area in Fig. 31.

7. SUBHARMONIC RESPONSE

We should not close this chapter without mentioning the subharmonic response to a given periodic excitation, since this phenomenon is very important in practical applications. It can be studied in the case of self-sustained oscillators by the method we have outlined in Chapter I, Section 7.6. Its theory has been studied by many research workers, chiefly by Mandelstam and Papalexi [7], by Van der Pol [5], and by Minorsky [32], who gives an extensive account of these methods in his book and analyzes the phenomenon from the viewpoint of the stroboscopic method.

BIBLIOGRAPHY

1. A. Lindstedt, Differentialgleichungen der Störungstheorie. *Mem. Acad. Imp. St. Petersburg* **31** (1883).
2. H. Poincaré, Les Méthodes nouvelles de la mécanique céleste." Gauthier-Villars, Paris, 1892.
3. J. Haag, Sur la synchronisation d'un système à plusieurs degrés de liberté. *Ann. l'Ecole Normale Supérieure,* 285–338 (1917).
4. E. V. Appleton and B. Van der Pol, On the Form of Free Triode Vibrations. *Phil. Mag.* **42**, 201–220 (1921).
5. B. Van der Pol, On a Type of Oscillation Hysteresis in a Simple Triode Generator. *Phil. Mag.* **43**, 177–193 (1922); On Oscillation Hysteresis in a Simple Triode Generator. *Phil. Mag.* **43**, 700–719 (1926); On Relaxation Oscillations. *Phil. Mag.* **2**, 978–992 (1926).
6. A. Lienard, Etude des oscillations entretenues. *Rev. Gen. l'Electricité* **23**, 901 (1928).
7. L. Mandelstam and N. Papalexi, Über Resonanzerscheinungen bei Frequenzteilung. *Z. Physik*, **73**, 223–248 (1932).
8. B. Van der Pol, The Nonlinear Theory of Electric Oscillations. *Proc. IRE* **22**, 1051–1086 (1934).
9. N. Krylov and N. Bogoliubov, "Introduction to Nonlinear Mechanics." Princeton Univ. Press, Princeton, N.J., 1943. (First published in Russian in 1937.)
10. V. S. Ivanov, Begründung einer Hypothese von Van der Pol aus der Theorie der Selbstschwingungen. *Uch. Zap. Leningr. Gos., Ser. Mat.* **40** (1940).
11. N. Levinson and O. K. Smith, A general equation for relaxation oscillations. *Duke Math. J.* **9** (1942).
12. K. O. Friedrichs, On nonlinear vibrations of third order. *J. Inst. Math. and Mechanics*, New York University, New York (1946).
13. J. Haag, Sur la stabilité des solutions de certains systèmes différentiels. *Bull. Sci. Math.* (1946).
14. A. D. Dragilev, *Prikl. Mat. i Mekhan.* **16** (1949).
15. G. Sansone, Sopra l'equazione di A. Liénard per le oscilliazioni di rilassamento. *Ann. Math. Pura e Appl.* **28**, 153–181 (1949).
16. J. Van Slooten, On Synchronisation of LC Oscillators. *Electronic Application Bull.* **12**, No. 617 (1951). N.V. Philips' Gloeilampenfabrieken Eindhoven. Electronic tube division.
17. N. Minorsky, Sur une équation différentielle de la physique. *Compt. Rend.* **232** (1951); *Rend. Accad. Sci. Bologna* (1952).
18. A. Blaquière, L'Effet du bruit de fond sur la fréquence d'un oscillateur à lampe. *Compt. Rend.* **234**, 419–421 (1952); L'Effet du bruit de fond sur l'amplitude d'un oscillateur à lampe. *Compt. Rend.* **234**, 710–712, 1140–1142 (1952).
19. R. Conti, Soluzioni periodiche dell'equazione di Liénard generalizzata. Esistenza ed Unicità. *Bull. Un. Mat. Italiana* **7**, 111–118 (1952).

20. J. Haag, "Les Mouvements vibratoires." Presses Universitaires de France, Paris, 1952.
21. A. Blaquière, Effet du bruit de fond sur la fréquence des auto-oscillateurs à lampes. Précision ultime des horloges radioélectriques. *Ann. Radioelec.* **8** (1953).
22. A. de Castro, Soluzioni periodiche di una equazione differenziale del secondo ordine. *Bull. Un. Mat. Italiana* **8**, 26–29 (1953).
23. R. Chaleat, Recherches sur la synchronisation. *Ann. Franc. Chronométrie* (1954); La théorie générale de la synchronisation du Professeur J. Haag. *Actes du Congrès Intern. de Chronométrie, Paris* **1**, 295–312 (1954).
24. S. Lefschetz, *Proc. Fifth Symp. Appl. Math.* **5** (1954).
25. N. Bogoliubov and Y. A. Mitropolsky, "Asymptotic Methods in the Theory of Nonlinear Oscillations," Moscow, 1958. (English transl., Gordon and Breach, New York, 1961).
26. F. N. H. Robinson, Nuclear Resonance Absorption Circuit. *J. Sci. Instr.* **36**, 481–487 (1959).
27. N. Minorsky, Méthode stroboscopique et ses applications. *Cahiers Phys.* No. **119**, Paris, 1960.
28. F. L. Stumpers, Balth. van der Pol's Work on Nonlinear Circuits. *IRE Trans. Circuit Theory* **7**, 366 (1960).
29. R. Chaleat, Synchronisation d'un oscillateur nonisochrone. *Symp. intern. sur les oscillations nonlinéaires, Kiev*, p. 458, 1961.
30. A. Blaquière and P. Grivet, L'Effet non linéaire du bruit blanc et du bruit de scintillation dans les spectromètres à résonance nucléaire, du type oscillateur marginal. *Arch. Sci. Geneva, Spec. Issue* (1961).
31. A. Blaquière, Une nouvelle méthode de linéarisation locale des opérateurs non-linéaires; approximation optimale. *Second Conf. Nonlinear Vibrations, Warsaw*, 1962.
32. N. Minorsky, "Nonlinear Oscillations." Van Nostrand, Princeton, N.J., 1962.
33. A. Blaquière and P. Grivet, Nonlinear Effects of Noise in Electronic Clocks. *Proc. IEEE, Spec. Intern. Issue*, p. 1606, Nov. 1963.
34. R. Reissig, G. Sansone, and R. Conti, "Qualitative Theorie Nichtlinearer Differentialgleichungen." Edizioni Cremonese, Rome, 1963.
35. J. Groszkowski, "Frequency of Self-Oscillations." Polish Scientific Publishers, Warsaw, 1964. (English trans., Pergamon Press, New York, 1964.)
36. Y. Nishikawa, "A Contribution to the Theory of Nonlinear Oscillations." Nippon Printing and Publishing Co., Osaka, 1964.
37. T. E. Stern, "Theory of Nonlinear Networks and Systems. An Introduction." Addison-Wesley, Reading, Massachusetts, 1965.

CHAPTER III

Classification of Singularities

1. SINGULAR POINTS

Consider an autonomous system of differential equations

$$\dot{x}_1 = f_1(x_1, x_2) \qquad \dot{x}_2 = f_2(x_1, x_2) \tag{1}$$

where f_1 and f_2 are continuously differentiable with respect to x_1, x_2. As pointed out earlier, it defines a vector field in the phase plane $R^2(x_1, x_2)$, a field of velocity vectors

$$\mathbf{f}(\mathbf{x})$$

the $\mathbf{x}$ vector with components x_1 and x_2 and the $\mathbf{f}$ vector with components f_1 and f_2.

The lines of force of this field are integral curves of system (1); they will be designated belows by Γ. At each point $P(\mathbf{x})$ of Γ, vector $\mathbf{f}(\mathbf{x})$ is tangent to Γ, except at a point for which

$$f_1(x_1, x_2) = f_2(x_1, x_2) = 0$$

Such a point, where $\mathbf{f}(\mathbf{x})$ vanishes, is called a *singular point*.

First of all we shall establish a classification of singular points based on the following: Characterize the position of a singular point P^0 by the vector $\mathbf{x}^0$ with components $x_1{}^0$ and $x_2{}^0$, and consider a point P in the neighborhood of P^0, namely $P^0 \triangleq P(\mathbf{x}^0)$ and $P \triangleq P(\mathbf{x}^0 + \epsilon\boldsymbol{\eta})$. $\boldsymbol{\eta}$ is a unit vector with components η_1 and η_2 and ϵ a positive parameter of first-order smallness. We have†

$$\begin{aligned} f_1(x_1{}^0 + \epsilon\eta_1, x_2{}^0 + \epsilon\eta_2) &= \epsilon\left(\frac{\partial f_1}{\partial x_1}\right)_0 \eta_1 + \epsilon\left(\frac{\partial f_1}{\partial x_2}\right)_0 \eta_2 + o(\epsilon) \\ f_2(x_1{}^0 + \epsilon\eta_1, x_2{}^0 + \epsilon\eta_2) &= \epsilon\left(\frac{\partial f_1}{\partial x_1}\right)_0 \eta_1 + \epsilon\left(\frac{\partial f_2}{\partial x_2}\right)_0 \eta_2 + o(\epsilon) \end{aligned} \tag{2}$$

where $[o(\epsilon)/\epsilon] \to 0$ uniformly as $\epsilon \to 0$.

† $(\partial f_1/\partial x_1)_0$, $(\partial f_1/\partial x_2)_0$, ... means that the partial derivatives are computed at the point $P^0(\mathbf{x}^0)$.

Then define vector $\mathbf{t}$:

$$\mathbf{t} = \lim_{\epsilon \to 0} \frac{1}{\epsilon} \mathbf{f}(\mathbf{x}^0 + \epsilon \boldsymbol{\eta})$$

From (2) it follows that

$$\mathbf{t} = A\boldsymbol{\eta}$$

where A is the linear operator defined by the matrix

$$\frac{D(f_1, f_2)}{D(x_1, x_2)} = \begin{pmatrix} \left(\dfrac{\partial f_1}{\partial x_1}\right)_0 & \left(\dfrac{\partial f_1}{\partial x_2}\right)_0 \\ \left(\dfrac{\partial f_2}{\partial x_1}\right)_0 & \left(\dfrac{\partial f_2}{\partial x_2}\right)_0 \end{pmatrix} \qquad (D, \text{jacobian}) \tag{3}$$

Henceforth, all vectors will be taken to be column vectors. For example,

$$\mathbf{t} = \begin{pmatrix} t_1 \\ t_2 \end{pmatrix} \qquad \boldsymbol{\eta} = \begin{pmatrix} \eta_1 \\ \eta_2 \end{pmatrix}$$

and we will identify the above linear operator with its representative matrix, which we will designate A. Its characteristic equation is

$$\lambda^2 - \lambda\Theta + \Delta = 0 \tag{4}$$

where

$$\Theta = \text{trace of } A = \left(\frac{\partial f_1}{\partial x_1}\right)_0 + \left(\frac{\partial f_2}{\partial x_2}\right)_0$$

$$\Delta = \text{determinant of } A = \left(\frac{\partial f_1}{\partial x_1}\right)_0 \left(\frac{\partial f_2}{\partial x_2}\right)_0 - \left(\frac{\partial f_1}{\partial x_2}\right)_0 \left(\frac{\partial f_2}{\partial x_1}\right)_0$$

The nature of the eigenvectors and eigenvalues at point P^0 will characterize the "structure" of the vector field *at that point*. Indeed in the neighborhood of P^0 the arrangement of the vectors and lines of force will retain the stamp of this structure.

The most typical cases are shown in Fig. 1. In Fig. 1a, b, and c the eigenvectors and the corresponding eigenvalues λ_1 and λ_2 are both real. Furthermore:

(a) In fig. 1a: $\lambda_1\lambda_2 < 0$; P^0 is called a *saddle point*.

(b) In figs. 1b and c: $\lambda_1\lambda_2 > 0$; P^0 is called a *nodal point*.

We will call a nodal point

convergent	if both $\lambda_1, \lambda_2 < 0$
divergent	if both $\lambda_1, \lambda_2 > 0$

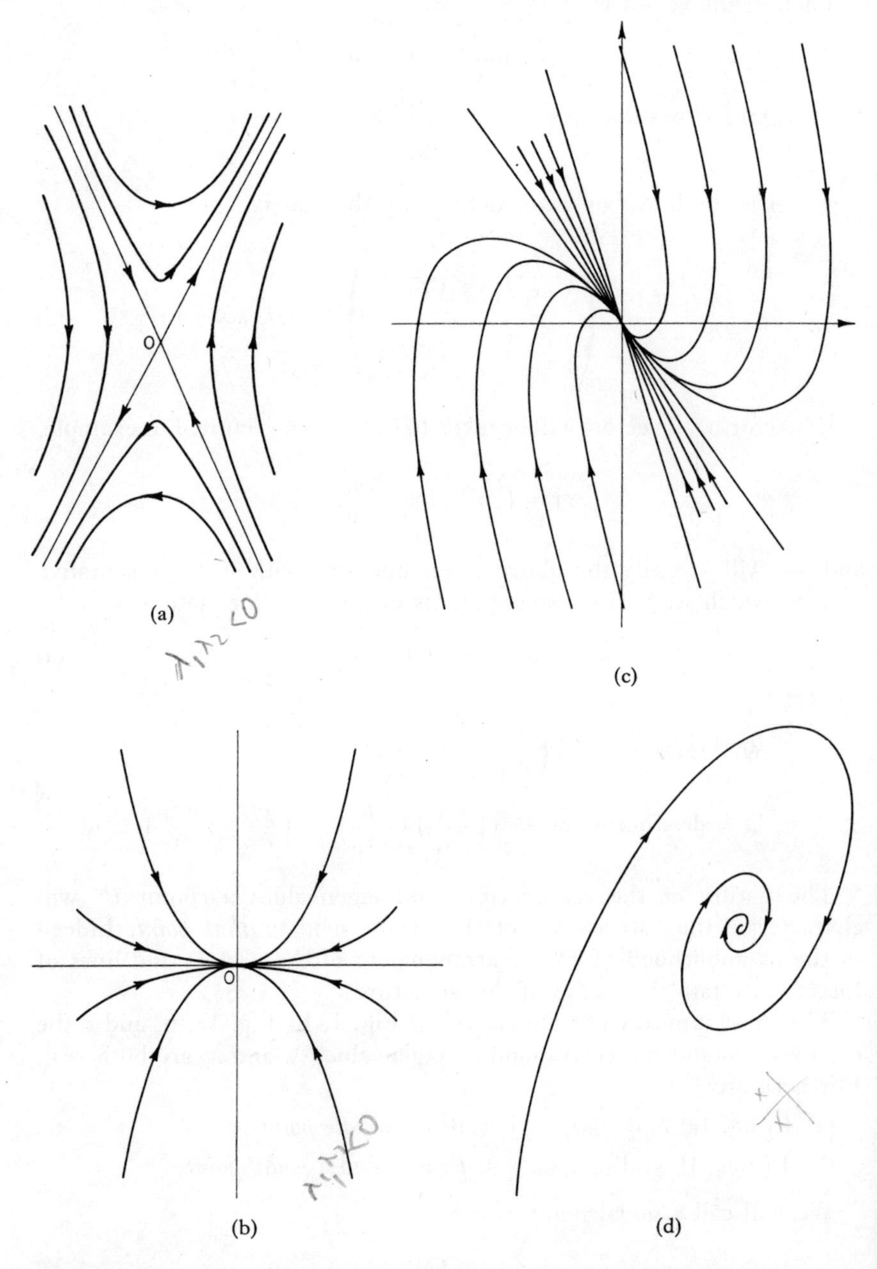

FIG. 1. Lines of force: (a) saddle point; (b) and (c) convergent nodal points; (d) convergent focal point.

Convergence or divergence is indicated by the arrows in Fig. 1b and c, whose directions are defined, on each trajectory, by the directions into which the tangent vectors $\mathbf{f}(\mathbf{x})$ are pointing. In Fig. 1d the eigenvectors are imaginary, and the eigenvalues λ_1 and λ_2 are complex conjugate; P^0 is called a *focal point*. In general, P^0 in this case is a spiral point surrounded by winding trajectories. If λ_1 and λ_2 have positive real parts, the spiral is divergent from P^0; then the singular point is called a *divergent focal point*. If their real parts are negative, the spiral is convergent to P^0, and the singular point is called a *convergent focal point*.

In the marginal case, where λ_1 and λ_2 are purely imaginary, trajectories are closed loops around P^0. Then P^0 belongs to the family of *centers* (Fig. 2).

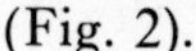

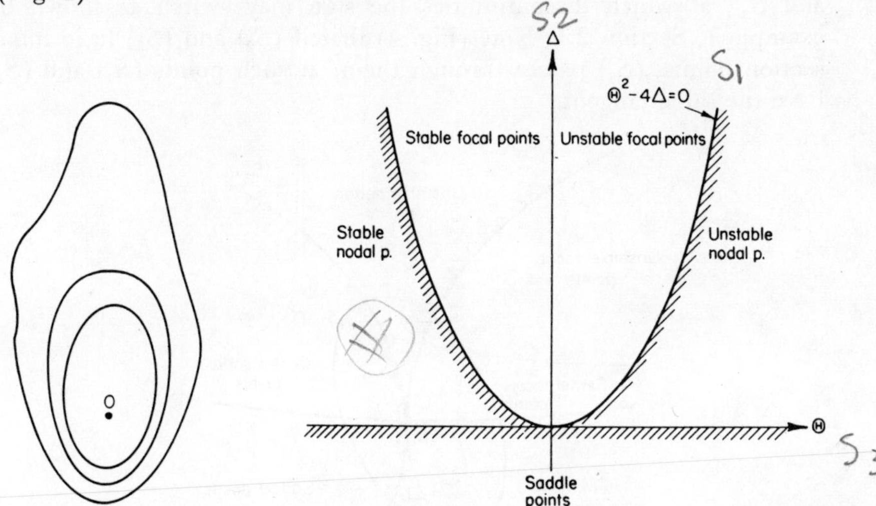

FIG. 2. A center.

FIG. 3. Classification of singularities in the plane Θ, Δ.

Leaving out the details of the discussion, we shall summarize these results.

Case 1. $\Theta^2 - 4\Delta > 0$:

(a) If $\Delta < 0$: P^0 is a *saddle point*.

(b) if $\Delta > 0$: P^0 is a *nodal point*: divergent if $\Theta > 0$, convergent if $\Theta < 0$.

Case 2. $\Theta^2 - 4\Delta < 0$: P^0 is a *focal point*: divergent if $\Theta > 0$, convergent if $\Theta < 0$.

The two cases are illustrated in Fig. 3, in which the different domains have been portrayed with respect to the axis Δ, Θ.

2. DISTRIBUTION OF SINGULAR POINTS IN PHASE PLANE R^2

From the above cases it follows that three separatrix lines, (S_1), (S_2), and (S_3), will play an important role:

$$(S_1): \qquad S_1(x_1^0, x_2^0) \triangleq \Theta^2 - 4\Delta = 0$$

$$(S_2): \qquad S_2(x_1^0, x_2^0) \triangleq \Theta = 0$$

$$(S_3): \qquad S_3(x_1^0, x_2^0) \triangleq \Delta = 0$$

It is also necessary to take account of possible discontinuities of S_1, S_2, and S_3, at which discontinuities the sign may switch as shown by example 1, Section 2.1. Note (Fig. 4) that if (S_2) and (S_3) have intersection points, (S_1) passes through them; at such points (S_1) and (S_3) have the same tangent.

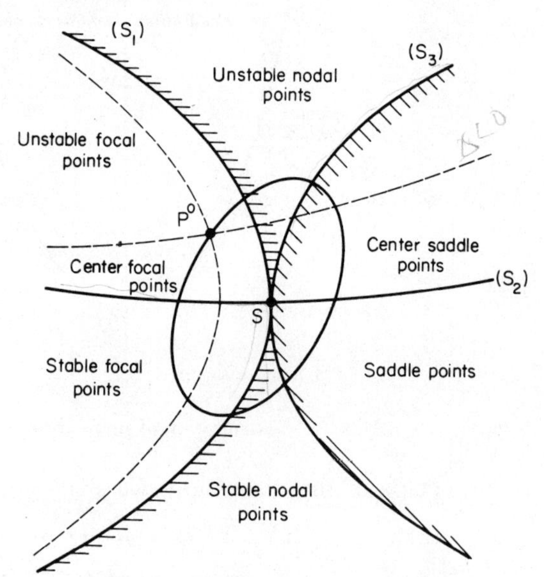

FIG. 4. Distribution of singular points in a phase plane.

Let S be an intersection point of (S_2) and (S_3); we have at that point

$$\frac{\partial S_1}{\partial x_1^0} = 2\Theta \frac{\partial \Theta}{\partial x_1^0} - 4 \frac{\partial \Delta}{\partial x_1^0} = -4 \frac{\partial \Delta}{\partial x_1^0}$$

$$\frac{\partial S_1}{\partial x_2^0} = 2\Theta \frac{\partial \Theta}{\partial x_2^0} - 4 \frac{\partial \Delta}{\partial x_2^0} = -4 \frac{\partial \Delta}{\partial x_2^0}$$

and, accordingly,

$$\left(\frac{\partial S_1}{\partial x_1{}^0} \Big/ \frac{\partial S_1}{\partial x_2{}^0}\right) = \left(\frac{\partial \Delta}{\partial x_1{}^0} \Big/ \frac{\partial \Delta}{\partial x_2{}^0}\right)$$

These lines (including the lines of discontinuity of S_1, S_2, and S_3) separate phase plane R^2 into different regions, each of which is the locus of singular points of a given kind.

Note that all points of (S_2) do not correspond to imaginary eigenvectors. Indeed, in the region of focal points the centers have the "structure" shown in Fig. 2. In this case they are called "*center-focal*" *points*. On the other hand, in the region of saddle points, singular points along (S_2) are defined by $\Delta < 0$, $\Theta = 0$. The eigenvalues λ_1 and λ_2, are both real:

$$\lambda_1 = +\sqrt{-\Delta} \qquad \lambda_2 = -\sqrt{-\Delta}$$

Such points are called "*center-saddle*" *points*. They are quite similar to ordinary saddle points. More generally, along each separatrix we find singular points which hold a plurality of structures.

Figure 4 is very interesting in that it gives an over-all picture of the geometric properties of the system under study. Then singular points can be examined in this context.

As a matter of fact, if the system has singular points and if we plot curves (F_1) and (F_2) separately,

$$(F_1): \quad f_1(x_1, x_2) = 0 \qquad (F_2): \quad f_2(x_1, x_2) = 0$$

we get intersection points, the singular points, which are located in some of the domains just defined. For example, in Fig. 4, P^0 is a divergent focal point.

Furthermore, if the system depends on a parameter which can be modified on demand, it will be beneficial to consider the corresponding path of P^0 on Fig. 4, by means of which the different behaviors can be easily analyzed. One can even modify the parameter to obtain an anticipated result.

This method was developed by T. Vogel and L. Sideriades [7, 9], whose work we shall follow in this chapter.

2.1. Example 1: Center-Saddle Points and Focal Points

The distribution of singular points in R^2 is fairly well illustrated by the following example [9]:

$$\dot{x}_1 = 1 - x_2 \log |x_1| \qquad \dot{x}_2 = x_1(x_1{}^2 + 2x_2 - a^2)$$

Then curves (F_1) and (F_2) pass through the origin 0, which is a singular point. It is possible to compute parameter a^2 in such a way that two other singular points, B and C, lie on separatrix (S_2), which is a parabola, in which case they are "*center-saddle*" *points*. There exist two other singular points, D and E, which are *focal points*. D is *convergent* and E is *divergent*.

In this example, the equations of (S_1), (S_2), (S_3) are

$$(S_1): \quad (x_2 - 2x_1^2)^2 - 4x_1^2[(3x_1^2 + 2x_2 - a^2) \log |x_1| - 2x_2] = 0$$

$$(S_2): \quad 2x_1 - x_2/x_1 = 0$$

$$(S_3): \quad (3x_1^2 + 2x_2 - a^2) \log |x_1| - 2x_2 = 0$$

with $a^2 = 8.8159$. $x_1 = 0$ is a line of discontinuity of S_2, namely the sign of $2x_1 - x_2/x_1$ switches on the parabola $2x_1^2 - x_2 = 0$ and on the axis $x_1 = 0$.

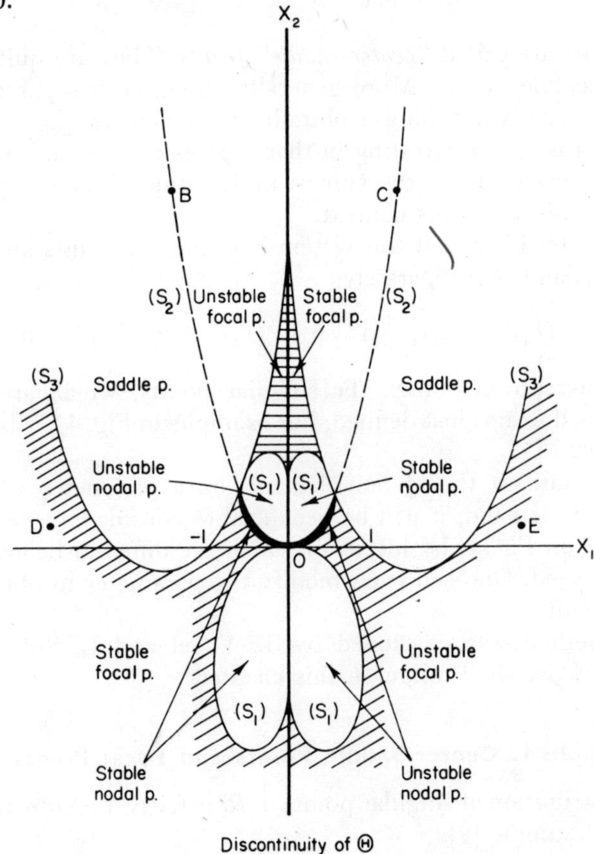

FIG. 5. Example 1: Center-saddle points and focal points.

The different domains are shown in Figs. 5 and 6. It is interesting to look at the connection between these diagrams and a set of trajectories obtained by the isoclynes' method (Fig. 7).

2.2. Example 2: Method of Andronov and Witt†

Here we shall return to the synchronization effect, which was discussed in Chapter II, Section 6.2, following the theory of Van der Pol, and we shall now apply to its analysis the method of Andronov and Witt, which also provides an illustration of the above ideas.

As pointed out in Chapter II, Section 6.2, a self-oscillator can be synchronized by a periodic excitation, provided that the following system of differential equations has a singular point, and provided that this point be a stable one:

$$\dot{a}_1 = f_1(a_1, a_2) \qquad \dot{a}_2 = f_2(a_1, a_2)$$

with

$$f_1(a_1, a_2) = \mu \left(1 - \frac{a_1^2 + a_2^2}{4V_0^2}\right) a_1 - a_2 \delta\omega$$

$$f_2(a_1, a_2) = -\frac{\omega_0 E_0}{2} + a_1 \delta\omega + \mu \left(1 - \frac{a_1^2 + a_2^2}{4V_0^2}\right) a_2$$

The notation is explained in Chapter II, Section 6.2, according to which we shall put

$$u = \frac{\delta\omega}{\mu} \qquad v = \frac{a_1^2 + a_2^2}{4V_0^2} \qquad F = \frac{E_0\omega_0}{4V_0\mu}$$

Furthermore, let $x = a_1/2V_0$, $y = a_2/2V_0$, and change variable t into $\theta = \mu t$.

Hence the basic equations are rewritten in the reduced form

$$\frac{dx}{d\theta} = \varphi_1(x, y) \qquad \frac{dy}{d\theta} = \varphi_2(x, y)$$

with

$$\varphi_1(x, y) = (1 - v)x - uy$$

$$\varphi_2(x, y) = ux + (1 - v)y - F$$

$$v = x^2 + y^2$$

Since the conditions for the existence and the stability of a singular point play the essential role, Andronov and Witt determine separatrices (S_1), (S_2), and (S_3), as outlined above.

† See [4].

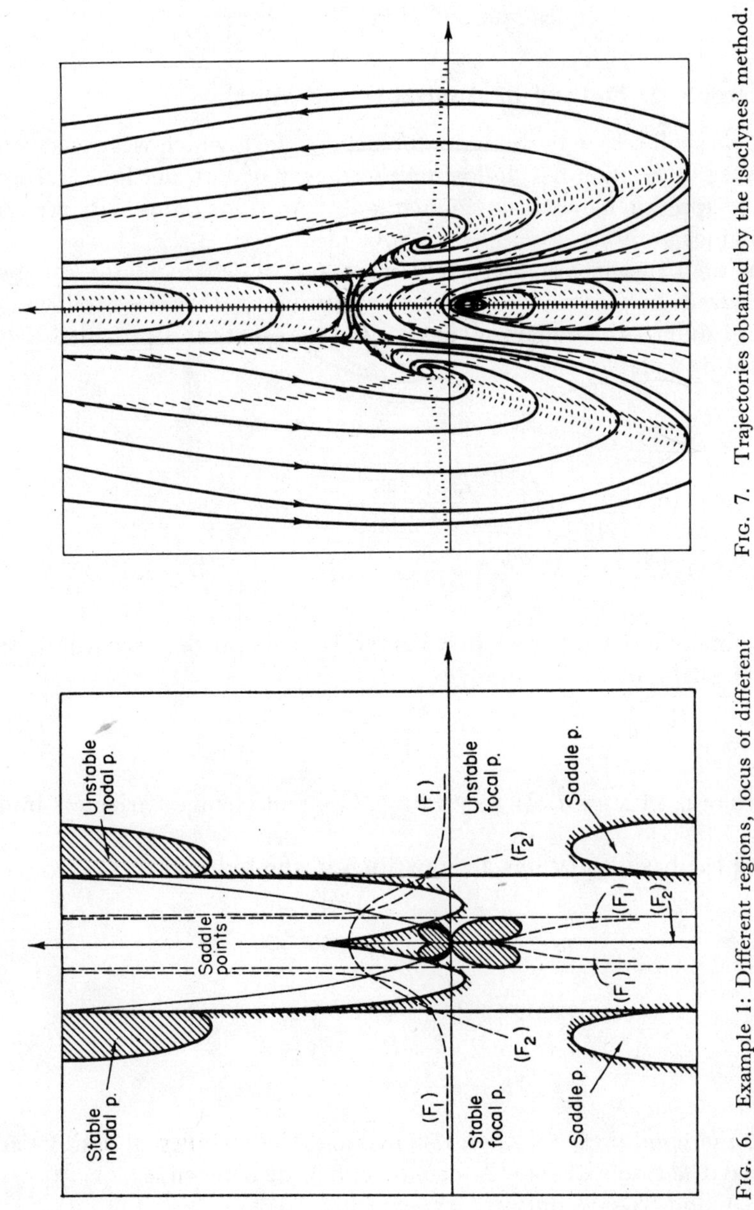

FIG. 6. Example 1: Different regions, locus of different kinds of singular points.

FIG. 7. Trajectories obtained by the isoclynes' method.

One finds

$$\frac{\partial \varphi_1}{\partial x} = 1 - v - 2x^2 \qquad \frac{\partial \varphi_1}{\partial y} = -u - 2xy$$

$$\frac{\partial \varphi_2}{\partial x} = u - 2xy \qquad \frac{\partial \varphi_2}{\partial y} = 1 - v - 2y^2$$

Then

$$\Theta = 2(1 - 2v)$$
$$\Delta = (1 - v)(1 - 3v) + u^2$$

Accordingly, the equations of the separatrices are

$$(S_1): \quad \Theta^2 - 4\Delta = 4(v^2 - u^2) = 0 \qquad (\text{say } u = \pm v)$$
$$(S_2): \quad \Theta = 2(1 - 2v) = 0$$
$$(S_3): \quad \Delta = (1 - v)(1 - 3v) + u^2 = 0$$

These separatrices are shown in Fig. 8 in the *uv* plane, which is more convenient in this problem than the *xy* representation. The nature of the different areas which they bound is also indicated, as well as the loci of the singular points for each given F, the curves represented by

$$v[u^2 + (1 - v)^2] = F^2$$

The conclusions are the same as those of Chapter II, Section 6.2.

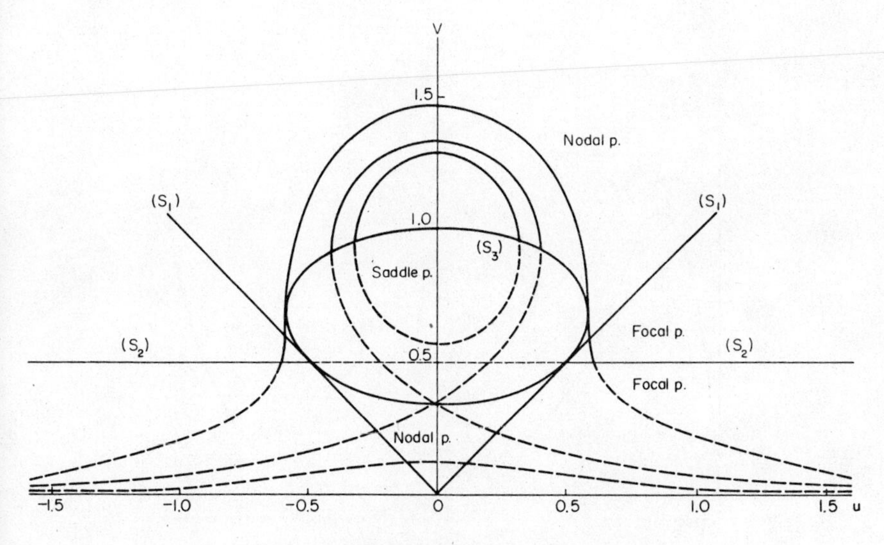

FIG. 8. Example 2: Separatrices in the *uv* plane.

3. STATIC AND DYNAMIC SYSTEMS

Consider Fig. 9, in which a tunnel diode (D) is connected to a generator with constant emf E, through a T network whose components are r, L, and C [17].

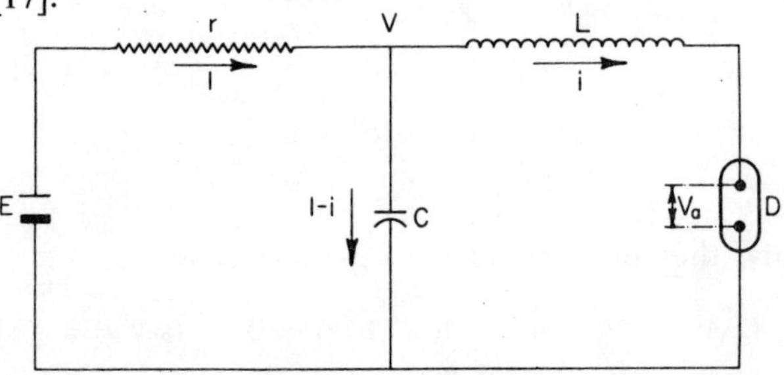

FIG. 9. Tunnel diode connected to a generator through a T network.

The characteristic curve of the diode is shown in Fig. 10, where V_a is the voltage across the diode and i the corresponding intensity. We shall write this relation

$$V_a = \psi(i)$$

Then, applying Kirchhoff's laws, we get, according to the notation shown in fig. 9,

$$E = rI + V$$

$$V = L\frac{di}{dt} + \psi(i)$$

$$I - i = C\frac{dV}{dt}$$

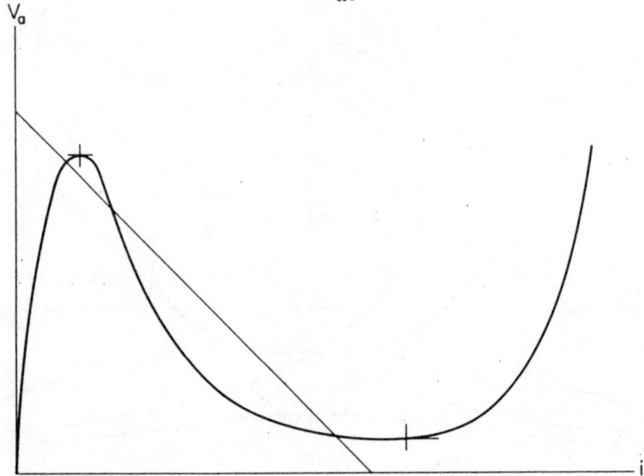

FIG. 10. Characteristic curve of a tunnel diode.

from which follows

$$\frac{dV}{dt} = \frac{E - V - ri}{rC} \qquad \frac{di}{dt} = \frac{V - \psi(i)}{L} \tag{5}$$

Now let us consider two different cases, which obviously are related to one another:

Case 1. The first one is the static system, in which

$$\frac{dV}{dt} \equiv 0 \qquad \frac{di}{dt} \equiv 0$$

and, accordingly,

$$V = E - ri \tag{6}$$

$$V = \psi(i) \tag{7}$$

As pointed out above, (7) is the equation of the characteristic curve of the tunnel diode, whereas (6) is commonly called the *load line.*

By superimposing the load line upon the volt-ampere characteristic, we get intersection points, each of which corresponds to a possible static equilibrium state.

Case 2. The other is the dynamic system, in which

$$\frac{dV}{dt} \not\equiv 0 \qquad \frac{di}{dt} \not\equiv 0$$

Now according to the definition of singular points, we see that:

Property 1. The singular points of the dynamic system are the same as the equilibrium points of the static one.

To make the present remarks more general, let

$$x_1 = V \qquad x_2 = i$$

Equation (5) is rewritten

$$\dot{x}_1 = f_1(x_1, x_2) \qquad \dot{x}_2 = f_2(x_1, x_2)$$

$$f_1(x_1, x_2) \triangleq \frac{E - V - rx_2}{rC} \tag{8}$$

$$f_2(x_1, x_2) \triangleq \frac{V - \psi(x_2)}{rC}$$

Singular points of the dynamic system and equilibrium points of the static one are defined by the same equation,

$$f_1(x_1, x_2) = f_2(x_1, x_2) = 0$$

Accordingly, we are faced with two topics to discuss: The first one concerns the static stability of the equilibrium points, the second one (which has already been studied) the character of singular points of the dynamic system.

In this section we shall accept the common definition of static stability. Considering a small deviation from the equilibrium position, the system is stable if the deviation tends to zero as $t \to \infty$.

Let $P(\mathbf{x}^0)$ be a *static equilibrium point*. Its stability will be discussed by considering a small deviation $\epsilon\boldsymbol{\eta}$ and going to variational equations†:

$$\begin{aligned} \dot{\eta}_1 &= \left(\frac{\partial f_1}{\partial x_1}\right)_0 \eta_1 + \left(\frac{\partial f_1}{\partial x_2}\right)_0 \eta_2 \\ \dot{\eta}_2 &= \left(\frac{\partial f_2}{\partial x_1}\right)_0 \eta_1 + \left(\frac{\partial f_2}{\partial x_2}\right)_0 \eta_2 \end{aligned} \tag{9}$$

where η_1 and η_2 are the components of $\boldsymbol{\eta}$. The index zero indicates that partial derivatives are computed at point P^0.

These variational equations are readily obtained in the same way as equations (2) by replacing x_1 in (8) by $x_1^0 + \epsilon\eta_1$ and x_2 by $x_2^0 + \epsilon\eta_2$. Then, since $f_1(x_1^0, x_2^0) = f_2(x_1^0, x_2^0) = 0$,

$$\begin{aligned} \epsilon\dot{\eta}_1 &= \left(\frac{\partial f_1}{\partial x_1}\right)_0 \epsilon\eta_1 + \left(\frac{\partial f_1}{\partial x_2}\right)_0 \epsilon\eta_2 + o(\epsilon) \\ \epsilon\dot{\eta}_2 &= \left(\frac{\partial f_2}{\partial x_1}\right)_0 \epsilon\eta_1 + \left(\frac{\partial f_2}{\partial x_2}\right)_0 \epsilon\eta_2 + o(\epsilon) \end{aligned}$$

Dividing both sides by ϵ and letting ϵ tend to zero, we get (9).

Then the stability of static equilibrium points depends on the behavior of the solution $\eta_1(t)$ and $\eta_2(t)$ of (9), say on the eigenvalues λ_1 and λ_2 of matrix A, since

$$\begin{aligned} \eta_1(t) &= K_1 \exp(\lambda_1 t) + K_2 \exp(\lambda_2 t) \\ \eta_2(t) &= K_3 \exp(\lambda_1 t) + K_4 \exp(\lambda_2 t) \end{aligned}$$

(K_1, K_2, K_3, K_4 are constants of integration). Again we find the same kind of problem as the one we met at the begining of the discussion of singular points of the dynamic system.

Considering the characteristic equation

$$\lambda^2 - \Theta\lambda + \Delta = 0$$

† Actually, the stability of a static equilibrium is discussed in terms of the dynamic behavior of the system.

we are led to the following conclusions:

(a) If $\Delta < 0$, eigenvalues λ_1 and λ_2 are both real and $\lambda_1\lambda_2 < 0$. Then since one of the eigenvalues is real positive, the static equilibrium is *unstable*.

(b) If $\Delta > 0$, the eigenvalues may be either real or complex-conjugate. In any case, the static system is *unstable* if $\Theta > 0$, *stable* if $\Theta < 0$.

(c) If $\Delta = 0$, one of the eigenvalues is $\lambda_1 = 0$, the other is $\lambda_2 = \Theta$. The static system is *unstable*.

By comparing this conclusion with the earlier one, we arrive at:

Property 2. Every equilibrium point $P(\mathbf{x}^0)$ of the static system is stable if and only if

$$\Theta < 0 \qquad \Delta > 0$$

Property 3. A stable equilibrium point of the static system is

(a) a convergent nodal point,
(b) a convergent focal point

of the dynamic system; and an unstable equilibrium point of the static system is

(a) a saddle point,
(b) a divergent nodal point
(c) a divergent focal point.

4. EXTENSION OF THE THEORY: SOURCES, SINKS, AND TRANSFORMATION POINTS†

The remarks in the preceding section provide a framework for a more general theory of transformations of static systems into dynamic ones. Following the ideas of Vogel and Sideriades, we will consider separately:

(a) A static system defined by the set of algebraic equations

$$f_1(x_1, x_2) = 0 \qquad f_2(x_1, x_2) = 0 \tag{10}$$

(b) A dynamic system governed by the differential equations

$$\begin{aligned} \alpha\dot{x}_1 + \beta\dot{x}_2 &= f_1(x_1, x_2) \\ \gamma\dot{x}_1 + \delta\dot{x}_2 &= f_2(x_1, x_2) \end{aligned} \tag{11}$$

† See [7, 9, 14].

where

$$M = \begin{pmatrix} \alpha & \beta \\ \gamma & \delta \end{pmatrix}$$

is a given matrix whose elements are analytic functions of $\mathbf{x}$, in which time t does not occur explicitely.

Next it will be convenient to write (11) in the matrix form

$$M\dot{\mathbf{x}} = \mathbf{f}(\mathbf{x})$$

We return to the case considered in Section 1 if $M = \mathrm{I}$ (I is the unity matrix). Also note that (11) can be rewritten

$$\frac{dx_1}{X_1} = \frac{dx_2}{X_2} = \frac{dt}{X_0} \tag{12}$$

where

$$\begin{aligned} X_1 &= \delta f_1 - \beta f_2 \\ X_2 &= -\gamma f_1 + \alpha f_2 \\ X_0 &= \alpha\delta - \beta\gamma \end{aligned} \tag{13}$$

or, in matrix form,

$$\mathbf{X} = T\mathbf{f} \qquad X_0 = \| T \| \tag{14}$$

where

$$T = \begin{pmatrix} \delta & -\beta \\ -\gamma & \alpha \end{pmatrix} \qquad \mathbf{X} = \begin{pmatrix} X_1 \\ X_2 \end{pmatrix}$$

Indeed,

$$T = X_0 M^{-1} \qquad \| T \| = \| M \|$$

Now, more generally, singular points of system (12) are defined by

$$X_1 = X_2 = 0$$

say by

$$T\mathbf{f} = 0 \tag{15}$$

Two cases need be considered:

Case 1. M, and accordingly T, are nonsingular matrices:

$$X_0 = \| T \| = \| M \| \neq 0$$

Then

$$T\mathbf{f} = 0 \quad \Leftrightarrow \quad \mathbf{f} = 0$$

At such singular points

$$\dot{x}_1 = (X_1/X_0) = 0 \qquad \dot{x}_2 = (X_2/X_0) = 0 \tag{16}$$

i.e., Property 1 is verified.

Case 2. M (and T) are singular matrices:

$$X_0 = \| T \| = \| M \| = 0 \tag{17}$$

Then (17) defines a curve at all of whose points velocity $\dot{\mathbf{x}}$ is either infinite or undetermined. The locus of all points of this kind is called a *shock curve.*

At last we are led to classifying singular points into two types:

Type 1. Points which are defined by the static system, i.e., solely by its resistive components (positive and negative resistances). Intuitively we can visualize the situation by considering trajectories along which the energy is transfered as lines of flow of a fluid. Energy is appearing and disappearing, respectively, at unstable and stable singular points of *this* kind, which play the same role as point *sources* and *sinks*. Accordingly, we shall call such points sources and sinks.

Type 2. Points which are originated by reactive elements, i.e., by the transformation of the static system into a dynamic one. These are called *transformation points.*

Remark. At a singular point of Type 1, say at a source or at a sink ($X_0 \neq 0$), all the arguments of Section 1 remain valid.

Equations

$$\dot{x}_1 = \frac{X_1}{X_0} \qquad \dot{x}_2 = \frac{X_2}{X_0}$$

are similar to (1).

Obviously, if $X_0 > 0$, we can use matrix $\mathscr{A}$ instead of A:

$$\mathscr{A} = \begin{pmatrix} \left(\dfrac{\partial X_1}{\partial x_1}\right)_0 & \left(\dfrac{\partial X_1}{\partial x_2}\right)_0 \\ \left(\dfrac{\partial X_2}{\partial x_1}\right)_0 & \left(\dfrac{\partial X_2}{\partial x_2}\right)_0 \end{pmatrix}$$

and the corresponding characteristic equation

$$\lambda^2 - \left[\left(\frac{\partial X_1}{\partial x_1}\right)_0 + \left(\frac{\partial X_2}{\partial x_2}\right)_0\right]\lambda + \left[\left(\frac{\partial X_1}{\partial x_1}\right)_0\left(\frac{\partial X_2}{\partial x_2}\right)_0 - \left(\frac{\partial X_1}{\partial x_2}\right)_0\left(\frac{\partial X_2}{\partial x_1}\right)_0\right] = 0$$

The discussion is the same.

5. TRANSFORMATIONS OF THE VECTOR FIELD

At each nonsingular point $P(\mathbf{x})$ of the phase plane R^2 the vector $\mathbf{X}(\mathbf{x})$ is tangent to the trajectory Γ which passes through that point. Let us designate by $\{\mathbf{X}(\mathbf{x})\}$ the vector field which is defined by the function $\mathbf{X}(\mathbf{x})$, and the corresponding set of trajectories by $\{\Gamma\}$.

To each given matrix $T(\mathbf{x})$ is associated a vector field, and accordingly a set of trajectories, which are, respectively, the transforms of the vector field $\{\mathbf{f}(\mathbf{x}))$ and the corresponding set of trajectories, say $\{\Gamma_f\}$. We shall write

$$\begin{gathered}\{\mathbf{X}(\mathbf{x})\} = T\{\mathbf{f}(\mathbf{x})\} \\ \{\Gamma\} = T\{\Gamma_f\}\end{gathered} \qquad T \triangleq T(\mathbf{x}) \tag{18}$$

We can apply different rules of transformation, defined by the matrices T, T', T'', ..., respectively, and consider the product of such transformations, which result in

$$\{\mathbf{X}'(\mathbf{x})\} = T'\{\mathbf{X}(\mathbf{x})\} = T'T\{\mathbf{f}(\mathbf{x})\} \tag{19}$$

$$\{\mathbf{X}''(\mathbf{x})\} = T''\{\mathbf{X}'(\mathbf{x})\} = T''T'T\{\mathbf{f}(\mathbf{x})\} \tag{20}$$

$$\vdots$$

We know that the coordinates of sources and sinks are not changed by transformations T, T', T'', ..., since they are defined only by the static system. However, in general the character of each singular point will change in a way that we shall wish to determine.

Consider a singular point $P(\mathbf{x}^0)$ of type 1 (a source or a sink) and a small deviation $\epsilon\boldsymbol{\eta}$. At point $P(\mathbf{x}^0 + \epsilon\boldsymbol{\eta})$ we have

$$X_1(\mathbf{x}^0 + \epsilon\boldsymbol{\eta}) = \left(\frac{\partial X_1}{\partial x_1}\right)_0 \epsilon\eta_1 + \left(\frac{\partial X_1}{\partial x_2}\right)_0 \epsilon\eta_2 + o(\epsilon)$$

$$X_2(\mathbf{x}^0 + \epsilon\boldsymbol{\eta}) = \left(\frac{\partial X_2}{\partial x_1}\right)_0 \epsilon\eta_1 + \left(\frac{\partial X_2}{\partial x_2}\right)_0 \epsilon\eta_2 + o(\epsilon)$$

$$X_1'(\mathbf{x}^0 + \epsilon\boldsymbol{\eta}) = \left(\frac{\partial X_1'}{\partial x_1}\right)_0 \epsilon\eta_1 + \left(\frac{\partial X_1'}{\partial x_2}\right)_0 \epsilon\eta_2 + o(\epsilon)$$

$$X_2'(\mathbf{x}^0 + \epsilon\boldsymbol{\eta}) = \left(\frac{\partial X_2'}{\partial x_1}\right)_0 \epsilon\eta_1 + \left(\frac{\partial X_2'}{\partial x_2}\right)_0 \epsilon\eta_2 + o(\epsilon)$$

and, accordingly, if for instance we refer to transformation (19):

$$\epsilon\mathscr{A}'\boldsymbol{\eta} = \epsilon T'(\mathbf{x}^0 + \epsilon\boldsymbol{\eta})\mathscr{A}\boldsymbol{\eta} + o(\epsilon)$$

with

$$\mathscr{A}' = \begin{bmatrix} \left(\dfrac{\partial X_1'}{\partial x_1}\right)_0 \left(\dfrac{\partial X_1'}{\partial x_2}\right)_0 \\ \left(\dfrac{\partial X_2'}{\partial x_1}\right)_0 \left(\dfrac{\partial X_2'}{\partial x_2}\right)_0 \end{bmatrix} \qquad \mathscr{A} = \begin{bmatrix} \left(\dfrac{\partial X_1}{\partial x_1}\right)_0 \left(\dfrac{\partial X_1}{\partial x_2}\right)_0 \\ \left(\dfrac{\partial X_2}{\partial x_1}\right)_0 \left(\dfrac{\partial X_2}{\partial x_2}\right)_0 \end{bmatrix}$$

from which, dividing by ϵ and letting $\epsilon \to 0$,

$$\mathscr{A}'\boldsymbol{\eta} = T'(\mathbf{x}^0)\mathscr{A}\boldsymbol{\eta}$$

or

$$\mathscr{A}' = T'\mathscr{A} \tag{21}$$

and similarly

$$\mathscr{A}'' = T''\mathscr{A}'$$

Now the problem is solved, since the characteristic equations of $\mathscr{A}'$, $\mathscr{A}''$, ... can be easily computed, starting from any given initial system, which can be, for example (but not necessarily), the static system.

Especially we deduce from (21)

$$\| \mathscr{A}' \| = \| T' \| \| \mathscr{A} \|$$

from which follows

$$\| \mathscr{A} \| = 0 \quad \Rightarrow \quad \| \mathscr{A}' \| = 0$$

We conclude that the boundary (S_3) of the saddle-points area remains unchanged.

5.1. Orthogonal Transformation

If T', T'', ... are orthogonal transformations

$$\| T' \| = \| T'' \| = \cdots = 1$$

it follows that

$$\| \mathscr{A}'' \| = \| \mathscr{A}' \| = \| \mathscr{A} \|$$

Thus the saddle-points area remains unchanged.

5.2. Pseudo-Orthogonal Transformation

On the other hand, a pseudo-orthogonal transformation is one in which the determinant is -1, or more generally $-k^2$, where k is any constant. For instance, if

$$\| T' \| = -k^2$$

then

$$\| \mathscr{A}' \| = -k^2 \| \mathscr{A} \|$$

Accordingly, when one applies a pseudo-orthogonal transformation, the saddle-points area and the union of the domains, which are assigned to focal and nodal points, respectively, switch together with respect to (S_3), which is unchanged, as pointed out above.

6. THREE-DIMENSIONAL SINGULARITIES†

The above representation has been extended by Poincaré to systems governed by the equations

$$\dot{x}_1 = f_1(x_1, x_2, x_3)$$
$$\dot{x}_2 = f_2(x_1, x_2, x_3)$$
$$\dot{x}_3 = f_3(x_1, x_2, x_3)$$

Then trajectories and singular points need to be considered in a three-dimensional phase space R^3 (x_1, x_2, x_3). In this space, basic surfaces (F_1), (F_2), and (F_3),

$$(F_1): \quad f_1(x_1, x_2, x_3) = 0$$
$$(F_2): \quad f_2(x_1, x_2, x_3) = 0$$
$$(F_3): \quad f_3(x_1, x_2, x_3) = 0$$

will play an important role, since their intersection, provided it is not empty, will define three-dimensional singularities.

Indeed matrix A (Section 1) will be replaced by

$$D(f_1, f_2, f_3)/D(x_1, x_2, x_3) \qquad (D, \text{jacobian})$$

whose characteristic equation, again, will be taken as the starting point of the discussion. Let

$$\lambda^3 + a\lambda^2 + b\lambda + c = 0 \tag{22}$$

be the characteristic equation.

† See [1, 14].

The condition which coefficients a, b, and c must fulfill in order that all the eigenvalues be real is

$$\Phi(a, b, c) \triangleq (9c - ab)^2 - (6b - 2a^2)(6ac - 2b^2) < 0$$

The general shape of the surface $\Phi(a, b, c) = 0$ is shown on Fig. 11.

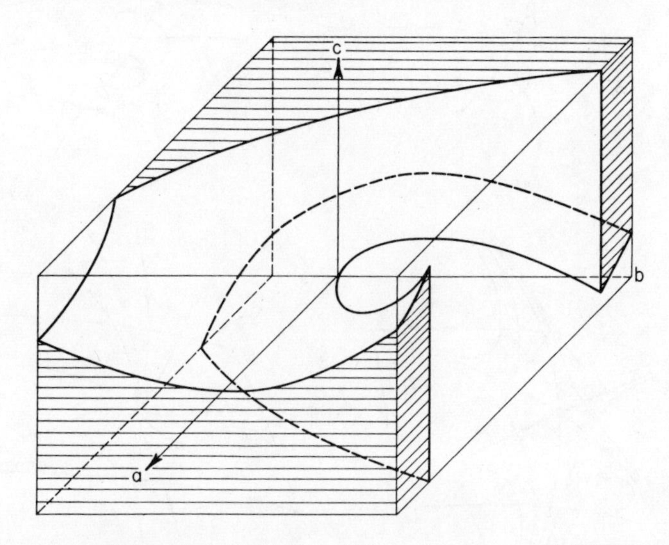

FIG. 11. Three-dimensional singularities: Locus of points at which the characteristic equation has a double root: $\Phi(a, b, c) = 0$.

This surface is the locus of all points at which the characteristic equation has a double root. On the other hand, the product of the eigenvalues is $-c$.

Accordingly, we are faced with the following different situations. In general we get singular points which hold a plurality of structures.

Case 1. $\Phi < 0$: Eigenvalues λ_1, λ_2, and λ_3 are all real.

(a) If $c < 0$ we have the following subcases: $\lambda_1, \lambda_2, \lambda_3 > 0$ is an instable nodal point, or $\lambda_1 > 0$; $\lambda_2, \lambda_3 < 0$ are stable saddle points.

(b) If $c > 0$: $\lambda_1, \lambda_2, \lambda_3 < 0$ is a stable nodal point, or $\lambda_1 < 0$; λ_2, $\lambda_3 > 0$ are instable saddle points.

(c) If $c = 0$: $\lambda_1 = 0$; $\lambda_2, \lambda_3 \neq 0$ are nodal points or saddle points.

Case 2. $\Phi = 0$: Eigenvalues λ_1, λ_2, and λ_3 are all real; furthermore, the characteristic equation has a double root. Then to each subcase (1a), (1b), there corresponds a subcase (2a) and (2b), respectively, which can be easily described, starting from (1a) and (1b), by bringing in

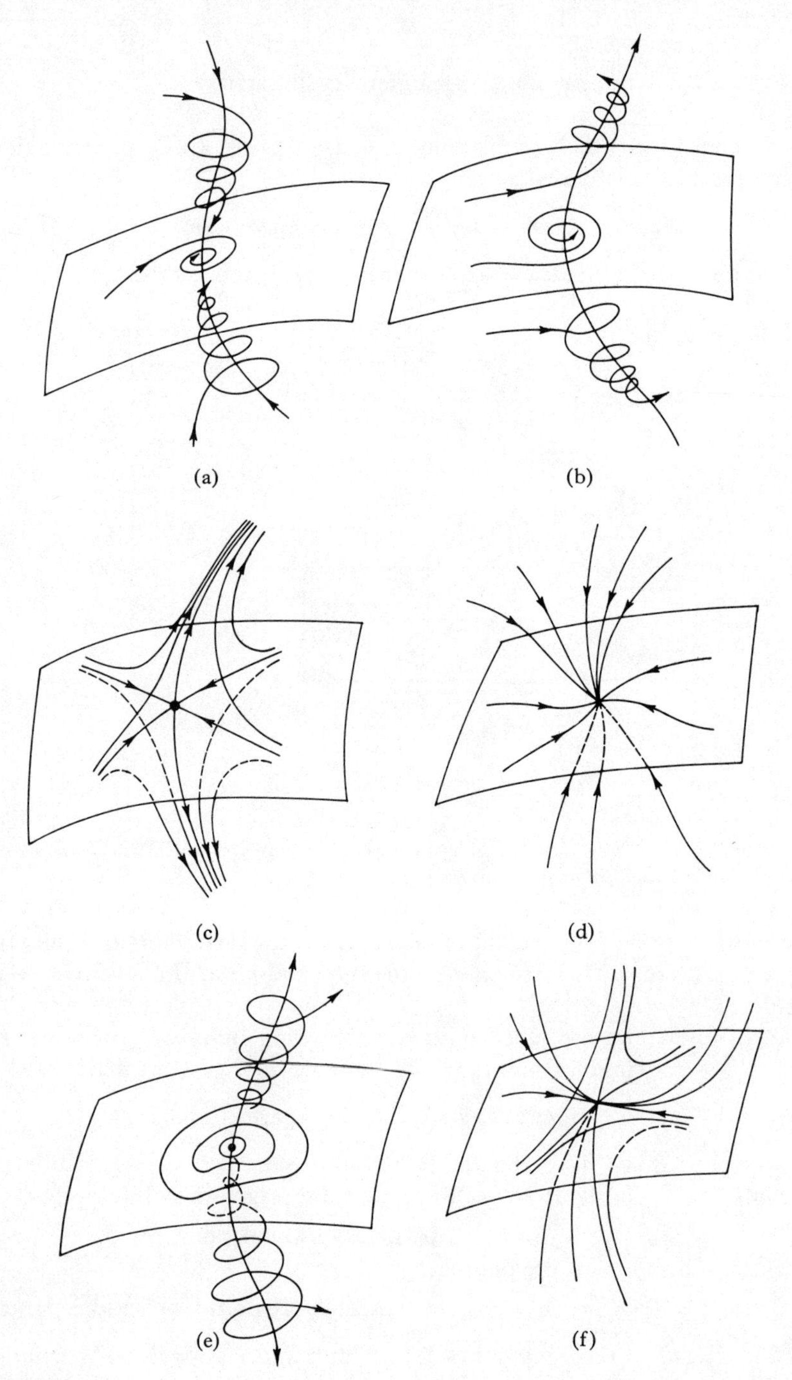

FIG. 12. Examples of three-dimensional singularities: (a) stable nodal-focal point; (b) saddle-focal point; (c) saddle point; (d) stable nodal point; (e) unstable center; (f) nodal-saddle point.

coincidence the eigenvectors which correspond to λ_2 and λ_3. For the third subcase, we have the following:

(c) When $c = 0$, (22) reduces to

$$\lambda(\lambda^2 + a\lambda + b) = 0$$

and $\Phi = 3b^2(4b - a^2) = 0$. If $b \neq 0$, $4b - a^2 = 0$, we get the solution

$$\lambda_1 = 0 \qquad \lambda_2 = \lambda_3 = -\frac{a}{2}$$

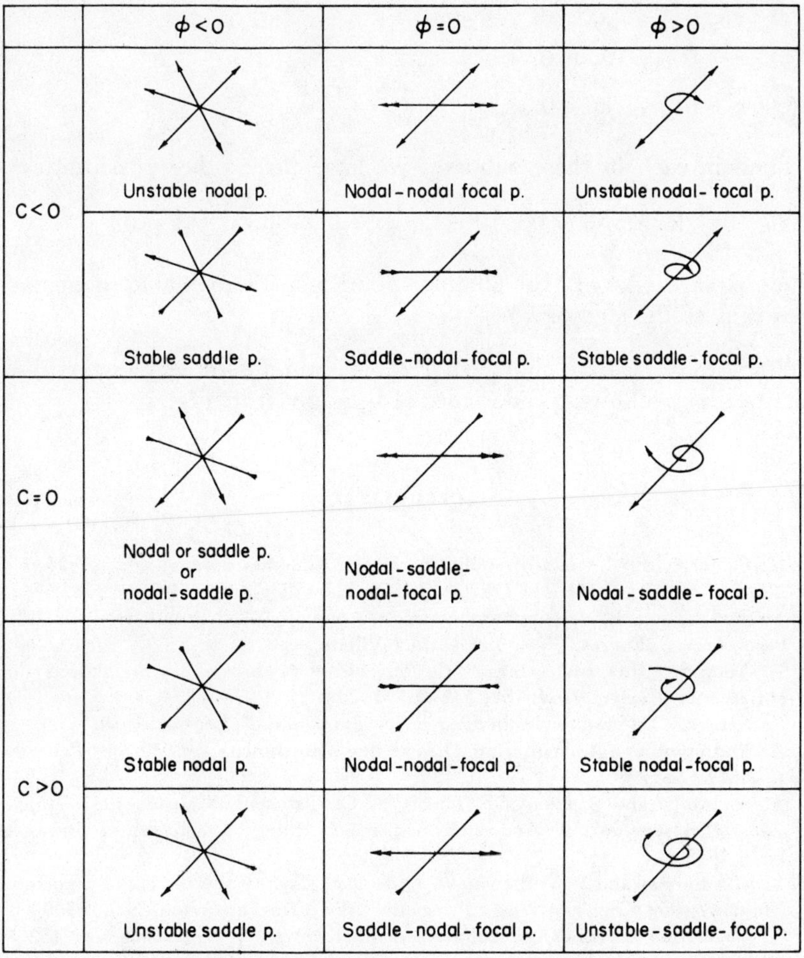

	$\phi < 0$	$\phi = 0$	$\phi > 0$
C<0	Unstable nodal p.	Nodal-nodal focal p.	Unstable nodal-focal p.
	Stable saddle p.	Saddle-nodal-focal p.	Stable saddle-focal p.
C=0	Nodal or saddle p. or nodal-saddle p.	Nodal-saddle-nodal-focal p.	Nodal-saddle-focal p.
C>0	Stable nodal p.	Nodal-nodal-focal p.	Stable nodal-focal p.
	Unstable saddle p.	Saddle-nodal-focal p.	Unstable-saddle-focal p.

FIG. 13. Summary of classification of three-dimensional singularities.

Then:

If $a > 0$, the singular point is stable; if $a < 0$, it is unstable.
If $b = 0$, $a \neq 0$, we have

$$\lambda_1 = \lambda_2 = 0 \qquad \lambda_3 = -a$$

The singular point is stable if $a > 0$, unstable if $a < 0$.

Case 3. $\Phi > 0$: One of the eigenvalues, say λ_1, is real, the other ones, λ_2 and λ_3, are complex conjugate.

(a) $c < 0 \Rightarrow \lambda_1 > 0$.

(b) $c = 0 \Rightarrow \lambda_1 = 0$.

(c) $c > 0 \Rightarrow \lambda_1 < 0$.

Then in each of these subcases we have three other possibilities:

$$\mathrm{Re}(\lambda_2, \lambda_3) > 0 \qquad \mathrm{Re}(\lambda_2, \lambda_3) = 0 \qquad \mathrm{Re}(\lambda_2, \lambda_3) < 0$$

When $\mathrm{Re}(\lambda_2, \lambda_3) = 0$, the singular point is a *center* (stable or unstable according to the sign of λ_1).

A few examples are shown in Fig. 12, and a summary of the above classification, following Sideriades [14], is given in Fig. 13.

REFERENCES

1. H. Poincaré, Sur les courbes définies par une équation différentielle. *J. Math.* **1**, 167–244 (1875); **2**, 151–217 (1876); **7**, 375–422 (1881); **8**, 251–296 (1882); "Les Méthodes nouvelles de la mécanique céleste" Vol. 1. Gauthier-Villars, Paris, 1892; Paris, 1928. "Œuvres," Vol. 1. Gauthier-Villars,
2. A. Andronov, Les cycles limites de Poincaré et la théorie des oscillations auto-entretenues. *Compt. Rend.* **189**, 559–561 (1929).
3. E. Kamke, "Differentialgleichungen reeler Functionen." Leipzig, 1930.
4. A. Andronov and A. Witt, Zur Theorie des Mitnehmens von van der Pol. *Arch. Electroth.* **24** (1930).
5. A. Andronov and S. Chaikin, "Theory of Oscillations." Moscow, 1937. (English transl., S. Lefschetz, A. Andronov, S. Chaikin, Princeton Univ. Press, Princeton, N.J., 1949.)
6. V. V. Nemitzky and V. V. Stepanov, "Qualitative Theory of Differential Equations." Moscow, 1949. (English transl., Princeton Univ. Press, Princeton, N.J., 1960.)
7. T. Vogel, Sur certaines oscillations à déferlement. *Ann. Telecommun.* **6**, 182–190 (1951); Les Méthodes topologiques de discussion des problèmes aux oscillations non linéaires. *Ann. Telecommun.* **6**, 1–9 (1951).

8. L. Sideriades, Systèmes couplés non linéaires. *Compt. Rend.* **242**, 1784–1787 (1956); Étude d'une bascule à quatre positions d'équilibre par les méthodes de l'analyse topologique. *Compt. Rend.* **242**, 1583–1586, 1704–1707 (1956).
9. L. Sideriades, Méthodes topologiques appliquées à l'électronique. Thèses, Faculté des Sciences de l'Université d'Aix-Marseille, Déc. 1956.
10. G. Sansone and R. Conti, "Equazioni Differenziali Nonlineari." Edizioni Cremonese, Rome, 1956.
11. S. Lefschetz, "Differential Equations: Geometric Theory." Wiley (Interscience), New York, 1957.
12. T. Vogel, Sur des systèmes dynamiques à hérédité non linéaire et à mémoire totale. *Compt. Rend.* **245**, 1224–1226 (1957); *Compt. Rend.* **246**, 59–61 (1958); Hérédité discontinue dans les systèmes dynamiques. *Compt. Rend.* 1379–1381 (1958).
13. L. Cesari, "Asymptotic Behaviour and Stability Problems." Springer, Berlin, 1959.
14. L. Sideriades, Méthodes topologiques et applications. *Ann. Télécommun.* **14**, 8 (1959).
15. L. Sideriades, Méthodes de topologie qualitative: Applications à l'étude des cheminées d'équilibre. *Symp. intern. sur les oscillations nonlinéaires, Kiev*, 1961.
16. T. Vogel, Systèmes déferlants, systèmes héréditaires, systèmes dynamiques. *Symp. intern. sur les oscillations nonlinéaires, Kiev*, 1961; Solutions périodiques des systèmes héréditaires. *Second Conf. Nonlinear Vibrations, Warsaw*, 1962.
17. N. Minorsky, "Nonlinear Oscillations." Van Nostrand, Princeton, N.J., 1962.

CHAPTER IV

Systems with Several Degrees of Freedom

1. INTRODUCTION

In the first three chapters we analyzed the motion of oscillators with one degree of freedom, governed by a differential equation of the second order. Now we shall consider more elaborate devices, in which a number of oscillators of this kind are linked together. The aim of this chapter will be to show how the theory of oscillations can be extended to systems with several degrees of freedom.

We shall start with systems with two degrees of freedom and, following our earlier classification, consider separately:

(a) An example of a conservative oscillator.

(b) A self-oscillatory system.

In the last part of this chapter we shall outline the general theory which was recently introduced and developed by Rosenberg, on the geometrization of normal vibrations of nonlinear systems having many degrees of freedom.

2. EXAMPLE OF A CONSERVATIVE OSCILLATOR

There are many examples of such oscillators; let us introduce them at the outset by considering a simple model, as shown in Fig. 1a. Two masses m_1 and m_2, mounted on horizontal string $x'x$, can execute vibrations; each mass, possessing a single degree of freedom of translation, is connected to a fixed point by an "anchor spring" and to its partner by a "coupling spring" Σ_0.

Let u_1 and u_2 be, respectively, the translations of masses m_1 and m_2 from their equilibrium positions, i.e., the elongations of anchor springs Σ_1 and Σ_2. Let $f_1(s)$, $f_2(s)$, and $f_{12}(s)$ be the relations between the restoring force and the elongation s for springs Σ_1, Σ_2, and Σ_0, respectively.

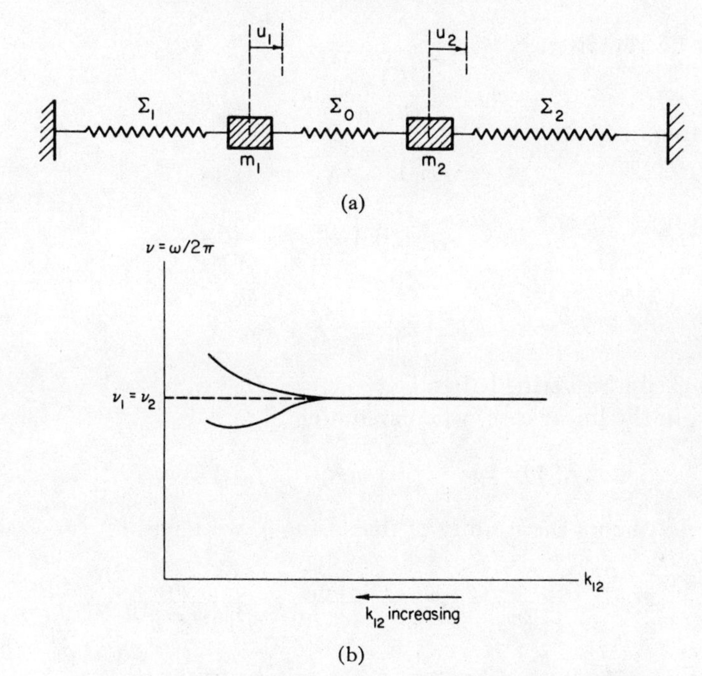

FIG. 1. Separation of frequencies due to coupling, two oscillators. (a) Coupled oscillators; (b) Frequency diagram.

Assuming that $f_i(s)$ $(i = 1, 2, 12)$ is an odd function, say

$$f_i(-s) = -f_i(s)$$

the system is governed by

$$\begin{aligned} m_1\ddot{u}_1 &= -f_1(u_1) - f_{12}(u_1 - u_2) \\ m_2\ddot{u}_2 &= -f_2(u_2) - f_{12}(u_2 - u_1) \end{aligned} \tag{1}$$

Note the well-known symmetry of the coupling terms, which is a consequence of the property of the coupling spring,

$$f_{12}(s) = f_{21}(s)$$

Next (Section 3) we shall also consider equations with dissymmetry in the coupling terms.

Since spring forces are derivable from a potential function $W(u_1, u_2)$ which verifies

$$\frac{\partial W}{\partial u_1} = -f_1(u_1) - f_{12}(u_1 - u_2)$$

$$\frac{\partial W}{\partial u_2} = -f_2(u_2) - f_{12}(u_2 - u_1)$$

(1) can be rewritten

$$m_1\ddot{u}_1 = \frac{\partial W}{\partial u_1} \qquad m_2\ddot{u}_2 = \frac{\partial W}{\partial u_2}$$

Or putting

$$x_1 = \sqrt{m_1}\, u_1 \qquad x_2 = \sqrt{m_2}\, u_2$$

$$U(x_1, x_2) = W\left[\frac{x_1}{\sqrt{m_1}}, \quad \frac{x_2}{\sqrt{m_2}}\right]$$

$$\ddot{x}_1 = \frac{\partial U}{\partial x_1} \qquad \ddot{x}_2 = \frac{\partial U}{\partial x_2} \tag{2}$$

It may easily be verified that $U(x_1, x_2) = U(-x_1, -x_2)$.

Now, in the linear case, when assuming

$$f_1(s) = k_1 s \qquad f_2(s) = k_2 s \qquad f_{12}(s) = k_{12} s$$

(k_1, k_2, k_{12} being the rigidity of the springs), we have

$$\begin{aligned} U &= -\frac{k_1 + k_{12}}{2m_1} x_1^2 - \frac{k_2 + k_{12}}{2m_2} x_2^2 + \frac{k_{12}}{(m_1 m_2)^{1/2}} x_1 x_2 \\ &= -\frac{k_1}{2m_1} x_1^2 - \frac{k_2}{2m_2} x_2^2 - \frac{k_{12}}{2}\left[\frac{x_1}{(m_1)^{1/2}} - \frac{x_2}{(m_2)^{1/2}}\right]^2 \end{aligned} \tag{3}$$

and the equations of motion become

$$\begin{aligned} \ddot{x}_1 + \frac{k_1}{m_1} x_1 &= -\frac{k_{12}}{(m_1)^{1/2}}\left[\frac{x_1}{(m_1)^{1/2}} - \frac{x_2}{(m_2)^{1/2}}\right] \\ \ddot{x}_2 + \frac{k_2}{m_2} x_2 &= \frac{k_{12}}{(m_2)^{1/2}}\left[\frac{x_1}{(m_1)^{1/2}} - \frac{x_2}{(m_2)^{1/2}}\right] \end{aligned} \tag{4}$$

Without coupling ($k_{12} = 0$) the angular frequencies of the oscillators ω_1 and ω_2 are

$$\omega_1 = \left(\frac{k_1}{m_1}\right)^{1/2} \qquad \omega_2 = \left(\frac{k_2}{m_2}\right)^{1/2}$$

Assume

$$\frac{k_1}{m_1} = \frac{k_2}{m_2} \qquad \omega_1 = \omega_2 = \omega_0$$

Then a well-known property of coupling is that it separates frequencies of the oscillators, which become Ω_1 and Ω_2:

$$\Omega_1 < \omega_0 < \Omega_2$$

This property is shown in Fig. 1b.

Indeed this property of coupling holds whatever the number of oscillators which are linked together. Let us refer for example to Fig. 2a or b, where each oscillator of a train of n oscillators is connected to the next one by a coupling spring Σ_0.

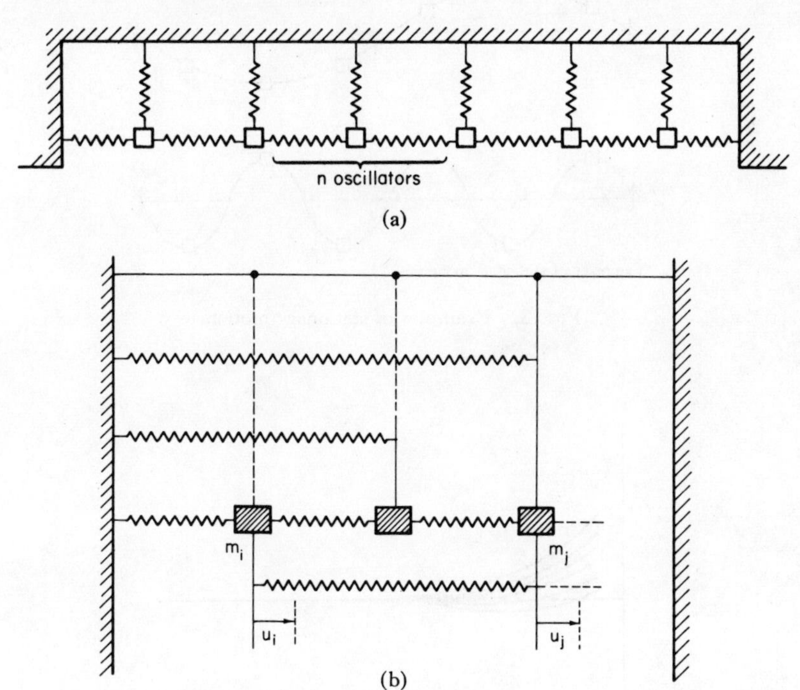

FIG. 2. Multiple oscillators. (a) Transverse vibrations; (b) Longitudinal vibrations.

For example, this periodic structure is used as a convenient model in quantum mechanics, for studying the motion of an electron in a one-dimensional potential lattice. When the coupling is removed, the n oscillators swing independent of one another, with their own frequency, which we shall assume to be the same for all of them.

When springs Σ_0 are introduced, the oscillators are no longer independent. Stationary waves settle down in the lattice. These stationary states depend on the boundaries of the lattice, and it may easily be verified that the number of such different states is equal to the number of independent oscillators, say n. Thus the number of degrees of freedom is independent of the coupling.

For example, in Figs. 2a and 3 there are six oscillators and six different stationary states. To each of these stationary states is associated a frequency. When the coupling is very weak, these frequencies are

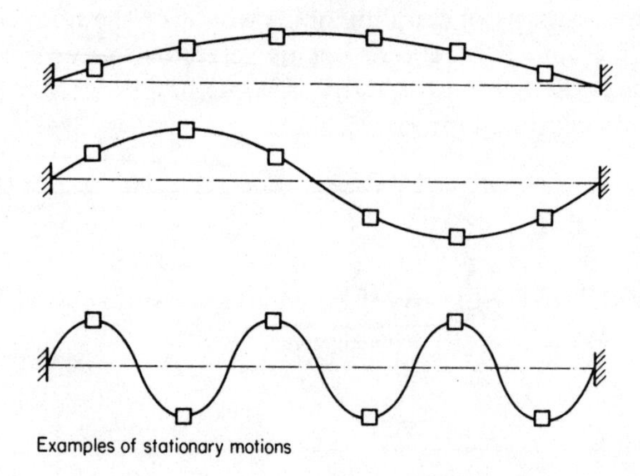

FIG. 3. Examples of stationary motion.

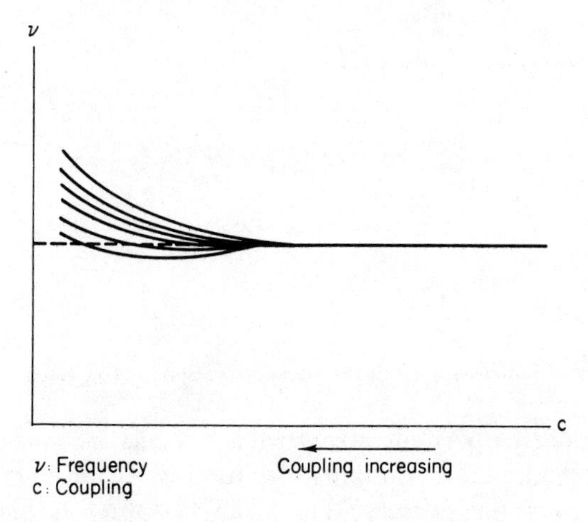

FIG. 4. Separation of frequencies due to coupling, n oscillators. ν, frequency; c, coupling.

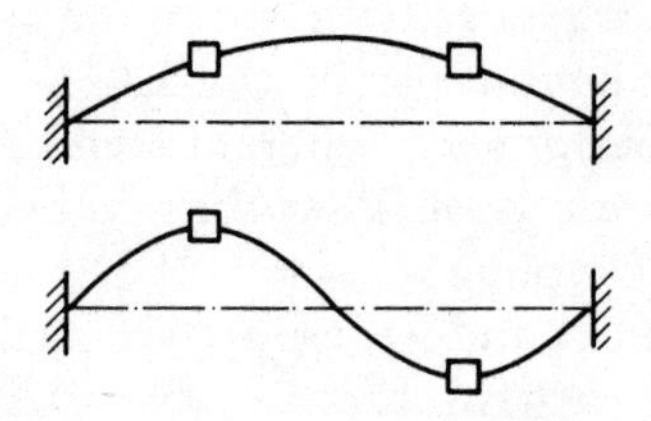

FIG. 5. Stationary states of two coupled oscillators.

approximately the same. When the coupling is increased, the frequencies become more and more separate, as shown in Fig. 4.

Note that in the case where $n = 2$, the number of stationary states is two (Fig. 5). In one of these stationary states, m_1 and m_2 are in phase, and in the other state the phase shift between the oscillators is π.

Now consider the function $U(x_1, x_2)$, and plot a level curve

$$U(x_1, x_2) = C$$

where C is any negative constant.

We shall assume that $k_1 = k_2 = k$, $m_1 = m_2 = m$. When the coupling is removed ($k_{12} = 0$), this curve is a circle,

$$x_1^2 + x_2^2 = R_c^2$$

with

$$R_c^2 = -\frac{2m}{k} C$$

On the other hand, when coupling spring Σ_0 is introduced, the equation of the level curve is

$$\left(1 + \frac{k_{12}}{k}\right)(x_1^2 + x_2^2) - 2\frac{k_{12}}{k} x_1 x_2 = R_c^2$$

It is an ellipse whose axis $A'A$ and $B'B$ are the bissectrices of $x_1 0 x_2$ (Fig. 6).

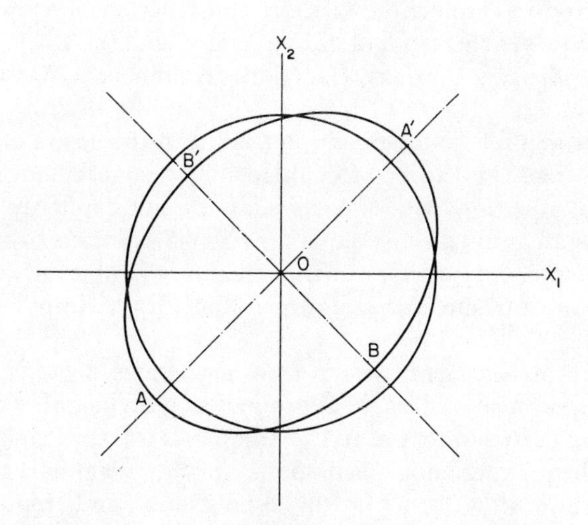

FIG. 6. The effect of coupling on equipotential curves.

At points A, A' and B, B', **grad** U has a supporting line which passes through 0, and since this is true whatever C (and R_c^2), we deduce from equations (2) the following properties:

(a) If the representative point P of the system in plane (x_1, x_2) is in coincidence with A or A' at the initial time t_0, and if its velocity is zero at that point, then it will move periodically along segment AA' for $t > t_0$.

(b) If it is in coincidence with B or B', at time t_0, and if its velocity is zero at that point, then it will move periodically along segment BB' for $t > t_0$.

These two special motions can be easily identified with the two stationary states mentioned above: Along AA', m_1 and m_2 are in phase; along BB', the phase shift is π.

We also note that the separation of frequencies due to the coupling is characterized by a flattening of the circle, which becomes an ellipse. To each axis of the ellipse is associated a stationary state and a frequency.

3. NONLINEAR OSCILLATIONS IN A PARTICLE ACCELERATOR

A few years ago, a number of problems in the field of nonlinear mechanics arose in connection with the construction of the alternating-gradient proton-synchrotron of CERN, at Geneva. They have been extensively studied by Courant, Hagedorn, Kolomenski, Moser, Schoch, Symon, and others [3, 5, 8–13].

The details of such problems do not fall into the scope of this book. Here, to illustrate the theory of coupled nonlinear oscillators governed by differential equations whose coefficients do not explicitly depend on time t, we shall only discuss equations which describe the so-called *betatron oscillations* of particles in the vacuum chamber at the "smooth approximation" introduced by Sigurgeirsson [3]. A simplified diagram is shown in fig. 7.

The alternating-gradient synchrotron accelerates protons around an orbit of constant mean radius, in a toroidal vacuum chamber of elliptical cross section. Particles rotate at the frequency of the applied electric field, and they experience oscillations in the neighborhood of the reference orbit: slow synchrotron oscillations, and high-frequency oscillations which are identical to those in a betatron.

The equations which govern the betatron oscillations are

$$x_1'' + [1 - n(\theta)]x_1 = \frac{\alpha(\theta)}{2}(x_1{}^2 - x_2{}^2)$$
$$x_2'' + n(\theta)x_2 = -\alpha(\theta)x_1x_2 \tag{5}$$

where x_1 and x_2 are, respectively, the radial and vertical deviations from the reference orbit; $\alpha(\theta)$ is a periodic function of the azimuthal

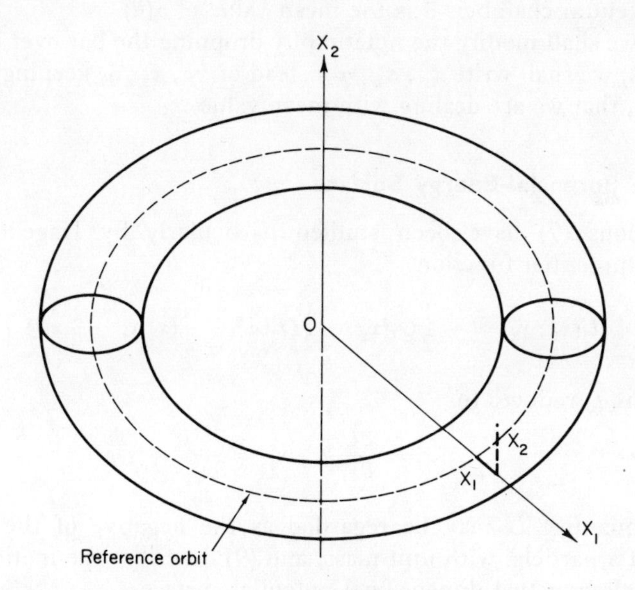

FIG. 7. Simplified diagram of a particle-accelerator vacuum chamber.

coordinate θ, which occurs in the coupling of radial and vertical deviations. The field index n defines the shaping of the magnetic field near the orbit. Since the gradient of the magnetic field is periodically alternated, $n(\theta)$ is a periodic function of θ. The double prime indicates $d^2/d\theta^2$.

If vertical deviations are temporarily disregarded, (5) reduces to

$$x_1'' + [1 - n(\theta)]x_1 - \frac{\alpha(\theta)}{2}x_1{}^2 = 0 \tag{6}$$

which is a kind of nonlinear Hill's equation, which we shall discuss later.

When both x_1 and x_2 are taken into account, a smoothing leads to replacing (5) by the equations for *mean* motion:

$$\begin{aligned} \bar{x}_1'' + Q_1{}^2\bar{x}_1 &= \frac{\bar{\alpha}}{2}(\bar{x}_1{}^2 - \bar{x}_2{}^2) \\ \bar{x}_2'' + Q_2{}^2\bar{x}_2^2 &= -\bar{\alpha}\bar{x}_1\bar{x}_2 \end{aligned} \tag{7}$$

These are the equations we shall discuss in this section.

Q_1 and Q_2 are, respectively, the numbers of betatron oscillations experienced by variables x_1 and x_2 during each revolution of the particles in the vacuum chamber. $\bar{\alpha}$ is the mean value of $\alpha(\theta)$.

Next we shall modify the notation by dropping the bar over the mean variables; we shall write x_1, x_2, α instead of $\bar{x}_1$, $\bar{x}_2$, $\bar{\alpha}$, keeping in mind, however, that we are dealing with mean values.

3.1. The Potential-Energy Surface

Equations (7) have been studied particularly by Hagedorn, who uses the potential function

$$U(x_1, x_2) = -\frac{1}{2}Q_1{}^2x_1{}^2 - \frac{1}{2}Q_2{}^2x_2{}^2 - \frac{\alpha}{2}(x_2{}^2x_1 - \frac{1}{3}x_1{}^3) \tag{8}$$

Equation (7) reduces to

$$x_1'' = \frac{\partial U}{\partial x_1} \qquad x_2'' = \frac{\partial U}{\partial x_2} \tag{9}$$

The quantity U may be regarded as the negative of the *potential energy* of a particle, with unit mass, and (9) describes the motion of this particle along a two-dimensional potential surface,

$$E_p(x_1, x_2) = -U(x_1, x_2)$$

The discussion is notably simplified by introducing new variables ξ and η:

$$\xi = -\frac{\alpha}{Q_2{}^2}x_1 \qquad \eta = -\frac{\alpha}{Q_2{}^2}x_2$$

and parameter ρ:

$$\rho = \frac{Q_1{}^2}{Q_2{}^2}$$

Then (8) is rewritten

$$\frac{2\alpha^2}{Q_2{}^6}E_p(x_1, x_2) = V(\xi, \eta) = \rho\xi^2 + \eta^2 - (\xi\eta^2 - \tfrac{1}{3}\xi^3) \tag{10}$$

A point at which the tangent plane to the surface $V(\xi, \eta)$ is horizontal corresponds to a stationary value of $V(\xi, \eta)$; i.e., at that point

$$\frac{\partial V}{\partial \xi} = \frac{\partial V}{\partial \eta} = 0$$

Then if the particle is located at that point with zero velocity, for $\theta = \theta_0$, it will remain at that point whatever $\theta > \theta_0$; this point is an equilibrium point.

Furthermore the equilibrium is:

(a) *Stable* if the potential energy is minimum at that point. Then it may be proved that, in the neighborhood of the equilibrium point, the equipotential curves

$$V(\xi, \eta) = C$$

where C is any positive constant, are closed loops which surround this point.

If the particle is released from a point of such a level curve, with zero velocity, it will oscillate in the region which is bounded by this curve.

(b) *Unstable* if the potential energy is maximum. In this case the level curves in the neighborhood of the equilibrium point are also closed loops surrounding this point.

(c) *Unstable* also at a saddle point of the potential surface, since there a small displacement can make the particle fall toward lower energies.

Now from (10) we deduce the stationary condition

$$\frac{\partial V}{\partial \xi} = 2\rho\xi - \eta^2 + \xi^2 = 0$$

$$\frac{\partial V}{\partial \eta} = 2\eta(1 - \xi) = 0$$

In Fig. 8 we have plotted the curves

$$\frac{\partial V}{\partial \xi} = 0 \qquad \text{and} \qquad \frac{\partial V}{\partial \eta} = 0$$

and we get the intersection points

B, C:	$\xi = 1$	$\eta = \pm\sqrt{1 + 2\rho}$	$V(\xi, \eta) = \rho + \frac{1}{3}$
0:	$\eta = 0$	$\xi = 0$	$V(\xi, \eta) = 0$
A:	$\eta = 0$	$\xi = -2\rho$	$V(\xi, \eta) = \frac{4}{3}\rho^3$

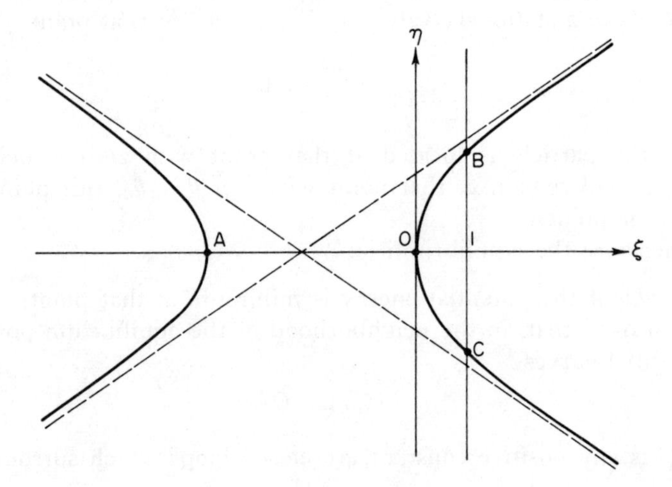

FIG. 8. Equilibrium points.

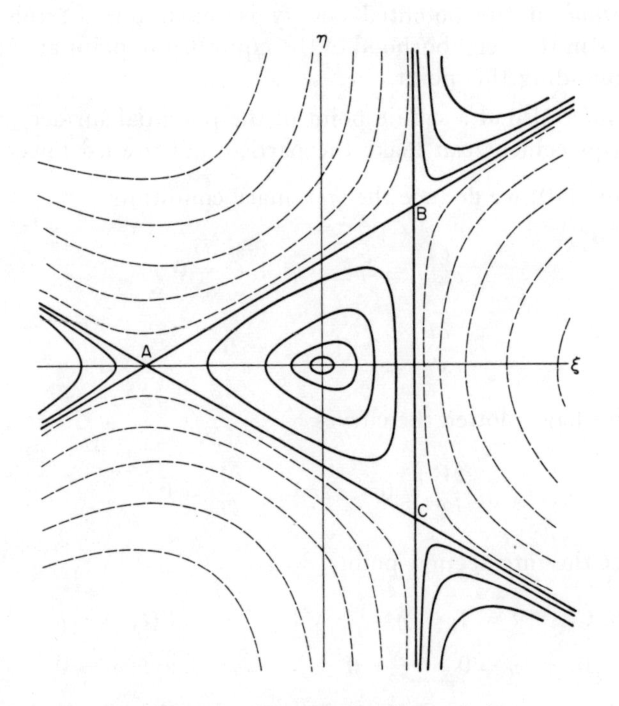

FIG. 9. Potential-energy surface: $\rho = 1$.

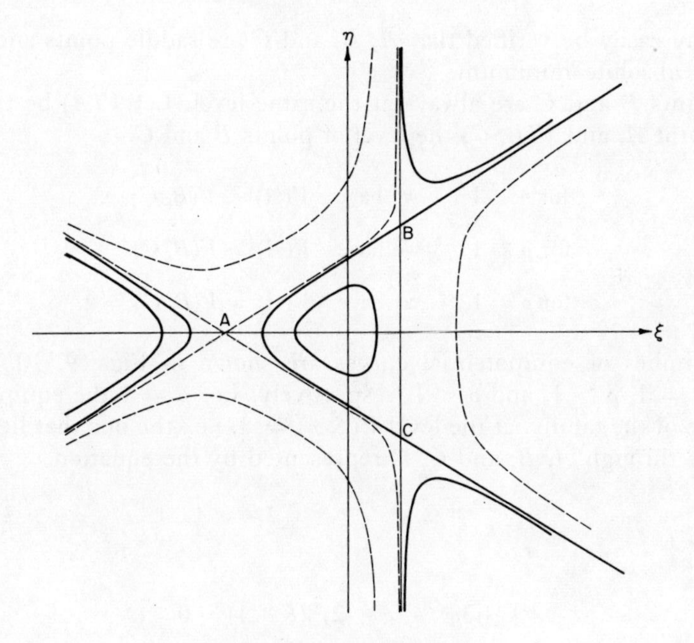

FIG. 10. Potential-energy surface: $\rho > 1$.

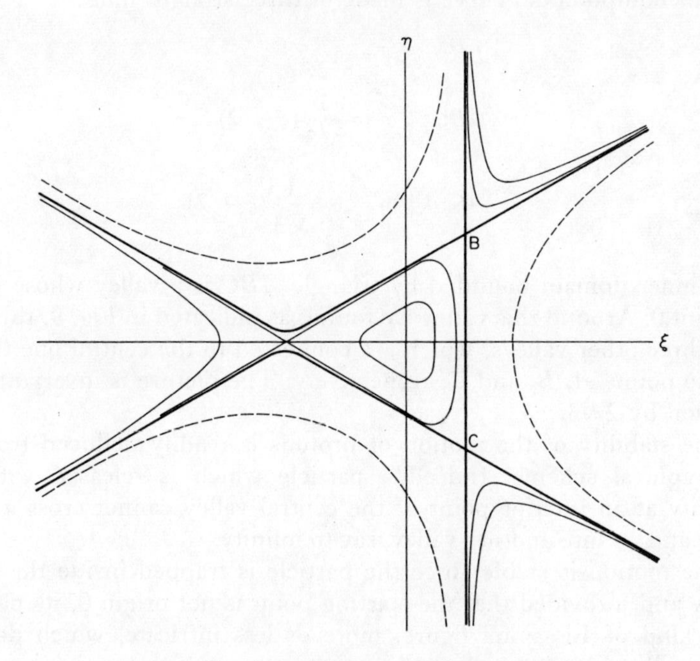

FIG. 11. Potential-energy surface: $\rho < 1$.

It may easily be verified that A, B, and C are saddle points and that 0 is an absolute minimum.

Points B and C are always at the same level. Let $V(A)$ be the level of point A, and $V(B, C)$ the level of points B and C.

$$\text{for } \rho < 1 \quad \text{we have} \quad V(A) < V(B, C)$$

$$\text{for } \rho > 1 \quad \text{we have} \quad V(A) > V(B, C)$$

$$\text{for } \rho = 1 \quad \text{we have} \quad V(A) = V(B, C) = \tfrac{4}{3}$$

Families of equipotential curves are shown in Figs. 9, 10, and 11 for $\rho = 1$, $\rho > 1$, and $\rho < 1$, respectively. For $\rho = 1$ the equipotential curve of the family, at the level $V(\xi, \eta) = \frac{4}{3}$, i.e., the one that lies in the plane through A, B, and C, is represented by the equation

$$\xi^2 + \eta^2 - \xi\eta^2 + \tfrac{1}{3}\xi^3 = \tfrac{4}{3}$$

or

$$[3\eta^2 - (\xi + 2)^2](\xi - 1) = 0$$

This equipotential curve is made of three straight lines:

$$BC: \quad \xi = 1$$

$$AB: \quad \eta = \frac{1}{\sqrt{3}}(\xi + 2)$$

$$AC: \quad \eta = -\frac{1}{\sqrt{3}}(\xi + 2)$$

The inner domain bounded by triangle ABC is a valley whose bottom is point 0. Around this valley we found, as indicated in Fig. 9, three hills and three other valleys, which are connected to the central one through saddle points A, B, and C, respectively. The picture is invariant under rotation by $2\pi/3$.

The stability of the motion of protons is readily deduced from this geographical scheme. Indeed a particle which is released with zero velocity at an interior point of the central valley cannot cross a saddle point and go into another valley, say to infinity.

The motion is stable since the particle is trapped inside the central valley and, provided that the starting point is not origin 0, its path will be a kind of Lissajoux figure, more or less intricate, which does not necessarily reduce to a closed loop.

For $\rho \neq 1$, the equation of the family of equipotential curves is

$$\rho\xi^2 + \tfrac{1}{3}\xi^3 + (1 - \xi)\eta^2 = C \tag{11}$$

Let us put

$$\rho = 1 + \delta$$

and consider the case $C = \rho + \frac{1}{3} = \frac{4}{3} + \delta$, which again corresponds to the horizontal plane through B and C. Equation (11) is rewritten

$$(\xi - 1)[(\xi + 1)\delta - \tfrac{1}{3}[3\eta^2 - (\xi + 2)^2]] = 0$$

Accordingly, straight line BC: $\xi = 1$ belongs to the equipotential curve at the level $V(\xi, \eta) = \frac{4}{3} + \delta$.

For $\xi \neq 1$, one gets the hyperbolas

$$\eta^2 = \tfrac{1}{3}(\xi + 2)^2 + (\xi + 1)\delta$$

These hyperbolas are shown in Figs. 10 and 11, which also portray the main topographic features of the potential-energy surface, in two interesting cases corresponding, respectively, to

$$\rho > 1 \qquad \text{i.e.,} \quad \delta > 0$$

and

$$\rho < 1 \qquad \text{i.e.,} \quad \delta < 0$$

When $\rho \neq 1$, the picture is no longer invariant under rotation by $2\pi/3$.

3.2. Admissible Energies for the Particles

It follows from the above discussion that the betatron oscillations are stable if the total energy E of the particles is less than the potential energy associated with the lower saddle point.

Indeed, for $\rho = 1$, the three saddle points A, B, and C are at the same level, and the stability condition is

$$V(\xi, \eta) < \tfrac{4}{3} \qquad \text{which implies} \qquad E < \frac{2}{3}\frac{Q_1^6}{\alpha^2} \tag{12}$$

Here

$$\rho = 1 \quad \Rightarrow \quad Q_1 = Q_2$$

For $\rho < 1$ we must have

$$V(\xi, \eta) < V(A) \qquad \text{say} \quad E < \frac{4}{3}\rho^3\frac{Q_2^6}{2\alpha^2}$$

and since $\rho = Q_1^2/Q_2^2$, the stability condition again reduces to (12):

$$E < \frac{2}{3}\frac{Q_1^2}{\alpha^2}$$

For $\rho > 1$ the condition is

$$V(\xi, \eta) < V(B, C) \qquad \text{say} \quad E < (\rho + \tfrac{1}{3})\frac{Q_2^6}{2\alpha^2}$$

or

$$E < \frac{Q_2^6 + 3Q_1^2Q_2^4}{6\alpha^2} \tag{13}$$

The admissible energy E increases when the coefficient α of the nonlinear coupling tends to zero. This conclusion is fairly intuitive, since then the equations tend to linear ones.

Another condition is provided by the geometry of the vacuum chamber, which introduces another critical energy, since a particle cannot touch the wall unless it is lost. The situation is illustrated by Fig. 12, which

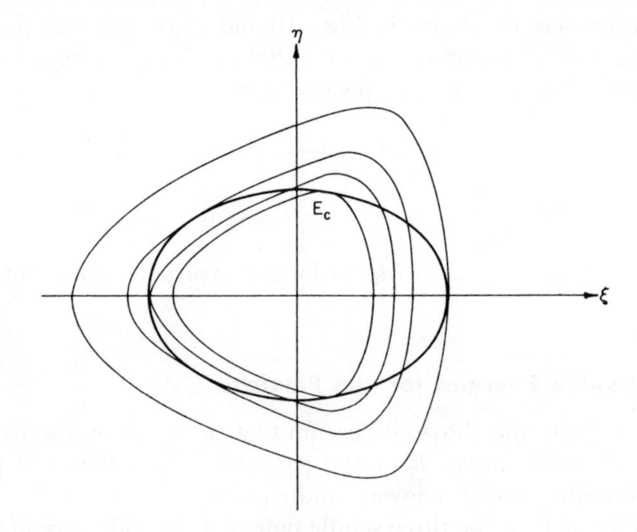

FIG. 12. Admissible energies for the particles, $E < E_c$. Elliptic cross section of the vacuum chamber and the just-touching equipotential line.

corresponds to the case $\rho = 1$, but the discussion is the same for $\rho \neq 1$. The elliptic cross section of the chamber is shown, together with the just-touching equipotential line.

A particle which is injected in the neighborhood of the reference orbit with a total energy E less than the level E_c of this critical equipotential curve is definitively trapped inside the potential valley which it bounds. Accordingly, it cannot touch the wall of the vacuum chamber.

3.3. Motion in the Near Neighborhood of the Reference Orbit. Analytic Method

Now we shall study more accurately the motion of a particle in the near neighborhood of the reference orbit, say in the central potential valley of Figs. 9–11.

We shall use an analytic method whose main idea is the following: As pointed out earlier, a particle which is trapped inside the central potential valley will in general execute a motion which looks like a Lissajoux figure but which does not reduce to a closed loop: The ratio of the frequencies in the x_1 and x_2 directions is not a rational number, in general.

Now we are led to this question: Is it possible to find periodic motions, with respect to time t, in such a potential valley?

We shall see that it is possible, then it is reasonable to compare such motions to the eigenstates of a linear system, with the proviso that in a nonlinear system the superposition principle does not hold. Accordingly, we shall have to consider these "pseudo eigenstates" *separately*.

In practice we shall start with the system

$$\begin{aligned} x_1'' + Q_1^2 x_1 &= \frac{\alpha}{2}(x_1^2 - x_2^2) \\ x_2'' + Q_2^2 x_2 &= -\alpha x_1 x_2 \end{aligned} \tag{14}$$

and we shall try to fit a solution of the form

$$\begin{aligned} x_1 &= a_1 \cos n_1 Q\theta + h_1 \qquad & a_1 > 0 \\ x_2 &= a_1 \cos(n_2 Q\theta - \varphi) + h_2 \qquad & a_2 > 0 \end{aligned} \tag{15}$$

where n_1 and n_2 are integers, φ an expected phase shift, and h_1 and h_2 two constants which will measure a small deviation between the center of the motion and the bottom of the valley due to nonlinearity. In general h_1 and h_2 will depend on amplitudes a_1 and a_2.

We shall assume that these quantities are of first-order smallness and will neglect αh_1, αh_2, h_1^2, and h_2^2, according to which, by substituting (15) in (14), we get

$$\begin{aligned} &(-n_1^2 Q^2 + Q_1^2) a_1 \cos n_1 Q\theta + Q_1^2 h_1 \\ &\qquad = \frac{\alpha}{2}\left[\frac{a_1^2 - a_2^2}{2} + \frac{a_1^2}{2}\cos 2n_1 Q\theta - \frac{a_2^2}{2}\cos 2(n_2 Q\theta - \varphi)\right] \\ &(-n_2^2 Q^2 + Q_2^2) a_2 \cos(n_2 Q\theta - \varphi) + Q_2^2 h_2 \\ &\qquad = -\frac{\alpha}{2} a_1 a_2 \{\cos[(n_1 + n_2)Q\theta - \varphi] + \cos[(n_2 - n_1)Q\theta - \varphi]\} \end{aligned} \tag{16}$$

A fairly simple answer to the question above is obtained without entering the details of a general discussion, by putting

$$n_1 = 2n_2 \qquad n_2 = 1$$

From (16) we get

$$h_1 = \frac{\alpha(a_1^2 - a_2^2)}{4Q_1^2} \qquad h_2 = 0 \tag{17}$$

and system (16) reduces to

$$\begin{aligned} (Q_1^2 - 4Q^2)a_1 \cos 2Q\theta &= -\frac{\alpha a_2^2}{4} \cos 2(Q\theta - \varphi) + \frac{\alpha a_1^2}{4} \cos 4Q\theta \\ (Q_2^2 - Q^2)a_2 \cos(Q\theta - \varphi) &= -\alpha \frac{a_1 a_2}{2} [\cos(3Q\theta - \varphi) + \cos(Q\theta + \varphi)] \end{aligned} \tag{18}$$

We shall also disregard harmonics 3 and 4, which do not play an important role. They only introduce a slight distortion, without modifying the over-all shape of the solution, provided that a_1 and a_2 are sufficiently small.

Finally, the motion in the central potential valley is represented by

$$\begin{aligned} (Q_1^2 - 4Q^2)a_1 \cos 2Q\theta &= -\frac{\alpha a_2^2}{4} \cos 2(Q\theta - \varphi) \\ (Q_2^2 - Q^2)a_2 \cos(Q\theta - \varphi) &= -\alpha \frac{a_1 a_2}{2} \cos(Q\theta + \varphi) \end{aligned} \tag{19}$$

Then we are faced with the following cases:

Case 1. $\varphi = 0$:

In Case 1 equations (19) reduce to

$$\begin{aligned} (Q_1^2 - 4Q^2)a_1 \cos 2Q\theta &= -\frac{\alpha a_2^2}{4} \cos 2Q\theta \\ (Q_2^2 - Q^2)a_2 \cos Q\theta &= -\alpha \frac{a_1 a_2}{2} \cos Q\theta \end{aligned} \tag{20}$$

from which, if $a_2 \neq 0$,

$$\begin{aligned} Q_1^2 - 4Q^2 &= -\frac{\alpha}{4} \frac{a_2^2}{a_1} \\ Q_2^2 - Q^2 &= -\frac{\alpha}{2} a_1 \end{aligned} \tag{21}$$

By eliminating Q between these two equations, we get a relation between a_1 and a_2 :

$$(Q_1^2 - 4Q_2^2)a_1 - 2\alpha a_1^2 = -\frac{\alpha}{4}a_2^2 \tag{22}$$

Now if we assume that a_1 is small and $Q_1 \neq 2Q_2$, we get

$$a_1 \simeq -\frac{\alpha a_2^2}{4(Q_1^2 - 4Q_2^2)} \tag{23}$$

Note that, since amplitude a_1 is positive, Case 1 is valid only if

$$\alpha(Q_1 - 2Q_2) < 0$$

Then the expected solution is

$$\begin{aligned} x_1 &= -\frac{\alpha a_2^2}{4(Q_1^2 - 4Q_2^2)} \cos\left(Q_1^2 + \frac{\alpha}{4}\frac{a_2^2}{a_1}\right)^{1/2} \theta + \alpha\frac{a_1^2 - a_2^2}{4Q_1^2} \\ x_2 &= a_2 \cos\left(Q_2^2 + \frac{\alpha a_1}{2}\right)^{1/2} \theta \end{aligned} \tag{24}$$

Case 2. $\varphi = \pi/2$:

By similar arguments we get, if $a_2 \neq 0$,

$$(Q_1^2 - 4Q_2^2)a_1 + 2\alpha a_1^2 = \frac{\alpha}{4}a_2^2 \tag{25}$$

Then, if a_1 is small and $Q_1 \neq 2Q_2$, the solution is

$$\begin{aligned} x_1 &= \frac{\alpha a_2^2}{4(Q_1^2 - 4Q_2^2)} \cos\left(Q_1^2 - \frac{\alpha}{4}\frac{a_2^2}{a_1}\right)^{1/2} \theta + \alpha\frac{a_1^2 - a_2^2}{4Q_1^2} \\ x_2 &= a_2 \sin\left(Q_2^2 - \frac{\alpha a_1}{2}\right)^{1/2} \theta \end{aligned} \tag{26}$$

As above, it may easily be seen that this solution is valid only if $\alpha(Q_1 - 2Q_2) > 0$.

Resonance. $Q_1 = 2Q_2$:

When $Q_1 = 2Q_2$, relations (22) and (25) become

$$2\alpha a_1^2 = \frac{\alpha}{4}a_2^2$$

from which follows

$$a_1 = \frac{a_2}{2\sqrt{2}} \tag{27}$$

Then the expected solution is

$$\begin{aligned} x_1 &= \frac{a_2}{2\sqrt{2}} \cos\left(Q_1^2 + \frac{\alpha}{\sqrt{2}} a_2\right)^{1/2} \theta - \frac{7\alpha}{32} \frac{a_2^2}{Q_1^2} \\ x_2 &= a_2 \cos\left(Q_2^2 + \frac{\alpha}{4\sqrt{2}} a_2\right)^{1/2} \theta \end{aligned} \tag{28}$$

if $Q_1 \to 2Q_2$, with $\alpha(Q_1 - 2Q_2) < 0$.

On the other hand, it is

$$\begin{aligned} x_1 &= \frac{a_2}{2\sqrt{2}} \cos\left(Q_1^2 - \frac{\alpha}{\sqrt{2}} a_2\right)^{1/2} \theta - \frac{7\alpha}{32} \frac{a_2^2}{Q_1^2} \\ x_2 &= a_2 \sin\left(Q_2^2 - \frac{\alpha}{4\sqrt{2}} a_2\right)^{1/2} \theta \end{aligned} \tag{29}$$

if $Q_1 \to 2Q_2$, with $\alpha(Q_1 - 2Q_2) > 0$.

Note that, given a_2, the value of a_1 at the resonance is very large as compared with the value given by formula (23), where a_1 is of the order

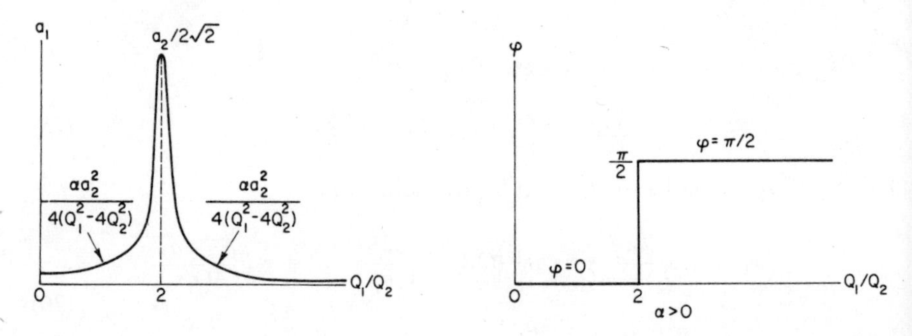

FIG. 13. Variations of amplitude a_1 and phase shift φ in the neighborhood of the resonance. $\alpha > 0$.

of α. In Fig. 13 amplitude a_1 and phase shift φ are plotted as functions of Q_1/Q_2, in the neighborhood of the resonance.

Remark. Note that when a_1 and a_2 are not sufficiently small, i.e., when the particle is not sufficiently close to the reference orbit, the first-harmonic approximation is not valid. By "sufficiently small" we mean that the moving point in the potential valley must execute its motion in the bottom of the valley. Indeed should the particle draw near a saddle point, the motion would become a relaxation oscillation, which is no longer relevant on the above approximate theory.

4. SELF-SUSTAINED OSCILLATORS WITH TWO DEGREES OF FREEDOM

In the above sections we have restricted the analysis to conservative systems; now we shall attack the case of systems with two degrees of freedom, in which regenerative couplings exist.

Such circuits usually consist of two coupled self-sustained oscillators which may be assumed to be identical except for a small difference between the frequencies of their free oscillations (when uncoupled). We shall see that under certain conditions the two oscillators will tend to synchronize mutually; i.e., their free oscillations will be extinguished by the coupling and they will vibrate at the same frequency.

This phenomenon was first described by Huyghens, who discovered the fact that two mechanical clocks hung on a thin wall tend to run at exactly the same speed. A similar observation in acoustics was made by Lord Rayleigh [1], who experimented on two organ pipes of slightly different frequencies coupled through a resonator.

Throughout this section we shall refer to the following two examples:

Example 1. The first one, which we borrow from Theodorchik [2], is shown in Fig. 14.

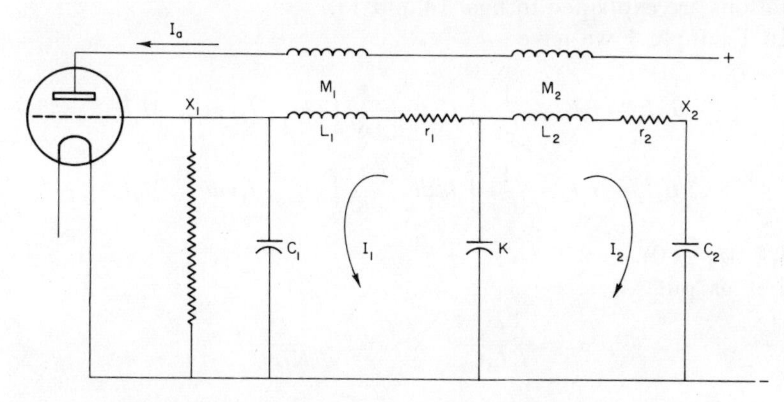

FIG. 14. Self-sustained oscillator with two degrees of freedom, first example.

Example 2. The second one, which is shown in Fig. 15, is similar to the first, but a difference is introduced by the fact that we have in this case two different regenerative circuits, whereas in Example 1 use of the same tube results in the maintenance of two coupled circuits.

From a physical viewpoint the difference is important. On the other hand, the second oscillator is basically similar to the coupled oscillators used in experiments by Huyghens and Lord Rayleigh.

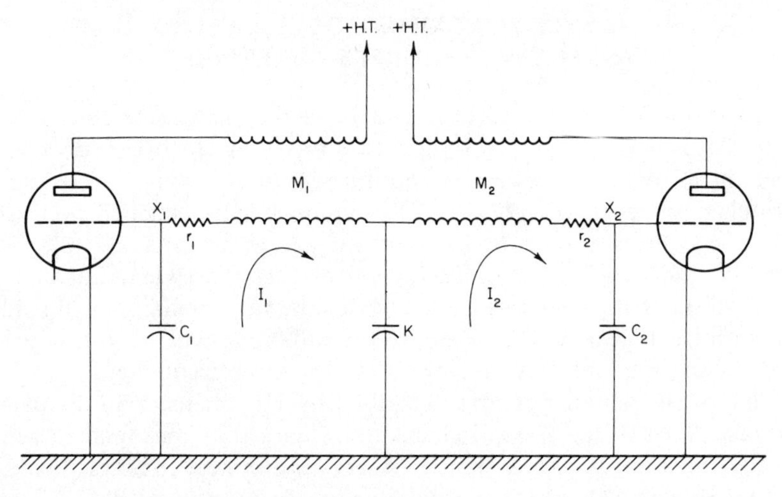

FIG. 15. Self-sustained oscillator with two degrees of freedom, second example.

4.1. Diagrams and Equations of Motion

The equations of motion of these oscillators are readly obtained. The notations are explained in figs. 14 and 15.

In Example 1 we have

$$\begin{aligned} L_1 \dot{I}_1 + r_1 I_1 + \frac{1}{C_1} \int I_1 \, dt + \frac{1}{K} \int (I_1 - I_2) \, dt = M_1 \dot{I}_a \\ L_2 \dot{I}_2 + r_2 I_2 + \frac{1}{C_2} \int I_2 \, dt + \frac{1}{K} \int (I_2 - I_1) \, dt = M_2 \dot{I}_a \end{aligned} \tag{30}$$

$(M_1, M_2 > 0)$.

Let us put

$$\alpha_1 = \frac{C_2}{KL_1C_1} \qquad \alpha_2 = \frac{C_1}{KL_2C_2}$$

$$x_1 = \frac{1}{C_1} \int I_1 \, dt \qquad x_2 = \frac{1}{C_2} \int I_2 \, dt$$

(α_1 and α_2 are positive).

Furthermore, since x_1 is the voltage which is applied by the leading tank circuit to the grid of the tube, assume

$$I_a = S_0 x_1 + S_2 x_1^3$$

$(S_0 > 0, S_2 < 0)$.

Then equations (30) are rewritten

$$\begin{aligned} \ddot{x}_1 + 2\delta_1\dot{x}_1 + \omega_1^2 x_1 &= \frac{M_1}{L_1C_1}(S_0 + 3S_2x_1^2)\dot{x}_1 + \alpha_1 x_2 \\ \ddot{x}_2 + 2\delta_2\dot{x}_2 + \omega_2^2 x_2 &= \frac{M_2}{L_2C_2}(S_0 + 3S_2x_1^2)\dot{x}_1 + \alpha_2 x_1 \end{aligned} \tag{31}$$

with

$$\delta_1 = \frac{r_1}{2L_1} \qquad \omega_1^2 = \frac{1}{L_1C_1}\left(1 + \frac{C_1}{K}\right)$$

$$\delta_2 = \frac{r_2}{2L_2} \qquad \omega_2^2 = \frac{1}{L_2C_2}\left(1 + \frac{C_2}{K}\right)$$

Similarly, Example 2, the equations of motion are

$$\begin{aligned} \ddot{x}_1 + 2\delta_1\dot{x}_1 + \omega_1^2 x_1 &= \frac{M_1}{L_1C_1}(S_0 + 3S_2x_1^2)\dot{x}_1 + \alpha_1 x_2 \\ \ddot{x}_2 + 2\delta_2\dot{x}_2 + \omega_2^2 x_2 &= \frac{M_2}{L_2C_2}(S_0 + 3S_2x_2^2)\dot{x}_2 + \alpha_2 x_1 \end{aligned} \tag{32}$$

Here we assume that the tubes are identical.

4.2. Synchronized Solution†

We shall follow the same method as in Section 3.3; we shall try to fit a solution such that the representative point in plane (x_1, x_2) moves along a periodic path (with respect to time t). Here it seems reasonable to try

$$\begin{aligned} x_1 &= a_1 \cos \omega t \\ x_2 &= a_2 \cos(\omega t - \varphi) \end{aligned} \tag{33}$$

$(a_1, a_2 > 0)$.

Such a solution, when it exists, is called a *synchronized solution.*

Let us consider the examples pointed out above.

4.2.1. Example 1

By making use of the quasi-linearization method, as explained in Chapter II, Section 1.4, we transform (31) into

$$\begin{aligned} \ddot{x}_1 + 2\delta_1\dot{x}_1 + \omega_1^2 x_1 &= \frac{M_1}{L_1C_1}\left(S_0 + \frac{3S_2}{4}a_1^2\right)\dot{x}_1 + \alpha_1 x_2 \\ \ddot{x}_2 + 2\delta_2\dot{x}_2 + \omega_2^2 x_2 &= \frac{M_2}{L_2C_2}\left(S_0 + \frac{3S_2}{4}a_1^2\right)\dot{x}_1 + \alpha_2 x_1 \end{aligned} \tag{34}$$

† See [1, 2, 4, 6, 7, 14, 18, 19].

Indeed it is a convenient form, since thereafter we shall be entitled to follow the usual linear techniques—we shall replace expressions (33) of x_1 and x_2 by

$$x_1 = a_1 e^{j\omega t} \qquad x_2 = a_2 e^{j(\omega t - \varphi)} \tag{35}$$

thereupon introducing a complex coupling coefficient between variables x_1 and x_2, say

$$\Lambda = \lambda e^{j\varphi} \qquad \text{with } \lambda = \frac{a_1}{a_2}, \qquad x_1 = \Lambda x_2 \tag{36}$$

By substituting (35) and (36) in (34) we get algebraic equations, from which ω, a_1, a_2, and φ can be computed:

$$\begin{aligned}\left\{(\omega_1^2 - \omega^2) + j\omega \left[2\delta_1 - \frac{M_1}{L_1 C_1}\left(S_0 + \frac{3S_2}{4} a_1^2\right)\right]\right\} \Lambda &= \alpha_1 \\ [(\omega_2^2 - \omega^2) + 2j\omega\delta_2] &= \left[j\omega \frac{M_2}{L_2 C_2}\left(S_0 + \frac{3S_2}{4} a_1^2\right) + \alpha_2\right] \Lambda\end{aligned} \tag{37}$$

By the way, this set is equivalent to

$$\begin{aligned}&(\omega_1^2 - \omega^2)\cos\varphi - \left[2\delta_1 - \frac{M_1}{L_1 C_1}\left(S_0 + \frac{3S_2}{4} a_1^2\right)\right] \omega \sin\varphi = \frac{\alpha_1}{\lambda} \\ &(\omega_1^2 - \omega^2)\sin\varphi + \left[2\delta_1 - \frac{M_1}{L_1 C_1}\left(S_0 + \frac{3S_2}{4} a_1^2\right)\right] \omega \cos\varphi = 0 \\ &\frac{\omega_2^2 - \omega^2}{\lambda} = \alpha_2 \cos\varphi - \frac{M_2}{L_2 C_2}\left(S_0 + \frac{3S_2}{4} a_1^2\right) \omega \sin\varphi \\ &\frac{2\omega\delta_2}{\lambda} = \alpha_2 \sin\varphi + \frac{M_2}{L_2 C_2}\left(S_0 + \frac{3S_2}{4} a_1^2\right) \omega \cos\varphi\end{aligned} \tag{38}$$

From (37) we deduce

$$\begin{aligned}&(\omega_1^2 - \omega^2)(\omega_2^2 - \omega^2) - 2\omega^2\delta_2 \left[2\delta_1 - \frac{M_1}{L_1 C_1}\left(S_0 + \frac{3S_2}{4} a_1^2\right)\right] \\ &\quad + j\omega \left\{2(\omega_1^2 - \omega^2)\delta_2 + (\omega_2^2 - \omega^2)\left[2\delta_1 - \frac{M_1}{L_1 C_1}\left(S_0 + \frac{3S_2}{4} a_1^2\right)\right]\right\} \\ &\qquad = \alpha_1\alpha_2 + j\omega\alpha_1 \frac{M_2}{L_2 C_2}\left(S_0 + \frac{3S_2}{4} a_1^2\right)\end{aligned} \tag{39}$$

which is equivalent to

$$(\omega_1^2 - \omega^2)(\omega_2^2 - \omega^2) - 2\omega^2\delta_2 \left[2\delta_1 - \frac{M_1}{L_1C_1}\left(S_0 + \frac{3S_2}{4} a_1^2\right)\right] = \alpha_1\alpha_2 \tag{40a}$$

$$2(\omega_1^2 - \omega^2)\delta_2 + (\omega_2^2 - \omega^2) \left[2\delta_1 - \frac{M_1}{L_1C_1}\left(S_0 + \frac{3S_2}{4} a_1^2\right)\right]$$

$$= \alpha_1 \frac{M_2}{L_2C_2}\left(S_0 + \frac{3S_2}{4} a_1^2\right) \tag{40b}$$

To each pair of solutions ω and a_1 there corresponds a synchronized solution. The corresponding values of a_2 and φ are obtained from (38).

4.2.2. Example 2

In the second example, the quasi-linear equations are also easily deduced from (32). They are

$$\begin{aligned} \ddot{x}_1 + \omega_1^2 x_1 &= (A_1 - B_1 a_1^2)\dot{x}_1 + \alpha_1 x_2 \\ \ddot{x}_2 + \omega_2^2 x_2 &= (A_2 - B_2 a_2^2)\dot{x}_2 + \alpha_2 x_1 \end{aligned} \tag{41}$$

when putting

$$A_1 = \frac{M_1S_0}{L_1C_1} - 2\delta_1 \qquad B_1 = -\frac{3M_1S_2}{4L_1C_1}$$

$$A_2 = \frac{M_2S_0}{L_2C_2} - 2\delta_2 \qquad B_2 = -\frac{3M_2S_2}{4L_2C_2}$$

As above we obtain

$$\begin{aligned} [(\omega_1^2 - \omega^2) - j\omega(A_1 - B_1a_1^2)]\Lambda &= \alpha_1 \\ (\omega_2^2 - \omega^2) - j\omega(A_2 - B_2a_2^2) &= \alpha_2\Lambda \end{aligned} \tag{42}$$

from which follows

$$\begin{aligned} A_1 - B_1a_1^2 &= \frac{\alpha_1}{\omega\lambda} \sin\varphi \\ A_2 - B_2a_2^2 &= -\frac{\alpha_2}{\omega} \lambda \sin\varphi \\ \omega_1^2 - \omega^2 &= \frac{\alpha_1}{\lambda} \cos\varphi \\ \omega_2^2 - \omega^2 &= \alpha_2\lambda \cos\varphi \end{aligned} \tag{43}$$

From (42) we also deduce

$$(\omega_1^2 - \omega^2)(\omega^2_2 - \omega^2) - (A_1 - B_1 a_1^2)(A_2 - B_2 a_2^2)\omega^2$$
$$- j\omega[(\omega_1^2 - \omega^2)(A_2 - B_2 a_2^2) + (\omega_2^2 - \omega^2)(A_1 - B_1 a_1^2)] = \alpha_1 \alpha_2 \quad (44)$$

or

$$\begin{aligned} (\omega_1^2 - \omega^2)(\omega_2^2 - \omega^2) - (A_1 - B_1 a_1^2)(A_2 - B_2 a_2^2)\omega^2 &= \alpha_1 \alpha_2 \\ (\omega_1^2 - \omega^2)(A_2 - B_2 a_2^2) + (\omega_2^2 - \omega^2)(A_1 - B_1 a_1^2) &= 0 \end{aligned} \quad (45)$$

4.3. Discussion. Existence of a Synchronized Solution

On the whole, (40) and (45) are complicated and do not lend themselves to easy discussion. However, many interesting results can be derived from them, provided they are simplified by some practical assumptions.

Since these equations (and the discussions) are rather different, we shall again consider separately Examples 1 and 2 of Section 4.2.

4.3.1. Example 1

Assumption 1. $M_1 = 0$, $M_2 \neq 0$:

If $M_1 = 0$, (40a) reduces to

$$(\omega_1^2 - \omega^2)(\omega_2^2 - \omega^2) - 4\delta_1\delta_2\omega^2 = \alpha_1 \alpha_2 \quad (46)$$

which is similar to the equation which gives the frequencies of a system consisting of two passive linear coupled circuits. Then from (40b) we get

$$a_1 = 2\left[\left(\frac{M_2 S_0}{L_2 C_2} - 2\delta_1 \frac{\omega_2^2 - \omega^2}{\alpha_1} - 2\delta_2 \frac{\omega_1^2 - \omega^2}{\alpha_1}\right) \Big/ - \frac{3 M_2 S_2}{L_2 C_2}\right]^{1/2} \quad (47)$$

with the condition

$$2\delta_1(\omega_2^2 - \omega^2) + 2\delta_2(\omega_1^2 - \omega^2) < \frac{M_2 S_0}{L_2 C_2}\alpha_1 \quad (48)$$

Assumption 2. $M_2 = 0$, $M_1 \neq 0$:

Assuming $M_2 = 0$, (40) becomes

$$(\omega_1^2 - \omega^2)(\omega_2^2 - \omega^2) - 4\delta_1\delta_2\omega^2 + 2\delta_2\omega^2 \frac{M_1}{L_1 C_1}\left(S_0 + \frac{3S_2}{4}a_1^2\right) = \alpha_1\alpha_2 \quad (49a)$$

$$2(\omega_1^2 - \omega^2)\delta_2 + 2(\omega_2^2 - \omega^2)\delta_1 - (\omega_2^2 - \omega^2)\frac{M_1}{L_1 C_1}\left(S_0 + \frac{3S_2}{4}a_1^2\right) = 0 \quad (49b)$$

From (49a) we get

$$a_1 = 2\left\{\left[\left(\frac{M_1 S_0}{L_1 C_1} - 2\delta_1\right) - \frac{\alpha_1\alpha_2 - (\omega_1{}^2 - \omega^2)(\omega_2{}^2 - \omega^2)}{2\delta_2\omega^2}\right] \Big/ -\frac{3M_1 S_2}{L_1 C_1}\right\}^{1/2} \tag{50}$$

and from (49b), by taking (49a) into consideration,

$$2(\omega_1{}^2 - \omega^2)\delta_2 - \frac{\omega_2{}^2 - \omega^2}{2\delta_2\omega^2}\left[\alpha_1\alpha_2 - (\omega_1{}^2 - \omega^2)(\omega_2{}^2 - \omega^2)\right] = 0 \tag{51}$$

We have to discuss the sign of the expression which occurs in the square root of the right side of (50). Note that if $\omega_1 = \omega_2$, then (51) reduces to

$$(\omega_1{}^2 - \omega^2)[(\omega_1{}^2 - \omega^2)^2 + 4\delta_2{}^2\omega^2 - \alpha_1\alpha_2] = 0 \tag{52}$$

Furthermore, if, the coupling is sufficiently loose, we have

$$(\omega_1{}^2 - \omega^2)^2 + 4\delta_2{}^2\omega^2 - \alpha_1\alpha_2 > 0 \tag{53}$$

and

$$\omega = \omega_1 = \omega_2 \tag{54}$$

Accordingly,

$$a_1 = a_m$$

$$a_m = 2\left\{\left[\left(\frac{M_1 S_0}{L_1 C_1} - 2\delta_1\right) - \frac{\alpha_1\alpha_2}{2\delta_2\omega_1{}^2}\right] \Big/ -\frac{3M_1 S_2}{L_1 C_1}\right\}^{1/2} \tag{55}$$

a_m is real only if

$$\frac{\alpha_1\alpha_2}{2\delta_2\omega_1{}^2} < \frac{M_1 S_0}{L_1 C_1} - 2\delta_1 \qquad \text{since} \quad \frac{M_1 S_2}{L_1 C_1} < 0 \tag{56}$$

On the other hand, it may easily be seen by comparing (50) and (55) that when a_m is real it is the minimum value of a_1, since for $\omega_1 \neq \omega_2$,

$$(\omega_1{}^2 - \omega^2)(\omega_2{}^2 - \omega^2) > 0 \tag{57}$$

As a matter of fact, this inequality is a consequence of (51), which can be rewritten

$$(\omega_1{}^2 - \omega^2)[(\omega_2{}^2 - \omega^2)^2 + 4\delta_2{}^2\omega^2] - (\omega_2{}^2 - \omega^2)\alpha_1\alpha_2 = 0 \tag{58}$$

Then

$$\omega_1{}^2 - \omega^2 = \frac{(\omega^2{}_2 - \omega^2)\alpha_1\alpha_2}{(\omega_2{}^2 - \omega^2)^2 + 4\delta_2{}^2\omega^2}$$

which proves (57). The discussion of the sign of the expression in the square root (50) follows in a straightforward manner.

On Fig. 16 a_1^2 is plotted as a function of ω_2^2/ω_1^2, with the assumptions that $M_2 = 0$ and that the coupling is loose.

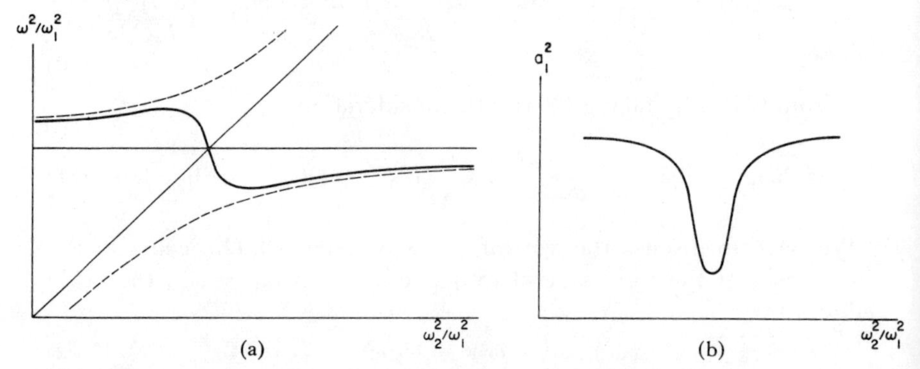

FIG. 16. Variations of (a) ω^2/ω_1^2 and (b) a_1^2 as functions of ω_2^2/ω_1^2; coupling very weak.

Two cases need to be considered:

Case 1. When the coupling is *very weak*, i.e.,

$$\frac{\alpha_1\alpha_2}{2\delta_2\omega_1^2} < \frac{M_1S_0}{L_1C_1} - 2\delta_1$$

a_m is real and, as shown in Fig. 16, the synchronized solution exists regardless of the value of ω_2/ω_1.

Case 2. When the coupling is weak but

$$\frac{\alpha_1\alpha_2}{2\delta_2\omega_1^2} > \frac{M_1S_0}{L_1C_1} - 2\delta_1$$

a_m is not real; accordingly, a_1 as given by (50) is not real between two values

$$\left(\frac{\omega_2^2}{\omega_1^2}\right)_1 \quad \text{and} \quad \left(\frac{\omega_2^2}{\omega_1^2}\right)_2$$

In Fig. 17 these boundaries are represented by points A and B, respectively. Hence the synchronized solution does not exist between A and B.

Note that Cases 1 and 2 correspond to the case where (58), whose degree is 3 with respect to ω^2, has one root which is real and two roots which are complex conjugate. This is a consequence of the fact that when $\omega_1 = \omega_2$, (52) has one root which is real, $\omega^2 = \omega_1^2 = \omega_2^2$, and two other roots which are complex conjugate according to (53). This situation is preserved by continuity when ω_1 becomes different from

ω_2. On Figs. 16 and 17, ω^2/ω_1^2 is plotted as a function of ω_2^2/ω_1^2, facing the curve which represents a_1^2 as a function of the same variable.

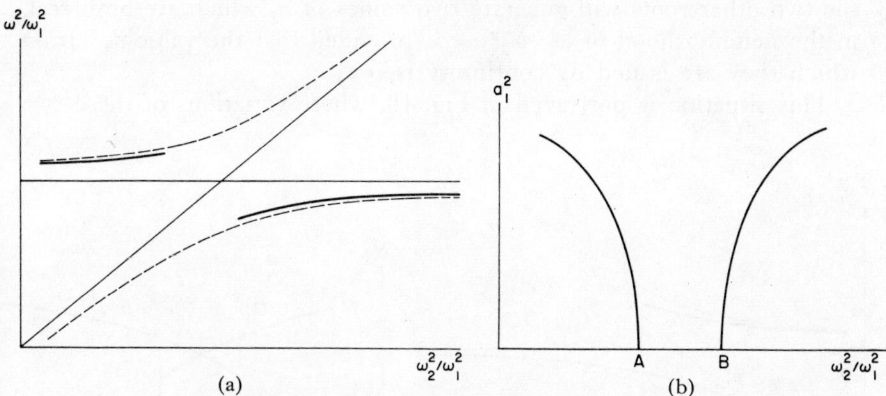

FIG. 17. Variations of (a) ω^2/ω_1^2 and (b) a_1^2 as functions of ω_2^2/ω_1^2; weak coupling, with $\alpha_1\alpha_2/2\delta_2\omega_1^2 > (M_1S_0/L_1C_1) - 2\delta_1$.

When the coupling is tight, say for

$$\alpha_1\alpha_2 > 4\delta_2^2\omega^2$$

(52) has three roots which are all real, $\omega^2 = \omega_1^2$, and the roots of the equation

$$(\omega_1^2 - \omega^2)^2 = \alpha_1\alpha_2 - 4\delta_2^2\omega^2 \tag{59}$$

This situation is preserved by continuity when ω_1 becomes different from ω_2.

In the case where $\omega_1 = \omega_2$, the value of a_1 which corresponds to the root $\omega = \omega_1 = \omega_2$ is not real, since (56) is not verified when the coupling is sufficiently tight; but the other roots that verify (59) lead to a value of a_1 which may be all right†:

$$a_1{}^* = 2\left[\left(\frac{M_1S_0}{L_1C_1} - 2\delta_1 - 2\delta_2\right)\Big/ - \frac{3M_1S_2}{L_1C_1}\right]^{1/2}$$

This value is real provided that

$$2\delta_2 < \frac{M_1S_0}{L_1C_1} - 2\delta_1$$

This is the condition for the synchronized solution to exist in the neighborhood of the point $\omega_2^2/\omega_1^2 = 1$.

† This value is readily obtained by replacing

$$\alpha_1\alpha_2 - (\omega_1^2 - \omega^2)(\omega_2^2 - \omega_2) \equiv \alpha_1\alpha_2 - (\omega_1^2 - \omega^2)^2$$

in (50) by $4\delta_2^2\omega^2$, according to (59).

When $\omega_1 \neq \omega_2$, the root which is deduced from $\omega = \omega_1 = \omega_2$ by continuity is not admissible, since it does not make a_1 real. However the two other roots will generate two values of a_1 which are both real, in the neighborhood of $\omega_2^2/\omega_1^2 = 1$, provided that the value $a_1{}^*$ from which they are issued by continuity is real.

This situation is portrayed in Fig. 18, where variations of these two

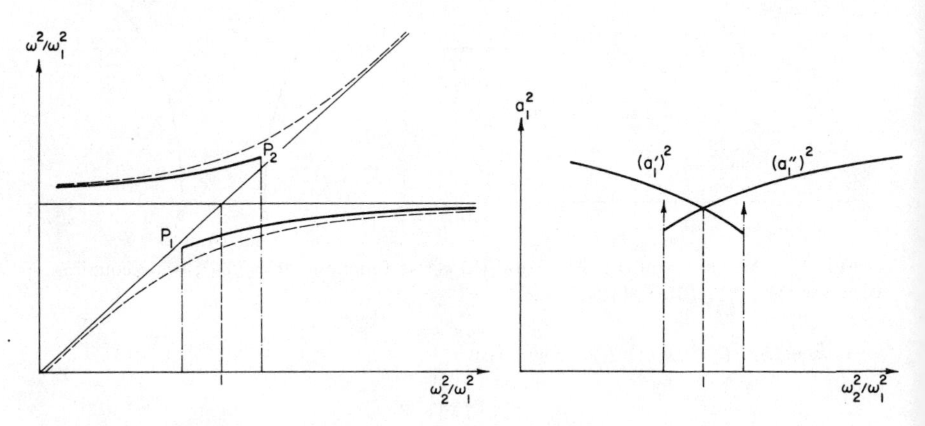

FIG. 18. Variations of (a) ω^2/ω_1^2 and (b) a_1^2 as functions of ω_2^2/ω_1^2; coupling tight. Jump phenomenon.

values, say a_1' and a_1'', and ω^2/ω_1^2 are plotted as functions of ω_2^2/ω_1^2. An example which illustrates the situation fairly well is the one in which $\delta_2 = 0$, i.e., circuit 2 is undamped. Then (58) reduces to

$$(\omega_2^2 - \omega^2)[\omega_1^2 - \omega^2)(\omega_2^2 - \omega^2) - \alpha_1\alpha_2] = 0$$

Since $\alpha_1\alpha_2/2\delta_2\omega_2^2 \to \infty$ as $\delta_2 \to 0$, however loose the coupling, the root $\omega = \omega_2$ gives no real value for a_1.

On the other hand, the equation

$$(\omega_1^2 - \omega^2)(\omega_2^2 - \omega^2) - \alpha_1\alpha_2 = 0$$

has two real roots:

$$\omega^2 = \frac{\omega_1^2 + \omega_2^2}{2} \pm \tfrac{1}{2}[(\omega_1^2 - \omega_2^2)^2 + 4\alpha_1\alpha_2]^{1/2}$$

Then, since $\omega \neq \omega_2$, (49b) reduces to

$$2\delta_1 - \frac{M_1}{L_1C_1}\left(S_0 + \frac{3S_2}{4}a_1^2\right) = 0$$

from which follows

$$a_1 = 2\left[\left(\frac{M_1 S_0}{L_1 C_1} - 2\delta_1\right) \Big/ - \frac{3M_1 S_2}{L_1 C_1}\right]^{1/2}$$

In this case both real values of ω^2 correspond to the same real value of a_1, which is constant whatever ω_2^2/ω_1^2. Variations of ω^2/ω_1^2 are plotted in Fig. 18 as dashed lines.

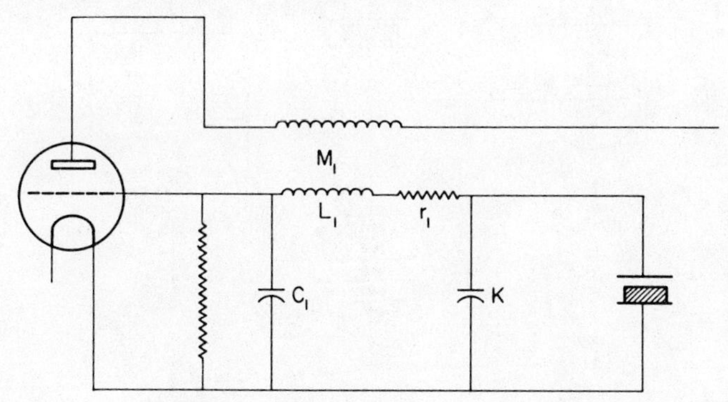

FIG. 19. Radioelectric oscillator stabilized by a quartz, example 1.

A more realistic case is the one in which δ_2 is very small but different from zero. Then we get from (58),

$$\frac{\alpha_1\alpha_2 - (\omega_1^2 - \omega^2)(\omega_2^2 - \omega^2)}{2\delta_2\omega^2} = \frac{\omega_1^2 - \omega^2}{\omega_2^2 - \omega^2} 2\delta_2$$

and

$$a = 2\left\{\left[\left(\frac{M_1 S_0}{L_1 C_1} - 2\delta_1\right) - \frac{\omega_1^2 - \omega^2}{\omega_2^2 - \omega^2} 2\delta_2\right] \Big/ - \frac{3M_1 S_2}{L_1 C_1}\right\}^{1/2}$$

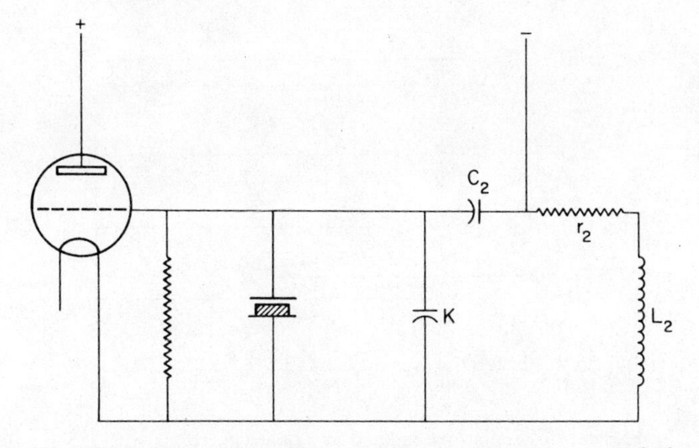

FIG. 20. Radioelectric oscillator stabilized by a quartz, example 2.

Obviously, for $\omega_1 = \omega_2$, a_1 is real (provided that δ_2 is sufficiently small). But:

(a) When $\omega \to \omega_2$, $(\omega_1^2 - \omega^2)/(\omega_2^2 - \omega^2)$ tends to ∞.

(b) When $\omega \to \omega_1$, $(\omega_1^2 - \omega^2)/(\omega_2^2 - \omega^2)$ becomes very small.

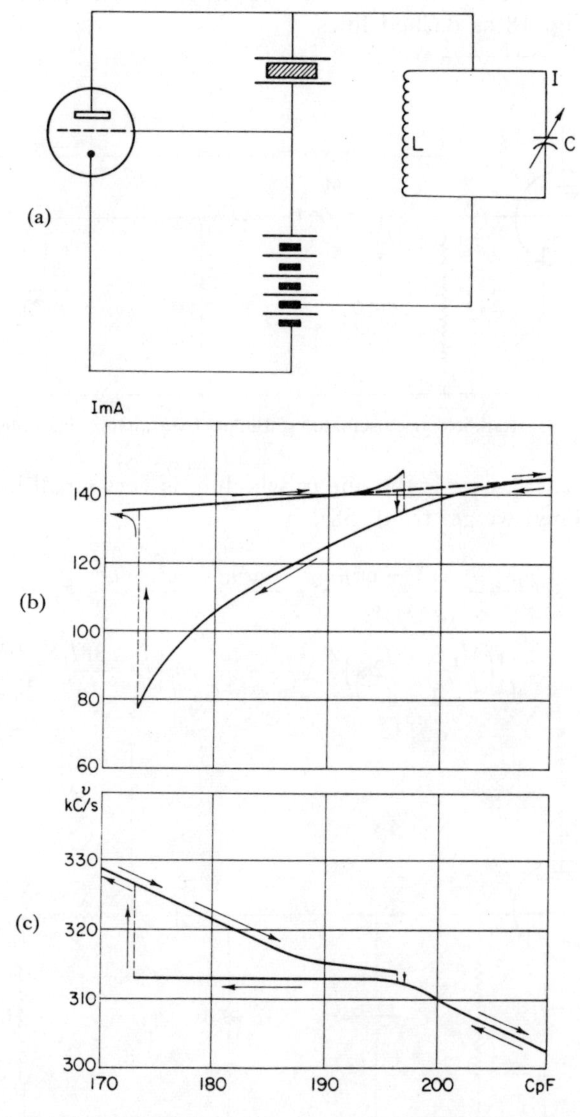

FIG. 21. (a) Dynatron oscillator stabilized by a quartz (Groszkowski [25]); (b) oscillating current I and (c) frequency ν as functions of circuit capacitance C.

Accordingly, we are led to the following conclusions:

If we plot ω^2/ω_1^2 as a function of ω_2^2/ω_1^2 (Fig. 18), we see that the curve which corresponds to $\delta_2 = 0$ is an hyperbola whose asymptotes have slopes zero and unity. On the other hand, the one which corresponds to $\delta_2 \neq 0$, very small, has only the asymptote

$$\frac{\omega^2}{\omega_1^2} = 1$$

since both branches have boundary points P_1 and P_2 according to (a). In this situation ω^2/ω_1^2 approaches unity for $\omega_2^2/\omega_1^2 \gg 1$ and for $\omega_2^2/\omega_1^2 \ll 1$. The oscillator may experience a jump at points P_1 and P_2.

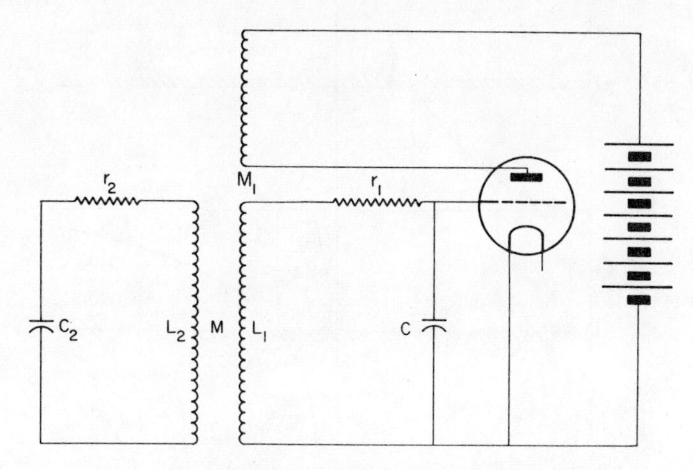

FIG. 22. Self-oscillator coupled with a resonant circuit.

The above discussion applies to the oscillator of Fig. 19, in which the frequency is stabilized by a quartz [2, 25]. On the other hand, the conclusions of Assumption 1 of Section 4.3.1, $M_1 = 0$, can be applied to the frequency-stabilized oscillator of Fig. 20. A typical example has been given by J. Groszkowski [25] and is shown in Fig. 21a. The oscillator shown in Fig. 22 is also relevant to the discussion of Assumption 2.

4.3.2. Example 2

From the last two equations of (43) we get

$$\cos \varphi = f(\lambda) = \frac{(\omega_1^2 - \omega_2^2)\lambda}{\alpha_1 - \alpha_2\lambda^2}$$

Accordingly, the synchronized solution can exist only if

$$-1 \leqslant \frac{(\omega_1^2 - \omega_2^2)\lambda}{\alpha_1 - \alpha_2\lambda^2} \leqslant +1$$

This condition is made more significant by plotting the curve $y = f(\lambda)$, for instance in the case $\omega_1 > \omega_2$ (Fig. 23).

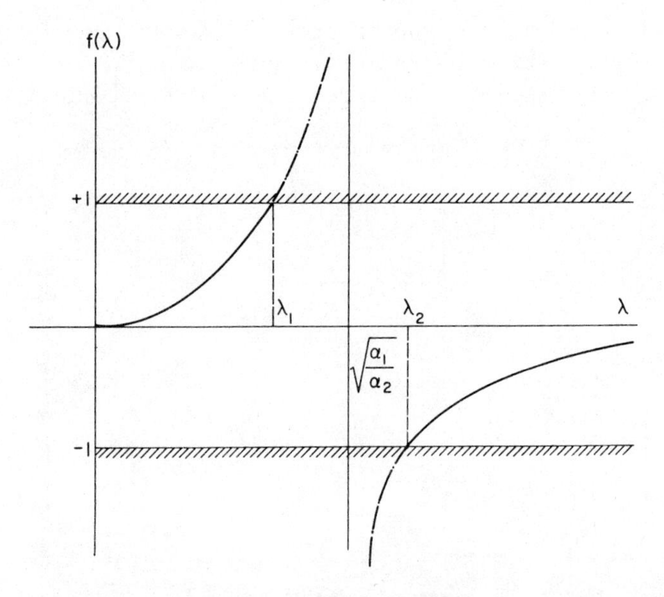

FIG. 23. Graphical discussion for the existence of the synchronized solution.

Because of the constraint conditions $-1 \leqslant y \leqslant +1$, its intersections with the straight lines $y = \pm 1$ define two values of λ, λ_1 and λ_2, such that the synchronization cannot exist in the interval (λ_1, λ_2):

$$\lambda_1 = -\frac{\omega_1^2 - \omega_2^2}{2\alpha_2} + \frac{1}{2\alpha_2}[(\omega_1^2 - \omega_2^2)^2 + 4\alpha_1\alpha_2]^{1/2}$$

$$\lambda_2 = \frac{\omega_1^2 - \omega_2^2}{2\alpha_2} + \frac{1}{2\alpha_2}[(\omega_1^2 - \omega_2^2)^2 + 4\alpha_1\alpha_2]^{1/2}$$

and

$$\Delta\lambda = \lambda_2 - \lambda_1 = \frac{\omega_1^2 - \omega_2^2}{\alpha_2}$$

We see that $\Delta\lambda$ decreases when the detuning $\omega_1 - \omega_2$ decreases or if the coupling coefficient α_2 increases.

The discussion is much simplified in the symmetric case, where the tubes are identical: $A_1 = A_2 = A$, $B_1 = B_2 = B$, and $a_1 = a_2$, say $\lambda = 1$.

From the first two equations of (43) we get

$$\frac{\alpha_1 + \alpha_2}{\omega} \sin \varphi = 0$$

from which follows

$$\varphi = 0 \qquad \text{or} \qquad \varphi = \pi$$

We have also, from the last two equations of (43),

$$\frac{\omega_1^2 - \omega_2^2}{\alpha_1 - \alpha_2} = \cos \varphi$$

Accordingly, we see that

(a) If $\omega_1 > \omega_2$ and $\alpha_1 > \alpha_2$, $\cos \varphi$ is positive, and $\varphi = 0$ is the only admissible solution, then $\cos \varphi = +1$.

(b) If $\omega_1 > \omega_2$ and $\alpha_1 < \alpha_2$, $\cos \varphi$ is negative, $\varphi = \pi$, and thus $\cos \varphi = -1$.

Converse conclusions hold for $\omega_1 < \omega_2$.

Then, returning to the first two equations of (43), we have

$$A - Ba_1^2 = A - Ba_2^2 = 0,$$

which means that the amplitude of the mutually synchronized oscillators is the same as if they where running freely without any interaction. But the coupling imposes the phase lag between them, $\varphi = 0$ or $\varphi = \pi$.

5. NORMAL VIBRATIONS ON NONLINEAR SYSTEMS†

5.1. The Physical System and the Pseudo System

Conservative systems with many degrees of freedom have been investigated from a geometric viewpoint by Rosenberg. Here we shall outline briefly some important remarks which provide a starting point for the theory, without going through the whole development.

The system under study is composed of n masses m_i $(i = 1, ..., n)$ which are linked together and to some fixed points by nonlinear springs,

† See [16, 20–23].

as shown in Fig. 2b. Each mass m_i possesses a single degree of freedom of translation u_i; it will be convenient to define the fixed points by letting $m_i \to \infty$ and $u_i \to 0$.

The equations of motion are obtained by assuming that the spring force between any two masses or between a mass and a fixed point, $f_{ij}(s)$, is a continuous odd function of spring deflection $s = u_i - u_j$ $(i, j = 1, ..., n)$, and can be represented by the series expansion

$$f_{ij}(s) = \sum_p k_{ij}^{(p)} s^p \tag{60}$$

in which each term is an odd function of s, i.e., $p = 1, 3, 5, ...$.

Accordingly, the potential function $U(x_1, ..., x_n)$ of the system is

$$U = -\sum_{i=1}^{n} \sum_{j=i+1}^{n} \sum_{p=1,3,5,...} \frac{k_{ij}^{(p)}}{p+1} \left[\frac{x_i}{(m_i)^{1/2}} - \frac{x_j}{(m_j)^{1/2}} \right]^{p+1} \tag{61}$$

with the proviso that we let $m_j \to \infty$ when we wish to express the fact that mass m_i is connected to fixed point j. Then the equations of motion are written

$$\ddot{x}_i = \frac{\partial U}{\partial x_i} \qquad (i = 1, ..., n) \tag{62}$$

Indeed, (62) and (61) generalize equations (2) and expression (3) which we first established in the two-dimensional case.

Note that if we consider a unit mass with potential energy

$$E_p = -U(x_1, ..., x_n)$$

equations (62) are also the ones which govern the motion of this particle in a potential-energy valley of n-dimensional phase space $R^n(x_1, ..., x_n)$. This establishes the bridge between the problems that we shall discuss in this section following Rosenberg's theory, and the ones which we have analyzed in Sections 3 and 4.

Note that in the present situation, equipotential surfaces possess symmetry with respect to the origin:

$$U(x_1, ..., x_n) = U(-x_1, ..., -x_n)$$

which was not the case, for instance, in the example of Section 3, the conclusions of which it might be interesting to compare to the ones of this section. The model consisting of this unit mass governed by (62) is called the *pseudo system*.

5.2. Normal Modes

Starting with this simple model, an interesting remark, which led Rosenberg to the concept of *normal modes*, will open the discussion:

The difference between the linear and the nonlinear cases is not at all obvious when one looks at the potential valley. In the linear case, the equipotential surfaces are ellipsoids; in the nonlinear case they are ovaloids. Moreover, in this problem and from a physical viewpoint, there is no discontinuity between the linear and the nonlinear case, that is:

(a) If one produces a slight deformation of the potential valley which transforms the ellipsoids of the linear case into ovaloids, provided that the starting point is the same in both cases, the motion of the particle will be only scarcely modified. As a matter of fact, it is in general impossible to infer at first sight whether the valley is elliptic or egg-shaped:

(b) The linear problem exhibits no particular feature amidst the set of nonlinear ones.

On the other hand, from a mathematical viewpoint, the difference between linear systems and nonlinear ones appears to be very important; indeed, they belong to two different kinds.

These points of view are irreconcilable, and this is very clear if one keeps in mind the fact that there is no relation between the physical properties of systems and the complexity of their equations of motion, which may be more or less difficult to solve.

If the starting position of the particle is arbitrary, whether the equations are linear or not, the trajectory in the potential valley will be a kind of Lissajoux figure whose shape is in general very intricate. However, in the linear case we know that there exist starting positions which result in motion along straight paths. These privileged starting positions lie along the axes of the family of ellipsoidal equipotential surfaces. To each of these starting points is associated one fundamental motion along a straight path which lies entirely along the corresponding axis. This path is a segment through origin 0, whose end points are the starting point itself and its symmetrical with respect to the origin. To each axis is associated a characteristic frequency, which does not depend on the amplitude of the motion. Indeed, we recognize the well-known eigenstates of the linear system.

In the nonlinear case, Rosenberg has also identified a discrete set of such privileged motions along curvilinear paths, each of which is represented by a single-valued function throughout the domain D of the potential valley in which it is defined. Each of these so-called *normal modes* may be described as a vibration in which all masses move periodically "in unison":

(a) The frequency of each component of the system is the same.

(b) Each mass passes through its equilibrium position 0 at the same time.

(c) Each mass reaches its maximum translation at the same time.

(d) The translation of each mass, at each given time t, is a single-valued function of the translation of one of them which can be arbitrarily chosen at that time.

The path which is associated with each normal mode is called a *modal line*. Modal lines are shown in Fig. 24a and b for the linear and nonlinear cases, respectively.

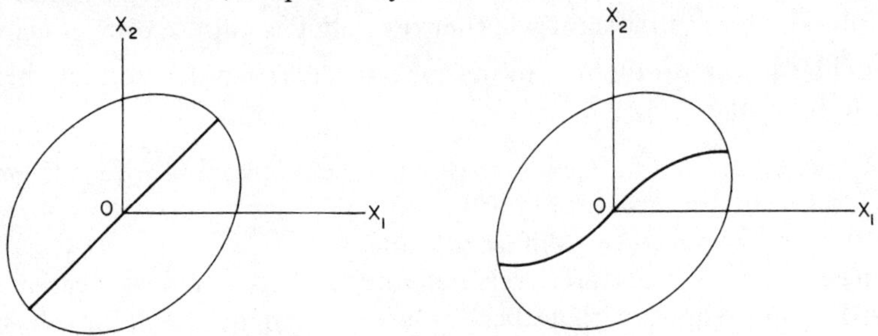

FIG. 24. Modal lines: (a) linear case; (b) nonlinear case.

From the analytical viewpoint, property (d) is really important, for if x_1 is the coordinate which defines the position of mass m_1 at each time t, it makes it possible to express the other coordinates as single-valued functions of x_1, say

$$x_2 = x_2(x_1) \cdots x_n = x_n(x_1) \tag{63}$$

Then by substituting (63) into the first equation of motion of the set (62) we get

$$\ddot{x}_1 = \frac{\partial}{\partial x_1} U[x_1, x_2(x_1), \ldots, x_n(x_1)] \tag{64}$$

which can be integrated by a single quadrature.

$x_2(t), \ldots, x_n(t)$ are computed by substituting in (63) the expression of $x_1(t)$ thus obtained. This is a basic property. Equation (64) is called *uncoupled* because it contains only the displacement variable x_1, whereas equations (62) are coupled ones. In general, for each normal mode the frequency will be a function of the amplitude, except in the linear case.

On the other hand, it may be that the modal lines are straight paths; then

$$\frac{x_j}{x_i} = C_{ij} \qquad (i, j = 1, ..., n)$$

where C_{ij} are constants. When this situation occurs, functions $x_i(t)$ $(i = 1, ..., n)$ are similar, and the corresponding normal mode is said to be *similar*. Then the following theorems hold:

THEOREM 1. *If a normal mode is similar, then this mode is independent of the amplitude of vibration.*

Conversely:

THEOREM 2. *If a normal mode is nonsimilar, then this mode depends on the amplitude of vibration.*

Now let us note that:

(a) In linear systems normal modes are similar and the frequencies associated with them do not depend on the amplitude of vibration.

(b) In nonlinear systems whose normal modes are similar, the frequency depends on the amplitude of vibration, whereas each normal mode does not depend on it.

(c) In nonlinear systems whose normal modes are nonsimilar, the frequency of each normal mode, as well as the normal mode itself, depends on the amplitude of vibration.

5.3. Representations H and S

Heretofore we have briefly outlined some qualitative features of the theory. When one wishes to state precisely the law of motion of the pseudo system, two kinds of representations can be used.

The first one is based on the Hamiltonian principle:

$$\delta \int_{t_0}^{t_1} (T + U)\, dt = 0 \tag{65}$$

where T is the kinetic energy of the system, and

$$L = T + U$$

is its Lagrangian.

Then from the Euler-Lagrange equations (65), the usual dynamic equations are obtained, namely (62):

$$\ddot{x}_1 = \frac{\partial U}{\partial x_i} \qquad (i = 1, ..., n)$$

This system is called the *H representation.*

The other representation is based on the Maupertuis least-action principle, according to which

$$\delta \int_{s_0}^{s_1} (U + h)^{1/2} \, ds = 0 \tag{66}$$

where

$$T - U = h \qquad \text{and} \qquad ds^2 = \sum_{i=1}^{n} dx_i^{\,2}$$

The Euler-Lagrange equations of (66) are

$$\begin{aligned} 2(U + h) \left(\sum_{j=1}^{n} x_i'' x_j'^2 - \sum_{j=1}^{n} x_i' x_j' x_j'' \right) \\ + \sum_{j=1}^{n} (x_j')^2 \left(\sum_{j=1}^{n} x_i' x_j' \frac{\partial U}{\partial x_j} - \sum_{j=1}^{n} x_j'^2 \frac{\partial U}{\partial x_i} \right) = 0 \qquad (i = 1, ..., n) \end{aligned} \tag{67}$$

where the primes denote differentiation with respect to s. System (67) is called the *S representation.*

It is worthwhile discussing the advantages and disadvantages of both representations. In the linear case it appears that (62) is in general more convenient than (67), whereas in nonlinear problems the simplicity of the H representation is fallacious and somewhat misleading. However, as was explained above, from the physical viewpoint the linear system does not deserve the very particular position which follows from mathematical considerations. Indeed this particularity should be emphasized by the simplicity of the H representation in the linear case, but at the same time we note that the S representation is not at all simplified by the assumption of linearity. It is *always* a nonlinear system of equations. It is tempting to conclude that the S representation reflects more accurately the physical properties of the oscillator.

Several important properties have been deduced from (66) and (67) by Rosenberg and Hsu [20], especially the following theorems:

THEOREM 3. *Every trajectory of the pseudo system, with U defined by* (61), *lies in an n-dimensional compact domain D of R^n whose boundary is the $(n-1)$-dimensional surface defined by*

$$U + h = 0$$

THEOREM 4. *Every trajectory of the pseudo system, with U defined by* (61), *intersects the bounding surface $U + h = 0$ orthogonally.*

Following these, special care is devoted to systems whose modal lines are straight paths.

5.4. Straight Modal Lines

Rosenberg and Hsu have studied the conditions under which a modal line reduces to a straight path. By using spherical coordinates in R^n,

$$x_i = r \sin \theta_{n+1-i} \prod_{j-1}^{n-i} \cos \theta_j \qquad \text{with} \quad \theta_n = \frac{\pi}{2} \qquad (i = 1 \; ..., n)$$

and by substituting in the potential function U, they arrive at the following necessary and sufficient condition for a modal line to be a straight path.

$\partial U/\partial \theta_i$ must be decomposed as follows:

$$\frac{\partial U}{\partial \theta_i} = \Theta_{1i}(r, \theta_1, ..., \theta_{n-1})\Theta_{2i}(\theta_1, ..., \theta_{n-1}) \qquad (i = 1, ..., n-1)$$

and the set of equations

$$\Theta_{2i}(\theta_1, ..., \theta_{n-1}) = 0 \qquad (i = 1, ..., n-1)$$

must have real roots θ_i^*. The roots θ_i^* determine the direction of the straight modal line.

Now it may be shown that there exist systems which have this property, among them:

System 1. The uniform system. A system in which all masses are equal, and all springs are identical. If one considers the linear system which is associated to it by limiting the series expansion (61) to its quadratic terms ($p = 1$), it may easily be verified that the equipotential ovaloids of the nonlinear system have the same symmetries as the equipotential ellipsoids of the associated linear system. Accordingly, the modal lines of the uniform system are straight lines whose directions coincide with those of the associated linear system.

System 2. The homogeneous system. A system whose potential energy has the form

$$U = -\sum_{i=1}^{n} \sum_{j=i+1}^{n} \frac{k_{ij}^{(p)}}{p+1} \left| \frac{x_i}{(m_i)^{1/2}} - \frac{x_j}{(m_j)^{1/2}} \right|^{p+1}$$

where p is any positive integer. The linear case, which corresponds to $p = 1$, is a member of this family.

From Theorem 4 follows another theorem:

THEOREM 5. *Every straight modal line intersects all equipotential surfaces orthogonally.*

Conversely:

THEOREM 6. *If a straight line intersects all equipotential surfaces orthogonally, it is a modal line.*

Theorem 6 is very useful, since it reduces the search for modal lines to the search for straight lines which intersect every equipotential surface orthogonally.

We can also prove the following:

THEOREM 7. *If a modal line intersects all equipotential surfaces orthogonally, it is a straight modal line.*

This is a consequence of (67), since

$$\sum_{j=1}^{n} x_i'' x_j'^2 - \sum_{j=1}^{n} x_i' x_j' x_j'' \equiv 0 \qquad (i = 1, ..., n)$$

must be verified everywhere in D.

5.5. Weak Superposition Principle

An interesting similarity between the normal modes of a nonlinear system and the eigenstates of a linear one is reported by Rosenberg based on a remark due to Caughey. A weak superposition principle is stated which applies to certain nonlinear systems, in which the general solution appears to be a linear combination of normal modes. This principle should not be identified with the strong superposition principle, which holds only in the case of linear systems and according to which *every* linear combination of particular solutions is also a solution.

The weak superposition principle can be illustrated by the following example. Consider a symmetric oscillator with two degrees of freedom, similar to the one which was described in Section 1, such that the anchor springs be identical and linear whereas the coupling spring is nonlinear:

$$m_1 = m_2 = m$$
$$f_1(s) \equiv f_2(s) \equiv ks$$
$$f_{12}(s) \equiv \sum_{p=1,3,5,\dots} k_p s^p \qquad (p \text{ an odd integer})$$

The equations of motion are

$$\begin{aligned} \ddot{x}_1 + \omega_0^2 x_1 &= -\sum_p \beta_p (x_1 - x_2)^p \\ \ddot{x}_2 + \omega_0^2 x_2 &= \sum_p \beta_p (x_1 - x_2)^p \end{aligned} \tag{68}$$

with

$$\omega_0^2 = \frac{k}{m} \qquad \beta_p = \frac{k_p}{m^{(p+1)/2}}$$

Let the initial conditions be

$$x_1(0) = x_1^0 \qquad x_2(0) = x_2^0$$
$$\dot{x}_1(0) = \dot{x}_1^0 \qquad \dot{x}_2(0) = \dot{x}_2^0$$

and replace x_1 and x_2 by the new variables u and v, such that

$$x_1 + x_2 = u \qquad x_1 - x_2 = v$$

Then equations (68) are rewritten

$$\ddot{u} + \omega_0^2 u = 0 \tag{69}$$
$$\ddot{v} + \omega_0^2 v = -2 \sum_p \beta_p v^p \tag{70}$$

from which we get by integration

$$u = (x_1^0 + x_2^0) \cos \omega_0 t + \frac{\dot{x}_1^0 + \dot{x}_2^0}{\omega_0} \sin \omega_0 t \tag{71}$$

$$v = v(x_1^0, x_2^0, \dot{x}_1^0, \dot{x}_2^0, t) \tag{72}$$

Function (72) is determined from (70) by a single quadrature; it is a periodic function whose period depends on the starting conditions $x_1{}^0$, $x_2{}^0$, $\dot{x}_1{}^0$, and $\dot{x}_2{}^0$. Accordingly, the general solution of (68) is

$$
\begin{aligned}
x_1 &= \frac{x_1{}^0 + x_2{}^0}{2} \cos \omega_0 t + \frac{\dot{x}_1{}^0 + \dot{x}_2{}^0}{2\omega_0} \sin \omega_0 t + \tfrac{1}{2} v(x_1{}^0, x_2{}^0, \dot{x}_1{}^0, \dot{x}_2{}^0, t) \\
x_2 &= \frac{x_1{}^0 + x_2{}^0}{2} \cos \omega_0 t + \frac{\dot{x}_1{}^0 + \dot{x}_2{}^0}{2\omega_0} \sin \omega_0 t - \tfrac{1}{2} v(x_1{}^0, x_2{}^0, \dot{x}_1{}^0, \dot{x}_2{}^0, t)
\end{aligned}
\tag{73}
$$

Since equations (69) and (70) are uncoupled, we see that the normal modes are defined, respectively, by $u = 0$ and $v = 0$, say $x_1 = x_2$ and $x_1 = -x_2$.

The conclusion follows directly from expressions (73). The general solution is a linear combination of the two normal modes:

$$
\begin{aligned}
x_1{}^* &= \frac{x_1{}^0 + x_2{}^0}{2} \cos \omega_0 t + \frac{\dot{x}_1{}^0 + \dot{x}_2{}^0}{2\omega_0} \sin \omega_0 t \\
x_2{}^* &= \frac{x_1{}^0 + x_2{}^0}{2} \cos \omega_0 t + \frac{x_1{}^0 + x_2{}^0}{2\omega_0} \sin \omega_0 t \\
x_1^{**} &= \tfrac{1}{2} v(x_1{}^0, x_2{}^0, \dot{x}_1{}^0, \dot{x}_2{}^0, t) \\
x_2^{**} &= -\tfrac{1}{2} v(x_1{}^0, x_2{}^0, \dot{x}_1{}^0, \dot{x}_2{}^0, t)
\end{aligned}
$$

We see that this nonlinear oscillator obeys the weak superposition principle.

5.6. Concluding Comments

Rosenberg's theory can be applied to a wide variety of systems. So it is very interesting from a practical viewpoint, so much the more so because it breaks up with the linearization methods (which are certainly an artificial way of approaching the problems of nonlinear mechanics), and it attacks nonlinearities head on. However, a number of difficulties occur which to be overcome seem to demand simplifying assumptions, such as the one that the system is conservative.†

Indeed, many practical problems deal with self-sustained oscillators with several degrees of freedom, which up to now have not entered the scope of this theory. However, perhaps there is some similarity between the synchronized solution of two coupled self-oscillators (as determined in Section 4) and a normal mode.

† In his more recent work Rosenberg has investigated the case of systems driven by a periodic force. His method gives interesting results in this nonconservative case.

On the other hand, for conservative systems the assumptions concerning the symmetries of the potential valley are not always met in practice, as shown by the example in Section 3. In this section the solution obtained exhibits similarities with a normal mode, although it does not exactly fit the definition.

REFERENCES

1. Lord Rayleigh (John William Strutt), "The Theory of Sound." Macmillan, London, 1894. (Reprint, Dover, New York, 1945.)
2. K. F. Theodorchik, "Autooscillating Systems." Moscow, 1948.
3. T. Sigurgeirsson, "Focusing in a Synchrotron with Periodic Field, Perturbation Treatment." CERN/T/TS-3, May 1953.
4. N. Minorsky, Sur les systèmes non-linéaires à deux degrés de liberté. *Rend. Seminario Mat. Fis. Milano* **13** (1953–54).
5. R. Hagedorn, "The Potential Energy Surface for Our Non-Linear Equations of Motion." CERN-PS/RH-6, Nov. 1954.
6. N. Rouche, Thèse, Liège, 1954.
7. N. Minorsky, Sur l'interaction des oscillations non-linéaires. *Rend. Seminario Mat. Fis. Milano* **25** (1955).
8. E. D. Courant, Non Linearities in the AG Synchrotron.†
9. R. Hagedorn, Note on an Instability on a Difference Resonance Line.†
10. R. Hagedorn, M. G. N. Hine, and A. Schoch, Non-Linear Orbit Problems in Synchrotrons.†
11. A. A. Kolomenski, On the Non-Linear Theory of Betratron Oscillations.†
12. L. J. Laslett and K. R. Symon, Particle Orbits in Fixed Field AG Accelerators.†
13. J. Moser, The Resonance Lines for the Synchrotron.†
14. L. Sideriades, Thèses: Méthodes topologiques appliquées à l'électronique. Faculté des Sciences de l'Université d'Aix-Marseille, 1956.
15. R. M. Rosenberg and C. P. Atkinson, On the Natural Modes and Their Stability in Nonlinear Two-Degree-of-Freedom Systems. *J. Appl. Mech.* **26**, 377–385 (1959).
16. R. M. Rosenberg, Normal Modes in Nonlinear Dual-Mode Systems. *J. Appl. Mech.* **27** (1960); *Inter. Congr. Appl. Mech., Stresa*, 1960.
17. C. Hayashi, H. Shibayama, and Y. Nishikawa, Frequency Entrainment in a Self-Oscillatory System with External Force. *IRE Trans. Circuit Theory* **7**, 413–422 (1960).
18. R. V. Khokhlov, A Method of Analysis in the Theory of Sinusoidal Self-Oscillations. *IRE Trans. Circuit Theory* **7**, 398–413 (1960).
19. N. Minorsky, On Synchronization. *Intern. Symp. Nonlinear Vibrations, Kiev*, 1961.
20. R. M. Rosenberg and C. S. Hsu, On the Geometrization of Normal Vibrations of Nonlinear Systems Having Many Degrees of Freedom. *Intern. Symp. Nonlinear Vibrations, Kiev*, 1961.
21. R. M. Rosenberg, The Normal Modes of Nonlinear n-Degree-of-Freedom Systems. *J. Appl. Mech.* **29** (1962).
22. R. M. Rosenberg, On Linearity and Nonlinearity. *Second Conf. Nonlinear Vibrations, Warsaw*, 1962.
23. W. Szemplinska-Stupnicka, Normal Modes of a Nonlinear Two-Degree-of-Freedom System and Their Properties. *Second Conf. Nonlinear Vibrations, Warsaw*, 1962.

† *CERN Symp. High Energy Accelerators Pion Phys.* **1956.**

24. F. Bertein, Sur le couplage entre modes voisins dans un résonateur électromagnétique. *Compt. Rend.* **258**, 123–126 (1964); Sur la synchronization entre modes voisins dans un résonateur électromagnétique. *Compt. Rend.* **258**, 1433–1436 (1964).
25. J. Groszkowski, "Frequencies of Self-Oscillations." Polish Scientific Publishers, Warsaw, 1964. (English trans., Pergamon Press, New York, 1964.)

CHAPTER V

Equivalent Linearization

1. STATING THE PROBLEM

The concept of frequency response is of such usefulness in the field of linear systems that it was tempting to try to investigate the properties of nonlinear systems by applying to them similar techniques of frequency analysis, and to extend to them the concept of transfer function. Although this concept is basically a linear one, it turns out that in many nonlinear cases of practical interest this method is fairly satisfactory, which explains that it has been extensively studied, developed, and applied to many engineering problems during the last decade. It became known as the *describing function method*.

So many publications have been devoted to this question that it is difficult to draw up an exhaustive bibliography and to determine the importance of the part taken by each of the authors who have contributed, in France and elsewhere, to the development and extension of the theory.

As far as the author knows, this method appeared in the work of Theodorchik [1] in the Soviet Union, of Kochenburger [2] in the United States—and in France at the same time, but independently and from different viewpoints, in the works of Loeb [4] and of Blaquière [3].

The extension of the concept of transfer function to nonlinear systems is based upon the principle of *equivalent linearization*, first introduced by Krylov and Bogoliubov.

In this chapter we shall examine the principle from different viewpoints, which will enable us to extend the concept of transfer function to nonlinear systems. Accordingly it may be considered an introduction to the describing function method, which we shall develop more completely in Chapter 6.

The concept of equivalent linearization embodies the following ideas: A linear passive quadrupole produces a *linear transformation* of any input $x(t)$ into an output $y(t)$:

$$y = \mathscr{A} x \tag{1}$$

For example, to each sinusoidal input will correspond a sinusoidal output, and $\mathscr{A}$ is a *linear functional operator*.

In view of the linearity of $\mathscr{A}$, the output response to *any* input function $x(t)$ will be readily obtained by Fourier's method, by expending the input function into a Fourier integral (or into a Fourier series if the function is a periodical one), and by determining the transform of each sinusoidal component. Then the output function $y(t)$ will be obtained by a summation over the whole spectrum.

Note that (1) can be thought of as a linear differential equation say, for instance, when the coefficients are constants,

$$\frac{d^n}{dt^n} y + \alpha_1 \frac{d^{n-1}}{dt^{n-1}} y + \cdots + \alpha_n y = \beta_1 \frac{d^{n-1}}{dt^{n-1}} x + \beta_2 \frac{d^{n-2}}{dt^{n-2}} x + \cdots + \beta_n x \tag{2}$$

Then $\mathscr{A}$ assumes the symbolic form

$$\mathscr{A} = \left(\beta_1 \frac{d^{n-1}}{dt^{n-1}} + \beta_2 \frac{d^{n-2}}{dt^{n-2}} + \cdots + \beta_n\right) \Big/ \left(\frac{d^n}{dt^n} + \alpha_1 \frac{d^{n-1}}{dt^{n-1}} + \cdots + \alpha_n\right) \tag{3}$$

On the other hand, when x and y assume the forms

$$x = Xe^{st} \qquad y = Ye^{st}$$

where X, Y, and s may be complex, (3) is replaced by

$$A(s) = \frac{\beta_1 s^{n-1} + \beta_2 s^{n-2} + \cdots + \beta_n}{s^n + \alpha_1 s^{n-1} + \cdots + \alpha_n} \tag{4}$$

$A(s)$ is called the *transfer function* for the linear element.

Of particular interest is the case where X and Y are real and $s = j\omega$, ω real. Then

$$A(s) = A(j\omega) = U_A(\omega) + jV_A(\omega) \tag{5}$$

where

$$U_A(\omega) = \mathrm{Re}[A(j\omega)] \qquad V_A(\omega) = \mathrm{Im}[A(j\omega)]$$

Such situations are sufficiently usual and well known in practice to need no further comment. As a matter of fact, the aim of this section is chiefly to point out the different ways that will enable us to extend the concept of transfer function to nonlinear systems.

In this respect, let us note that the definition of linear operator $\mathscr{A}$ and of function $A(s)$ associated with it is not restricted to cases where the linear element is a quadrupole with input x and output y. Indeed, if we consider a physical system governed by a linear differential equation of the form

$$\beta_0 \frac{d^n}{dt^n} x + \beta_1 \frac{d^{n-1}}{dt^{n-1}} x + \cdots + \beta_n x = 0 \tag{6}$$

we can always analyze the properties of the linear operator

$$\mathscr{H} = \beta_0 \frac{d^n}{dt^n} + \beta_1 \frac{d^{n-1}}{dt^{n-1}} + \cdots + \beta_n \tag{7}$$

independent of the fact that we are faced with a differential equation that we wish to solve.

More precisely, we can let $\mathscr{H}$ operate upon *any* differentiable function, for instance, upon functions of the form

$$x = Xe^{st}$$

Then $\mathscr{H}$ is replaced by

$$H(s) = \beta_0 s^n + \beta_1 s^{n-1} + \cdots + \beta_n \tag{8}$$

When X is real and $s = j\omega$, ω real,

$$H(s) = H(j\omega) = U(\omega) + jV(\omega) \tag{9}$$

In this case $H(s)$ is not a transfer function in the usual sense, since the transform of $x(t)$ is not the output of a quadrupole.

The nonlinear counterpart of the concept of transfer function became known as the describing function. This extension is based on the fact that the response of a nonlinear system to a sinusoidal excitation is sometimes very similar to a sinusoidal signal. This is true in particular when a filtering effect takes place in some linear component of the system. Then if one considers the first harmonic of the response, it is possible to extend the concept of amplitude ratio. In general this ratio will depend on the amplitude of the input signal, which is not the case in a linear system. Since the higher harmonics are disregarded, the method must be discussed carefully in each application.

However, the above remarks show clearly that we shall have to consider two such kinds of functions, in accordance with whether we are dealing with quadrupoles or with differential equations. This difference was also pointed out by Klotter [9], who noted that in most of the applications which have been treated in view of this extension—dead band, saturation, limiting, linkages, contactors, Coulomb friction, ... —the relationship between input and output is given by a nonlinear expression for the variables, in contrast with a differential equation.

In many practical problems the function $H(s)$ of a system is defined starting with the transfer functions of its components. For example,

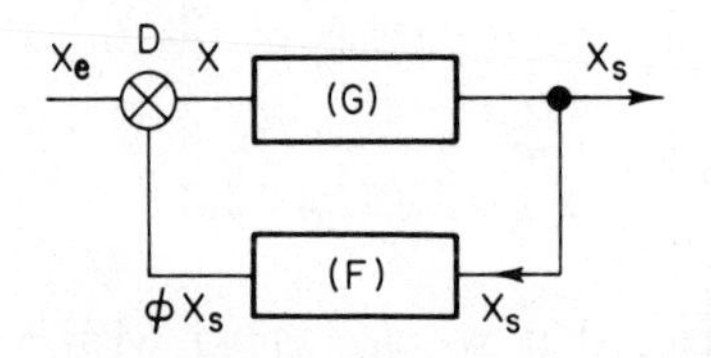

FIG. 1. Feedback loop.

if the system is a feedback loop, as shown in Fig. 1, the well-known expression for $H(s)$ is

$$H(s) = 1 - \Gamma(s)\Phi(s) \tag{10}$$

where $\Gamma(s)$ and $\Phi(s)$ are the transfer functions of amplifier (G) and passive quadrupole (F), respectively. If instead of $\Gamma(s)$, $\Phi(s)$, and $H(s)$ we consider the corresponding differential operators $\mathscr{G}$, $\mathscr{F}$, and $\mathscr{H}$, the differential equation of motion of the loop, when the input signal x_e is identically zero, is

$$(1 - \mathscr{G}\mathscr{F})x = 0 \qquad \text{or} \qquad \mathscr{H}x = 0 \tag{11}$$

where $x(t)$ is the variable under consideration. Next we shall define in this way a function $H(s)$ for nonlinear systems.

On the other hand, from a geometric viewpoint, we can consider functions $x(t)$ and $y(t)$ as generalized *vectors* of a function space, namely *Hilbert space*.

We shall denote these vectors by $| x\rangle$ and $| y\rangle$, following Dirac's notation, and represent the above transformations by

$$\tilde{\mathscr{A}}\,| x\rangle \qquad \text{and} \qquad \tilde{\mathscr{H}}\,| x\rangle$$

In linear systems $\tilde{\mathscr{A}}$ and $\tilde{\mathscr{H}}$ can be written in matrix form. Bertein [10] and, independently, Clauser [7] have developed the extension of such matrix representations to nonlinear systems. Other treatments can be found in the references at the end of this chapter.

2. A MODEL IN CLASSICAL OPTICS

In the theory of electrical systems it is well known that quadrupolar networks exhibit similarities with some optical instruments, for instance, with the lenses of classical optics, or, reversing the statement, any lens can be conveniently represented by a quadrupolar model.

Furthermore, let us visualize the generalized vector space of the input signals by considering in a plane a regular square lattice, in which the junction points (or nodes) are intended to represent a set of vectors. Each vector represent an admissible input function (Fig. 2a).

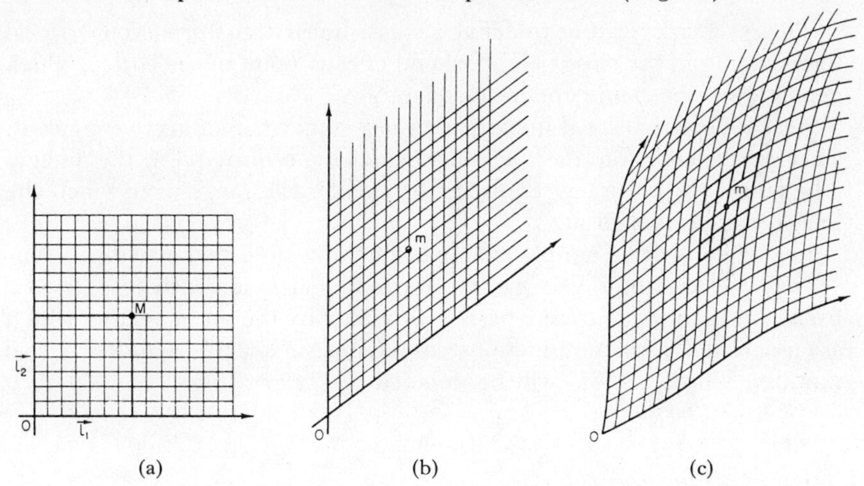

FIG. 2. A square lattice (a), with (b) linear and (c) nonlinear transformations.

If the instrument has linear properties, the image of the lattice is a plane regular lattice—all its meshes are similar parallelograms in a plane (Fig. 2b). This property can be expressed in the following way:

Let $| x_1\rangle$ and $| x_2\rangle$ be any two vectors and $| y_1\rangle$ and $| y_2\rangle$ their transforms

$$| y_1\rangle = \tilde{\mathscr{A}}|x_1\rangle \qquad | y_2\rangle = \tilde{\mathscr{A}}| x_2\rangle$$

If λ_1 and λ_2 are any two scalars, the transform of $\lambda_1 | x_1\rangle + \lambda_2 | x_2\rangle$ is $\lambda_1 | y_1\rangle + \lambda_2 | y_2\rangle$, say

$$\lambda_1 | y_1\rangle + \lambda_2 | y_2\rangle = \tilde{\mathscr{A}}\{\lambda_1 | x_1\rangle + \lambda_2 | x_2\rangle\}$$

On the other hand, if the instrument is nonlinear, the image of the lattice is an irregular lattice which exhibits distortion (Fig. 2c), since

$$\lambda_1 | y_1\rangle + \lambda_2 | y_2\rangle \neq \tilde{\mathscr{A}}\{\lambda_1 | x_1\rangle + \lambda_2 | x_2\rangle\}$$

Then:

(a) The distortion of the image may be looked upon as a characteristic of the nonlinearity of the instrument, namely of the operator which determines the transformation.

(b) If the nonlinearity of the instrument is weak, it is possible to define a *local linearization*, which means that if we consider the image over a small range, it will not be possible to perceive the distortion at first sight, and this piece of lattice will resemble a piece of regular lattice. The distortion is visible only in a more general view.

These remarks lead us to define a quasi-linear transformation or local linearization in the close neighborhood of each point of the lattice, which depends on the point under consideration.

Also we see that local linearization is a somewhat subjective concept, since it depends on the accuracy which is required for the image: The better the accuracy required, the smaller the range over which the image can be looked at.

Let us consider a simple example, but first note that it will be convenient to determine vectors of the functional space with respect to a basis. An example of such a basis is provided by the set of vectors which are associated with the functions $\cos \omega t$ and $\sin \omega t$, where ω is any real number. These vectors will be denoted

$$| \xi_\omega \rangle \qquad \text{and} \qquad | \eta_\omega \rangle$$

respectively.

In these problems we shall consider only discrete sets, such as $\cos \omega t$, $\cos 2\omega t$, $\cos 3\omega t$, etc. ...; then the basic vectors will be denoted by

$$| \xi_1 \rangle \qquad | \xi_2 \rangle \qquad | \xi_3 \rangle \quad \cdots$$

Similarly, the vectors

$$| \eta_1 \rangle \qquad | \eta_2 \rangle \qquad | \eta_3 \rangle$$

will be associated with $\sin \omega t$, $\sin 2\omega t$, $\sin 3\omega t$. ... .

Now consider a quadrupole in which the input-output relation is

$$y = S_0 x + S_2 x^3 \qquad \text{with} \quad S_0 > 0 \tag{12}$$

To every $x(t)$ and $y(t)$ we shall associate vectors $| x \rangle$ and $| y \rangle$. Further, the set of functions

$$a_0 \cos \omega t$$

where ω is given and a_0 is an arbitrary positive number will define the axis L_1 (Fig. 3). Namely,

$$|\xi_1\rangle \in L_1 \quad \text{and} \quad a_0 |\xi_1\rangle \in L_1 \quad \forall a_0 > 0$$

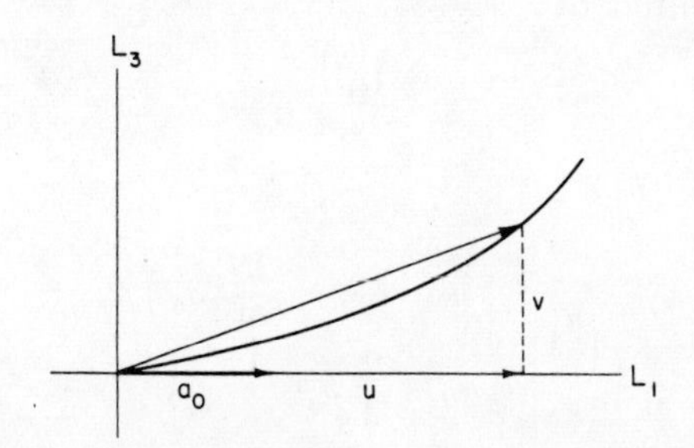

FIG. 3. Nonlinear transformation of axis L_1.

If $S_2 = 0$, the quadrupole is linear, and the transform of any vector of L_1 lies along L_1; L_1 is invariant by the transformation. On the other hand, if $S_2 \neq 0$, we have

$$S_0 a_0 \cos \omega t + S_2 a_0^3 \cos^3 \omega t = \left(S_0 + \frac{3S_2}{4} a_0^2\right) a_0 \cos \omega t + \frac{S_2}{4} a_0^3 \cos 3\omega t \tag{13}$$

This means that the transform of any vector $a_0 |\xi_1\rangle$ does not lie along L_1.

This transform has a component along L_1,

$$u = \left(S_0 + \frac{3S_2}{4} a_0^2\right) a_0 \tag{14}$$

and a component

$$v = \frac{S_2}{4} a_0^3 \tag{15}$$

along axis L_3 which is associated with the vector $|\xi_3\rangle$ in the same way L_1 is associated with $|\xi_1\rangle$.

If we think of the quadrupole as a kind of optical instrument, we see that the image of axis L_1 is not a straight line but a distorted line whose parametric equations are (14) and (15). Then if coefficient S_2 is small, this line is close to the axis L_1 for a wide range of values of amplitude a_0; accordingly, the nonlinear transformation (12) can be approximated in this range by the transformation

$$y = \left(S_0 + \frac{3S_2}{4} a_0^2\right) x \tag{16}$$

This means that we neglect the small deviations from axis L_1. As a matter of fact, this is also a first-harmonic approximation, since that amounts to neglecting the third harmonic in expression (13).

Now, in matrix form, in the reference defined by $|\xi_1\rangle$ and $|\xi_3\rangle$, $a_0 |\xi_1\rangle$ is represented by

$$\begin{pmatrix} a_0 \\ 0 \end{pmatrix}$$

and its transform by

$$\begin{pmatrix} u \\ v \end{pmatrix} \qquad \begin{aligned} u &= \left(S_0 + \frac{3S_2}{4} a_0^2\right) a_0 \\ v &= \frac{S_2}{4} a_0^3 \end{aligned}$$

Accordingly, we see that the nonlinear transformation which is introduced by the quadrupole can be represented by the matrix

$$\tilde{M} = \begin{pmatrix} S_0 + \dfrac{3S_2}{4} a_0^2 & 0 \\ \dfrac{S_2}{4} a_0^2 & 0 \end{pmatrix} \tag{17}$$

and the above first-harmonic approximation is obtained by using, instead of $\tilde{M}$, the matrix

$$\tilde{N} = \begin{pmatrix} S_0 + \dfrac{3S_2}{4} a_0^2 & 0 \\ 0 & 0 \end{pmatrix} \tag{18}$$

This approximation is deduced from the nonlinear transformation $\tilde{M}$ by a *projection* on axis L_1.

As a matter of fact, *if we freeze parameter* a_0, (17) defines a linear transformation, which can be used, instead of the nonlinear one, *in the close neighborhood of given* a_0. It will be denoted $\tilde{M}^*$.

To each a_0 there corresponds a new linearized transformation, which can be called a *tangent linear approximation* to express the fact that the nonlinear transformation looks like a kind of "envelope" of such linear approximations.

As concerns (18), it is also not a linear transformation although it let be L_1 invariant, since it also depends on a_0. If we freeze a_0, it becomes a linear approximation and can be used in the close neighborhood of the given value of a_0. This linearized transformation is sometimes more convenient than $\tilde{M}^*$, although it is not so good.

It will be denoted

$$\tilde{N}^* = \operatorname*{proj}_{L_1} \tilde{M}^*$$

since it is deduced from $\tilde{M}^*$ by a projection on axis L_1.

This simple example is interesting because it shows how different viewpoints—first-harmonic approximation, matrix representation, and tangent linear approximation are related to one another and help give a better understanding of nonlinear behavior.

As a matter of fact, it will also help to introduce the idea of optimal linearization in a very simple way.

3. INTRODUCTION TO THE OPTIMAL LINEARIZATION METHOD†

Suppose that, *at the outset*, we choose to replace (12) by linear approximation (19):

$$y^* = \lambda x \tag{19}$$

Then at once we have to choose the parameter λ to approximate the nonlinear transformation as closely as possible.

We have to define a criterion of optimality. Let us consider the difference

$$y - y^* = (S_0 - \lambda)x + S_2 x^3$$

and minimize the integral

$$\int_{t_0}^{t_1} (y - y^*)^2 \, dt$$

over some conveniently chosen interval t_0, t_1.

Let

$$\overline{x^2} = \int_{t_0}^{t_1} x^2(t) \, dt \qquad \overline{x^4} = \int_{t_0}^{t_1} x^4(t) \, dt \qquad \overline{x^6} = \int_{t_0}^{t_1} x^6(t) \, dt$$

We have

$$\int_{t_0}^{t_1} (y - y^*)^2 \, dt = (S_0 - \lambda)^2 \overline{x^2} + 2S_2(S_0 - \lambda)\overline{x^4} + S_2{}^2 \overline{x^6}$$

and the optimal choice of λ is determined by

$$\frac{\partial}{\partial \lambda} \int_{t_0}^{t_1} (y - y^*)^2 \, dt = 0 \tag{20}$$

† See [14].

say

$$2(S_0 - \lambda)\overline{x^2} + 2S_2\overline{x^4} = 0$$

from which follows

$$\lambda = S_0 + S_2 \frac{\overline{x^4}}{\overline{x^2}} \tag{21}$$

When

$$x = a_0 \cos \omega t$$

we get, by performing the integrations over a period $2\pi/\omega$,

$$\lambda = S_0 + \frac{3S_2}{4} a_0^2 \tag{22}$$

Again we obtain the approximation of Section 2, which is a linear approximation *for each given* a_0, and also is the best linear approximation *with respect to the above choice* (19) *and to the above criterion of optimality* (20).

We shall consider (22) as a *generalized transfer function* for the quadrupole, *with respect to a sinusoidal input with amplitude* a_0.

It is interesting to note that (21) enables us to define the generalized transfer functions with respect to various input functions. For instance, instead of a sinusoidal input, let us apply at the input of the quadrupole a gaussian noise $x(t)$, whose probability distribution is

$$p(x) = \frac{1}{\sigma_0(2\pi)^{1/2}} \exp\left(-\frac{x^2}{2\sigma_0^2}\right) \tag{23}$$

and let

$$\overline{x^2} = \int_0^\infty x^2 p(x)\, dx \qquad \overline{x^4} = \int_0^\infty x^4 p(x)\, dx \qquad \overline{x^6} = \int_0^\infty x^6 p(x)\, dx \tag{24}$$

These functions occur in expression (21) if we define the optimality criterion as

$$\frac{\partial}{\partial\lambda} \int_0^\infty (y - y^*)^2 p(x)\, dx = 0 \tag{25}$$

Then by computing integrals (24) we get

$$\overline{x^2} = \frac{\sigma_0^2}{2} \qquad \overline{x^4} = \frac{3\sigma_0^4}{2}$$

and, finally,

$$\lambda = S_0 + 3S_2\sigma_0^2 \tag{26}$$

Equation (26) is a *generalized transfer function for the quadrupole with respect to the gaussian input.* In this case it depends on the parameter σ_0.

Before proceeding, let us investigate the meaning of (19). As a matter of fact, $x(t)$ can be *any* input: a sinusoidal input, a gaussian noise, etc. However, (19) is a rather restrictive assumption since, λ being a scalar, it means that the transform of any vector $| x\rangle$, namely $| y^*\rangle$, has the same *direction* as the vector $| x\rangle$ itself, and we know that this is not true in general for the given nonlinear transformation.

Giving function $x(t)$ is the equivalent of giving the direction of $| x\rangle$ in the functional space, namely axis L_x. Then the question is: What, *on this axis,* is the transformation which most resembles the given nonlinear transformation?

The answer is (19) [with (21)] and it may easily be seen that it is *the projection of the actual nonlinear transformation on* L_x, which generalizes the remarks of Section 2. That is,

$$\forall | x\rangle, \ | x\rangle \in L_x$$

$$| y^*\rangle = \operatorname*{proj}_{L_x} | y\rangle$$

As pointed out above, this is an easy way to extend the concept of transfer function to nonlinear quadrupoles, and it will also prove useful later for defining the describing function associated with a nonlinear differential equation.

4. SIMILARITY WITH FOURIER'S METHOD

The optimal linearization method amounts to substituting a linear operator for a nonlinear operator, and will depend in general on a set of parameters $\lambda_1, \lambda_2, ..., \lambda_n$.

Indeed the choice of the form of the linear operator, for instance, the choice of the number of parameters $\lambda_1, \lambda_2, ..., \lambda_n$, introduces some restrictive conditions. Then the problem is the following: *These restrictive conditions being taken into consideration, what is the best linear approximation?*

This problem is similar to the one in which one tries to approach a periodic function $f(t)$ which contains an infinite number of harmonics by another periodic function $s(t)$, subject to some restrictive condition. This restrictive condition is introduced in general by limiting the number of harmonics to be taken into consideration. Indeed, by decreasing this number one increases the discrepancy between the two functions; i.e., if $s(t)$ is considered an approximation of $f(t)$, one reduces the accuracy of the representation with the benefit of a noteworthy simplification.

Then, the number of harmonics being given (the restrictive condition), Fourier's criterion defines an optimal choice for the function $s(t)$. From a geometric viewpoint, if $|f\rangle$ and $|s\rangle$ are vectors of the functional space which represent $f(t)$ and $s(t)$, Fourier's criterion leads to minimizing the norm of the difference

$$|f\rangle - |s\rangle$$

That is the criterion we have discussed in the previous section. Indeed,

$$\int_{t_0}^{t_1} (y - y^*)^2 \, dt \qquad \text{and} \qquad \int_0^{\infty} (y - y^*)^2 p(x) \, dx$$

are definitions of the norm of $|y\rangle - |y^*\rangle$ in the cases where $y(t)$ are continuous and random functions, respectively.

As pointed out previously, this criterion is also the one used in the Ritz-Galërkin approximation (Chapter I, Section 3.4).

5. OPTIMAL LINEAR OPERATOR

In this section we shall consider a nonlinear operator $\mathscr{H}$. It may be the functional which defines the input-output relation for a quadrupole; i.e., to each input function $x(t)$ output $y(t)$ is associated such that

$$y = \mathscr{H} x$$

Also it may be the funtional which is joint to a differential equation. For instance, the functional which is joint to Van der Pol's equation

$$\ddot{x} + \mu(x^2 - 1)\dot{x} + x = 0$$

is

$$\mathscr{H} \triangleq \frac{d^2}{dt^2} + \mu(x^2 - 1)\frac{d}{dt} + 1$$

i.e., to each differentiable function $x(t)$ is associated the transform

$$y(t) = \ddot{x} + \mu(x^2 - 1)\dot{x} + x$$

However, $y(t)$ is not the output of a quadrupole.

The optimal linearization method can be applied to both cases, and later it will lead to a general method for defining and producing the describing function associated with $\mathscr{H}$.

As shown by the simple examples of Section 3, this method amounts to approaching the given nonlinear operator $\mathscr{H}$ by linear operator $\mathscr{H}^*$, under the following conditions:

Condition 1. The approximation will be defined for $x(t) \equiv x_0(t)$, where $x_0(t)$ is a given function, namely at (or in the close neighborhood of) a given point $| x_0\rangle$ of the functional space.

Condition 2. In the general case we shall let $\mathscr{H}^*$ depend on n parameters $\lambda_1, \lambda_2, ..., \lambda_n$. For instance, we may assume

$$\mathscr{H}^*(\lambda_1, \lambda_2, ..., \lambda_n) \triangleq \lambda_1 \frac{d^{n-1}}{dt^{n-1}} + \lambda_2 \frac{d^{n-2}}{dt^{n-2}} + \cdots + \lambda_n$$

In some problems it may be convenient to use only one or two parameters.

Condition 3. The optimal $\mathscr{H}^*$ will be the one which minimizes the norm of the difference

$$| \epsilon\rangle \triangleq \mathscr{H} | x_0\rangle - \mathscr{H}^* | x_0\rangle \triangleq (\mathscr{H} - \mathscr{H}^*) | x_0\rangle$$

We shall also write

$$\epsilon(t) \triangleq (\mathscr{H} - \mathscr{H}^*) x_0(t)$$

Norm of $| \epsilon\rangle$:

$$N | \epsilon\rangle = \overline{\epsilon^2} \qquad \overline{\epsilon^2} = \int_{t_0}^{t_1} \epsilon^2(t)\, dt \qquad \text{(for a continuous function)}$$

$$\overline{\epsilon^2} = \int_0^{\infty} \epsilon p(x)\, dx \qquad \text{(for a random function)}$$

The optimality criterion will be written

$$\frac{\partial}{\partial \lambda_1} N | \epsilon\rangle = \frac{\partial}{\partial \lambda_2} N | \epsilon\rangle = \cdots = \frac{\partial}{\partial \lambda_n} N | \epsilon\rangle = 0 \tag{27}$$

It provides n equations from which coefficients $\lambda_1, \lambda_2, ..., \lambda_n$ are to be computed.

We shall now apply this method to a few additional examples. (Also refer to Chapter I, Section 3.5, and Chapter II, Section 5.4.)

5.1. Example 1

Let

$$y \triangleq \mathscr{H}x \triangleq \ddot{x} + \beta\dot{x} + \omega_0^2[x + \epsilon g(x)]$$

where $g(x)$ is a given function. Coefficients β and ω_0^2 are constant as well as ϵ, which is assumed to be small.

$x_0(t)$ being any given differentiable function defined on the interval $[0, T]$, we shall wish to approach this nonlinear relation by the linear one,

$$y^* \triangleq \mathscr{H}^*x \triangleq \ddot{x} + \beta\dot{x} + \lambda x \qquad \text{for} \quad x \equiv x_0(t)$$

We have

$$\epsilon = (\mathscr{H} - \mathscr{H}^*)x_0 = (\omega_0^2 - \lambda)x_0 + \epsilon\omega_0^2 g(x_0)$$

$$N \mid \epsilon\rangle = \frac{1}{T}\int_0^T \epsilon^2\, dt = (\omega_0^2 - \lambda)^2\overline{x_0^2} + 2\epsilon\omega_0^2(\omega_0^2 - \lambda)\overline{x_0 g(x_0)} + \epsilon^2\omega_0^4\overline{g^2(x_0)}$$

$$\frac{\partial}{\partial\lambda} N \mid \epsilon\rangle = -2(\omega_0^2 - \lambda)\overline{x_0^2} - 2\epsilon\omega_0^2\overline{x_0 g(x_0)} = 0$$

from which follows

$$\lambda = \omega_0^2\left[1 + \epsilon\frac{\overline{x_0 g(x_0)}}{\overline{x_0^2}}\right] \tag{28}$$

When $g(x) \equiv x$ we find $\lambda = \omega_0^2(1 + \epsilon)$, which is a consequence of the linearity of the given equation in this case. Then the optimal linear law is identical to the given equation.

5.2. Example 2

Let

$$y \triangleq \mathscr{H}x \triangleq \ddot{x} + \mu(x^2 - 1)\dot{x} + x$$

μ being assumed to be small, approach this nonlinear relation by the equation

$$y^* \triangleq \mathscr{H}^*x \triangleq \ddot{x} + \lambda\dot{x} + x \qquad \text{for} \quad x \equiv x_0(t)$$

We have

$$\epsilon = \mu x_0^2\dot{x}_0 - (\mu + \lambda)\dot{x}_0$$

$$N \mid \epsilon\rangle \simeq \lambda^2\overline{\dot{x}_0^2} - 2\lambda\mu\overline{(x_0^2 - 1)\dot{x}_0^2}$$

(when neglecting μ^2) and

$$\frac{\partial}{\partial\lambda} N \mid \epsilon\rangle = 2\lambda\overline{\dot{x}_0^2} - 2\mu\overline{(x_0^2 - 1)\dot{x}_0^2} = 0$$

from which follows

$$\lambda = \mu \frac{\overline{(x_0^2 - 1)\dot{x}_0^2}}{\overline{\dot{x}_0^2}} \tag{29}$$

When $x_0 = a_0 \cos \omega t$ $(0 < t < T)$, we get

$$\lambda = \mu \left(\frac{a_0^2}{4} - 1\right) \tag{30}$$

This is also the approximation obtained by the first-harmonic method, which appears from this viewpoint as an optimal approximation, in the neighborhood of the function $a_0 \cos \omega t$.

5.3. Example 3: Application to a Nonautonomous System

The optimal linearization method can also be applied to nonautonomous systems, i.e., systems whose equations depend explicitly on time t.

Let us consider as an example a system whose equation of motion is

$$\ddot{x} + \beta\dot{x} + (1 + \alpha \cos 2t)x + \gamma x^3 = 0 \tag{31}$$

α, β, and γ are constants; the coefficient γ is assumed to be small.

We shall try to approach the nonlinear transformation

$$y \triangleq \mathscr{H}x \triangleq \ddot{x} + \beta\dot{x} + (1 + \alpha \cos 2t)x + \gamma x^3$$

by the linear one

$$y^* \triangleq \mathscr{H}^*x \triangleq \ddot{x} + \lambda_1\dot{x} + \lambda_2 x \qquad \text{for} \quad x \equiv x_0(t)$$

We have

$$\epsilon = (\beta - \lambda_1)\dot{x}_0 + (1 + \alpha \cos 2t - \lambda_2)x_0 + \gamma x_0^3$$

$$\begin{aligned} N \mid \epsilon\rangle = {} & (\beta - \lambda_1)^2\overline{\dot{x}_0^2} + \overline{(1 + \alpha \cos 2t - \lambda_2)^2 x_0^2} \\ & + 2(\beta - \lambda_1)\gamma\overline{\dot{x}_0 x_0^3} + 2(\beta - \lambda_1)\overline{(1 + \alpha \cos 2t - \lambda_2)\dot{x}_0 x_0} \\ & + 2\overline{(1 + \alpha \cos 2t - \lambda_2)\gamma x_0^4} \end{aligned}$$

(when neglecting γ^2).

$$\frac{\partial N \mid \epsilon\rangle}{\partial \lambda_1} = -2(\beta - \lambda_1)\overline{\dot{x}_0^{\,2}} - 2\gamma\overline{\dot{x}_0 x_0^{\,3}} - 2\overline{(1 + \alpha \cos 2t - \lambda^2)\dot{x}_0 x_0} = 0$$

$$\frac{\partial N \mid \epsilon\rangle}{\partial \lambda_2} = 2\overline{(1 + \alpha \cos 2t - \lambda_2)x_0^{\,2}} - 2(\beta - \lambda_1)\overline{\dot{x}_0 x_0} - 2\gamma\overline{x_0^{\,4}} = 0$$

Accordingly, we get λ_1 and λ_2:

$$\lambda_1 = \beta + \gamma\frac{\overline{x_0^{\,3}\dot{x}_0}}{\overline{\dot{x}_0^{\,2}}} + \frac{\overline{(1 + \alpha \cos 2t - \lambda_2)\dot{x}_0 x_0}}{\overline{\dot{x}_0^{\,2}}}$$

$$\lambda_2 = 1 + \gamma\frac{\overline{x_0^{\,4}}}{\overline{x_0^{\,2}}} + \alpha\frac{\overline{x_0^{\,2}\cos 2t}}{\overline{x_0^{\,2}}} + (\beta - \lambda_1)\frac{\overline{\dot{x}_0 x_0}}{\overline{x_0^{\,2}}}$$

Finally, if we replace $x_0(t)$ by a first approximate solution of (31), of the form

$$x_0 = a_0 \cos (t - \varphi)$$

we find well-known expressions for λ_1, and λ_2, which we shall also obtain later by other methods—the stroboscopic method and the describing function method.

$$\begin{aligned} \lambda_1 &= \beta + \frac{\alpha}{2}\sin 2\varphi \\ \lambda_2 &= 1 + \frac{3\gamma}{4}a_0^{\,2} + \frac{\alpha}{2}\cos 2\varphi \end{aligned} \tag{32}$$

6. ITERATION OF THE PROCEDURE

The method for producing the optimal linear operator associated with a given nonlinear operator requires knowledge of the function $x = x_0(t)$, the point of the functional space at which the linearization is performed. However, in many practical problems, the function which is known is $y = y_0(t)$, and $x = x_0(t)$ is unknown. It needs to be determined, which means that the problem has to be solved before linearization can be performed. Thus the method we have described above seems to be of no help in this circumstance.

Fortunately this is not exactly true, since the method can be complemented by an iterative procedure which brings an approximate solution for this kind of problem, provided the nonlinearity is sufficiently weak.

We shall explain this procedure by a simple example. Consider the equation

$$y = \ddot{x} + \omega_0^2(x + \epsilon x^3) \tag{33}$$

where ϵ is small and $y = F \cos \omega t$ is the given function.

Following the optimal linearization method (Section 5.1) we will approach (33), in the neighborhood of the *unknown* $x_0(t)$, by the linear equation $y = \ddot{x} + \lambda x$, which gives, by applying the optimization procedure,

$$\lambda = \omega_0^2 \left(1 + \epsilon \frac{\overline{x_0^4}}{\overline{x_0^2}}\right) \tag{34}$$

as above.

6.1. First Approximation

Now, since ϵ is small, let $\lambda = \omega_0^2$ *as a first approximation.* We get $x(t) = a_0 \cos \omega t$ with

$$a_0 = \frac{F}{\lambda - \omega^2} = \frac{F}{\omega_0^2 - \omega^2} \tag{35}$$

6.2. Second Approximation

From the first approximation we deduce

$$\overline{x_0^4} = \frac{3a_0^4}{8} \qquad \overline{x_0^2} = \frac{a_0^2}{2}$$

Then, by substituting in (34),

$$\lambda = \omega_0^2 \left(1 + \frac{3\epsilon}{4} a_0^2\right) \tag{36}$$

which is *the second approximation* for λ. Finally, substituting (36) in (35), we get

$$a_0 = \frac{F}{\omega_0^2 - \omega^2 + \frac{3\epsilon}{4} \omega_0^2 a_0^2}$$

This is the well-known relation, obtained in Chapter I, Section 7, between amplitude and frequency for a nonlinear pendulum driven by a sinusoidal external force.

It should be pointed out that this iterative procedure can also be applied when the excitation is not a sinusoidal function. For instance, it has been applied by Crandall [13] in the case where the excitation is a gaussian noise.

7. THE DESCRIBING FUNCTION†

Summarizing the results of the previous sections, we see that in many cases of practical interest, the behavior of a nonlinear system can be *approximately* described by a linear equation, and we have given a method for producing this equation in such a way that the approximation is as good as possible with respect to some criterion.

If the system is a nonlinear quadrupole, this approximate linear equation enables us to extend the classical concept of transfer function, by assuming that the input is a sinusoidal signal and by using formulas (4) and (5).

For instance, in the example of Section 3 the generalized transfer function is

$$A(a_0) \triangleq S_0 + \frac{3S_2}{4} a_0^2$$

It depends on amplitude a_0 of the sinusoidal input signal, but not on the frequency. Note that in more general examples it will depend on a_0 and ω. In this example we also defined a generalized transfer function with respect to a gaussian random input:

$$A(\sigma_0) \triangleq S_0 + 3S_2\sigma_0^2$$

It depends on the parameter σ_0, which characterizes the spread of the gaussian distribution.

If the given nonlinear equation is the differential equation of motion of a nonlinear system, then, again, we have approximated this equation by a linear one, which enables us to extend formulas (8) and (9) by assuming a quasi-sinusoidal behavior. Indeed, such an assumption must be discussed in each practical example. In general it is valid when the system has a linear component which is sufficiently selective, and when the nonlinearities are weak.

For instance, using the examples of Section 5, we are led to introduce the following functions:

† See [1–6, 12].

In Example 1, for $g(x) \equiv x^3$:

$$H(a_0, j\omega) \triangleq -\omega^2 + \omega_0^2 \left(1 + \frac{3\epsilon}{4} a_0^2\right) + j\omega\beta$$

which is associated with the linearized operator (linear *for each given* a_0)

$$\mathscr{H}^* \triangleq \ddot{x} + \beta\dot{x} + \omega_0^2 \left(1 + \frac{3\epsilon}{4} a_0^2\right)$$

In Example 2:

$$H(a_0, j\omega) \triangleq -\omega^2 + 1 + j\omega\mu \left(\frac{a_0^2}{4} - 1\right)$$

In example 3:

$$H(a_0, \varphi, j\omega) \triangleq -\omega^2 + \left(1 + \frac{3\gamma}{4} a_0^2 + \frac{\alpha}{2} \cos 2\varphi\right) + j\omega \left(\beta + \frac{\alpha}{2} \sin 2\varphi\right)$$

As a matter of fact, we see that such functions depend in general on ω, which also occurs in the function H of linear systems, but also on amplitude a_0 of the quasi-sinusoidal oscillation.

Frequently, as will be seen later, a_0 is a slowly varying function of time, in contrast with the case of quadrupoles where a_0 is fixed. In the present case a_0 is the amplitude of the oscillation during a certain interval of time.

Furthermore, we see, Example 3, that the function H can also depend on other parameters, such as phase angle φ. This generalized function is called the *describing function* for the nonlinear system. It appears that in most of the papers which have been published on this question no difference is made, as concerns the name, between generalized transfer functions and functions associated with differential equations of motion. As a matter of fact there is no trouble in doing so, and in the following we shall represent the describing function in both cases by $H(a_0, j\omega)$, by $H(a, j\omega)$, or by H, $\mathscr{H}$ being the nonlinear operator from which the describing function is generated. Indeed, it may depend on other parameters, beside a and ω.

From the equivalence between the first-harmonic approximation and the optimal linearization method when applied to input $x = a_0 \cos \omega t$, it follows that the describing function can also be determined by picking out the first harmonic of the transform y when x is a sinusoidal function:

$$x = a_0 \cos \omega t$$

That is,

$$y = \mathscr{H}(a_0 \cos \omega t)$$

is decomposed into a Fourier series

$$y = b_0 + b_1 \cos(\omega t - \varphi_1) + b_2 \cos(2\omega t - \varphi_2) + \cdots + b_N \cos(N\omega t - \varphi_N) + \cdots$$

and only the first harmonic is taken into consideration.

Hower, since the definition of the describing function involves the use of the complex exponential function $e^{j\omega t}$, we are led to replace $x(t)$ and $y(t)$ by

$$\begin{aligned} \hat{x}(t) &= a_0 e^{j\omega t} \\ \hat{y}(t) &= Y_0 + Y_1 e^{j\omega t} + Y_2 e^{2j\omega t} + \cdots + Y_N e^{Nj\omega t} + \cdots \end{aligned} \tag{37}$$

where

$$Y_0 = b_0 \qquad Y_k = b_k e^{-j\varphi_k} \qquad (k = 1, 2, ..., N)$$

$$x(t) = \text{Re}[\hat{x}(t)] \qquad y(t) = \text{Re}[\hat{y}(t)]$$

Then the describing function is the amplitude ratio

$$H = \frac{Y_1}{a_0} \tag{38}$$

As a matter of fact, it is the way in which we had introduced the describing function earlier [5], and only afterward did we established the connection with the optimal linearization method.

Equation (38) can be expressed by making use of Fourier's formulas:

$$Y_k = \frac{\omega}{\pi} \int_0^{2\pi/\omega} [\mathscr{H}(a_0 \cos \omega t)] e^{-jk\omega t}\, dt \qquad (k = 1, 2, ..., N)$$

from which follows

$$H = \frac{\omega}{\pi a_0} \int_0^{2\pi/\omega} [\mathscr{H}(a_0 \cos \omega t)] e^{-j\omega t}\, dt \tag{39}$$

For example, let us apply this formula to a nonlinear element with saturation, whose input-output characteristic is shown in Fig. 4a:

$$\mathscr{H}x = S_0 x \quad \text{if} \quad -m \leqslant x \leqslant +m$$

$$\mathscr{H}x = mS_0 \quad \text{if} \quad \begin{cases} x \geqslant +m \\ x \leqslant -m \end{cases}$$

(S_0, constant slope). If $a_0 < m$, we get

$$H = S_0$$

If $a_0 > m$, one finds easily by performing the integration that

$$H = \frac{2S_0 m}{\pi a_0} \left\{ \frac{a_0}{m} \sin^{-1}\left(\frac{m}{a_0}\right) + \left[1 - \left(\frac{m}{a_0}\right)^2\right]^{1/2} \right\}$$

Function $H(a_0)$ is shown in Fig. 4b.

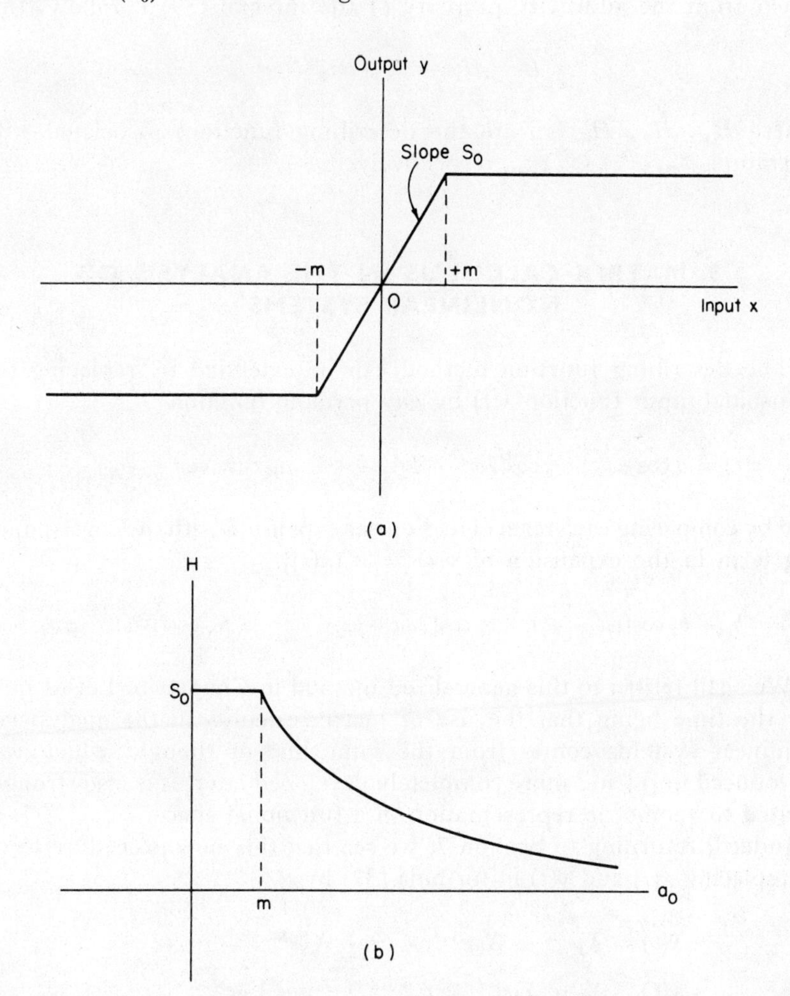

FIG. 4. (a) Input-output characteristic of a nonlinear element with saturation; (b) describing function.

8. ADDITIVE PROPERTY OF THE DESCRIBING FUNCTION

In practical applications it may happen that nonlinearities are split into a number of terms which can be considered independent of one another in the resolution of the describing function. Thus $\mathscr{H}x$ can be written

$$\mathscr{H}x = (\mathscr{H}_1 + \mathscr{H}_2 + \mathscr{H}_3 + \cdots)x$$

Then from the additivity property of the integral (39) it follows that

$$H = H_1 + H_2 + H_3 + \cdots$$

where H_1, H_2, H_3, ... are the describing functions associated with operators $\mathscr{H}_1$, $\mathscr{H}_2$, $\mathscr{H}_3$, ..., respectively.

9. MATRIX CALCULUS IN THE ANALYSIS OF NONLINEAR SYSTEMS†

The describing function method can be extended by replacing the sinusoidal input function $x(t)$ by *any* periodic function

$$x(t) = a \cos \omega t + a_2 \cos(2\omega t - \psi_2) + \cdots + a_N \cos(N\omega t - \psi_N) + \cdots$$

and by comparing each term of its Fourier expansion with the corresponding term in the expansion of $y(t) = \mathscr{H}(x(t)]$:

$$y(t) = b_0 + b_1 \cos(\omega t - \varphi_1) + b_2 \cos(2\omega t - \varphi_2) + \cdots + b_N \cos(N\omega t - \varphi_N) + \cdots$$

We shall return to this generalized method in Chapter 6. Let us note for the time being that the use of matrix calculus in the analysis of nonlinear systems comes from the same line of thought which was introduced in [8] and more completely developed later. It is also strongly related to geometric representation in a functional space.

Indeed, returning to Section 7, we see that this new procedure leads to replacing $\hat{x}(t)$ and $\hat{y}(t)$ in formula (37) by

$$\begin{aligned} \hat{x}(t) &= X_1 e^{j\omega t} + X_2 e^{2j\omega t} + \cdots + X_N e^{Nj\omega t} + \cdots \\ \hat{y}(t) &= Y_0 + Y_1 e^{j\omega t} + Y_2 e^{2j\omega t} + \cdots + Y_N e^{Nj\omega t} + \cdots \end{aligned} \tag{40}$$

† See [7, 8, 10].

with

$$X_1 = a \qquad X_k = a_k \exp(-j\psi_k) \qquad (k = 1, 2, ..., N)$$
$$Y_0 = b_0 \qquad Y_k = b_k \exp(-j\varphi_k)$$
$$x(t) = \mathrm{Re}[\hat{x}(t)] \qquad y(t) = \mathrm{Re}[\hat{y}(t)]$$

Then, if we disregard the constant Y_0, which can in general be ignored in practice, the relation between $\hat{x}(t)$ and $\hat{y}(t)$ can be expressed either by making explicit the set of equations

$$\begin{aligned} Y_1 &= H_1(\omega, a, a_2, ..., a_N, ...)X_1 \\ Y_2 &= H_2(2\omega, a, a_2, ..., a_N, ...)X_2 \\ &\vdots \\ Y_N &= H_N(N\omega, a, a_2, ..., a_N, ...)X_N \\ &\vdots \end{aligned} \tag{41}$$

or by introducing the matrix relation

$$\begin{pmatrix} Y_1 \\ Y_2 \\ \vdots \\ Y_N \\ \vdots \end{pmatrix} = \begin{pmatrix} H_{11} & H_{12} & \cdots & H_{1N} & \cdots \\ H_{21} & H_{22} & \cdots & H_{2N} & \cdots \\ \vdots \\ H_{N1} & H_{N2} & \cdots & H_{NN} & \cdots \\ \vdots \end{pmatrix} \begin{pmatrix} X_1 \\ X_2 \\ \vdots \\ X_N \\ \vdots \end{pmatrix} \tag{42}$$

In the first alternative we are led to replace the describing function of Section 7 with a *set of describing functions*:

$$H_1, H_2, ..., H_N$$

which makes the description of the behavior of the nonlinear system more rigorous, and therefore more complexe. The matrix formulation will also enable us to investigate properties which were masked by the excessive simplicity of the first-harmonic approximation.

Note that (42) makes the geometric representation in a Hilbert space more precise. That is, the basis is now defined by vectors which are associated with functions

$$e^{j\omega t} \qquad e^{2j\omega t} \cdots e^{Nj\omega t}$$

say

$$| \zeta_1\rangle \qquad | \zeta_2\rangle \cdots | \zeta_N\rangle$$

and $\tilde{H}$ is the operator which transforms the vector $| X\rangle$, whose components are X_1, X_2, ..., X_N, into the vector $| Y\rangle$, whose components are Y_1, Y_2, ..., Y_N:

$$\tilde{H} = \begin{pmatrix} H_{11} & H_{12} \cdots H_{1N} & \cdots \\ H_{21} & H_{22} \cdots H_{2N} & \cdots \\ \vdots & & \\ H_{N1} & H_{N2} \cdots H_{NN} & \cdots \\ \vdots & & \end{pmatrix}$$

In general computation of the elements of the matrix H is toilsome. However, in many practical applications it is possible to take advantage of some simplifying assumptions, and to determine these elements easily.

9.1. Connection with the Optimal Linearization Method†

Assume, for instance, that $| X\rangle$ has been chosen in the plane $\prod_{1,3}$ defined by the vectors $| \zeta_1\rangle$ and $| \zeta_3\rangle$. In general its transform $| Y\rangle$ does not belong to this plane. From the functional viewpoint this means that we have considered a periodic signal $x(t)$ whose fundamental frequency ω is accompanied by a third harmonic 3ω. The transformation usually generates higher-order harmonics.

A simplifying assumption consists of neglecting higher-order harmonics and taking into consideration only the fundamental term and the third harmonic of the transform. From a geometric viewpoint this means that we are interested in *the projection* $| Y^*\rangle$ *of the transform* $| Y\rangle$ *on the plane* $\prod_{1,3}$:

$$| Y^*\rangle = \operatorname*{proj}_{\Pi_{1,3}} | Y\rangle$$

If instead of analyzing the transformation due to the actual nonlinear operator, we only consider the projection of this transformation on plane $\prod_{1,3}$, we shall have an imperfect picture of the process. However, the lack of rigor will again be counterpoised by the greater simplicity of the computations; the exact representation will be replaced by the approximate one:

$$| Y^*\rangle = \tilde{\Lambda} \,| X\rangle \tag{43}$$

with

$$| X\rangle = \begin{pmatrix} X_1 \\ X_3 \end{pmatrix} \qquad | Y^*\rangle = \begin{pmatrix} Y_1 \\ Y_3 \end{pmatrix}$$

$$\tilde{\Lambda} = \begin{pmatrix} H_{11} & H_{13} \\ H_{31} & H_{33} \end{pmatrix}$$

† See [14].

In general, the elements H_{11}, H_{13}, H_{31}, and H_{33} of this simplified matrix will be functions of ω, a, and a_3, say of the amplitudes of the fundamental component of $x(t)$ and of its third harmonic.

As a matter of fact, (43) is a generalization of (19). It becomes a *linear approximation* in the close neighborhood of a given $| X\rangle$ as soon as amplitudes a and a_3 are "frozen" at that point. Furthermore, it may easily be seen that it is an *optimal* approximation with respect to the prescribed restrictive conditions and to our earlier criterion of optimality, since it is the *orthogonal projection* of the actual transformation on the plane $\prod_{1,3}$.

Obviously, instead of two vectors $| \zeta_1\rangle$ and $| \zeta_3\rangle$, we might consider a set consisting of a larger number of vectors:

$$| \zeta_1\rangle \qquad | \zeta_2\rangle \cdots | \zeta_k\rangle$$

and, again, define an optimal approximation by projecting the actual transformation on the plane $\prod_{1,2,\ldots,k}$ defined by these vectors. Namely, $| X\rangle \in \prod_{1,2\ldots,k}$ being given, we would replace its transform $| Y\rangle$ by

$$|Y^*\rangle = \operatorname*{proj}_{\Pi_{1,2,\ldots,k}} | Y\rangle$$

The arguments are similar: The larger the number of components, the better the approximation, but the complexity of the computations is increased.

9.2. Example

To illustrate the above outline of the theory, let us carry out the computations in a simple example in which the nonlinear operator $\mathscr{H}$ is defined by the relation $y = x^3$. Let

$$x(t) = a\cos(\omega t - \psi_1) + a_3\cos(3\omega t - \psi_3)$$

We find

$$\begin{aligned} y(t) = {} & \frac{3a}{4}(a^2 + 2a_3^2)\cos(\omega t - \psi_1) + \frac{3a^2a_3}{4}\cos(\omega t - \psi_3 + 2\psi_1) \\ & + \frac{a^3}{4}\cos(3\omega t - 3\psi_1) + \frac{3a_3}{4}(a_3^2 + 2a^2)\cos(3\omega t - \psi_3) \\ & + \frac{3a^2a_3}{4}\cos(5\omega t - 2\psi_1 - \psi_3) + \frac{3aa_3^2}{4}\cos(5\omega t - 2\psi_3 + \psi_1) \\ & + \frac{3aa_3^2}{4}\cos(7\omega t - 2\psi_3 - \psi_1) + \frac{a_3^3}{4}\cos(9\omega t - 3\psi_3) \end{aligned}$$

Accordingly, we have

$$\hat{x}(t) = a \exp(-j\psi_1) \exp(j\omega t) + a_3 \exp(-j\psi_3) \exp(3j\omega t)$$

$$\begin{aligned}\hat{y}(t) = &\left\{\frac{3a}{4}(a^2 + 2a_3^2) \exp(-j\psi_1) + \frac{3a^2a_3}{4} \exp[-j(\psi_3 - 2\psi_1)]\right\} \exp(j\omega t)\\ &+ \left[\frac{a^3}{4} \exp(-3j\psi_1) + \frac{3a_3}{4}(a_3^2 + 2a^2) \exp(-j\psi_3)\right] \exp(3j\omega t)\\ &+ \left\{\frac{3a^2a_3}{4} \exp[-j(2\psi_1 + \psi_3)] + \frac{3aa_3^2}{4} \exp[-j(2\psi_3 - \psi_1)]\right\} \exp(5j\omega t)\\ &+ \frac{3aa_3^2}{4} \exp[-j(2\psi_3 + \psi_1)] \exp(7j\omega t) + \frac{a_3^3}{4} \exp(-3j\psi_3) \exp(9j\omega t)\end{aligned} \tag{44}$$

say

$$X_1 = a \exp(-j\psi_1) \qquad X_3 = a_3 \exp(-j\psi_3) \qquad X_5 = X_7 = X_9 = 0$$

$$Y_1 = \frac{3a}{4}(a^2 + 2a_3^2) \exp(-j\psi_1) + \frac{3a^2a_3}{4} \exp[-j(\psi_3 - 2\psi_1)]$$

$$Y_3 = \frac{a^3}{4} \exp(-3j\psi_1) + \frac{3a_3}{4}(a_3^2 + 2a^2) \exp(-j\psi_3)$$

$$Y_5 = \frac{3a^2a_3}{4} \exp[-j(2\psi_1 + \psi_3)] + \frac{3aa_3^2}{4} \exp[-j(2\psi_3 - \psi_1)]$$

$$Y_7 = \frac{3aa_3^2}{4} \exp[-j(2\psi_3 + \psi_1)]$$

$$Y_9 = \frac{a_3^3}{4} \exp(-3j\psi_3)$$

Here we can define

$$\frac{Y_1}{X_1} = H_1(a, a_3, \psi_1, \psi_3) = \frac{3}{4}(a^2 + 2a_3^2) + \frac{3aa_3}{4} \exp[-j(\psi_3 - 3\psi_1)]$$

$$\frac{Y_3}{X_3} = H_3(a, a_3, \psi_1, \psi_3) = \frac{a^3}{4a_3} \exp[-j(3\psi_1 - \psi_3)] + \frac{3}{4}(a_3^2 + 2a^2)$$

On the other hand, the matrix relation (42) can be written

$$\begin{pmatrix} Y_1 \\ Y_3 \\ Y_5 \\ Y_7 \\ Y_9 \end{pmatrix} = \begin{pmatrix} H_{11} & 0 & 0 & 0 & 0 \\ 0 & H_{33} & 0 & 0 & 0 \\ H_{51} & H_{53} & 0 & 0 & 0 \\ H_{71} & 0 & 0 & 0 & 0 \\ 0 & H_{93} & 0 & 0 & 0 \end{pmatrix} \begin{pmatrix} X_1 \\ X_3 \\ 0 \\ 0 \\ 0 \end{pmatrix}$$

with

$$H_{11} = \frac{3}{4}(a^2 + 2a_3^2) + \frac{3aa_3}{4}\exp[-j(\psi_3 - 3\psi_1)]$$

$$H_{33} = \frac{a^3}{4a_3}\exp[-j(3\psi_1 - \psi_3)] + \frac{3}{4}(a_3^2 + 2a^2)$$

$$H_{51} = \frac{3aa_3}{4}\exp[-j(\psi_1 + \psi_3)]$$

$$H_{53} = \frac{3aa_3}{4}\exp[-j(\psi_3 - \psi_1)]$$

$$H_{71} = \frac{3a_3^2}{4}\exp(-2j\psi_3)$$

$$H_{93} = \frac{a_3^2}{4}\exp(-2j\psi_3)$$

Obviously that is not the only admissible representation for operator $\tilde{H}$. Other possibilities could be easily found.

Now if we take into consideration only the fundamental term and the third harmonic of the transform, say if we project the actual transformation on the plane $\prod_{1.3}$, $\tilde{H}$ reduces to the matrix $\tilde{\Lambda}$:

$$\tilde{\Lambda} = \begin{pmatrix} H_{11} & 0 \\ 0 & H_{33} \end{pmatrix}$$

where

$$H_{11} = H_1(a, a_3, \psi_1, \psi_3) \qquad H_{33} = H_3(a, a_3, \psi_1, \psi_3)$$

Then the connection between the matrix representation and the generalized describing function method is pretty clear, since the two components of the generalized describing function are the diagonal terms of the matrix $\tilde{\Lambda}$.

Indeed other representations can be used for the operator $\tilde{\Lambda}$; this is quite obvious, since (44) is rewritten

$$\hat{x}(t) = a\exp(-j\psi_1)\exp(j\omega t) + a_3\exp(-j\psi_3)\exp(3j\omega t)$$

$$\hat{y}(t) = \left\{\frac{3a}{4}(a^2 + 2a_3^2)\exp(-j\psi_1) + \frac{3a^2a_3}{4}\exp[-j(\psi_3 - 2\psi_1)]\right\}\exp(j\omega t)$$
$$+ \left[\frac{a^3}{4}\exp(-3j\psi_1) + \frac{3a_3}{4}(a_3^2 + 2a^2)\exp(-j\psi_3)\right]\exp(3j\omega t) + \cdots$$

For instance one can use

$$\tilde{\Lambda} = \begin{pmatrix} \frac{3}{4}(a^2 + 2a_3{}^2) & \frac{3a^2}{4}\exp(2j\psi_1) \\ \frac{a^2}{4}\exp(-2j\psi_1) & \frac{3}{4}(a_3{}^2 + 2a^2) \end{pmatrix}$$

9.3. Different Kinds of Matrices†

If $x(t)$ is the input of a quadrupole whose output is $y(t)$, and if $x(t)$ and $y(t)$ are physical variables of the same kind, $\tilde{H}$ is a *transfer matrix*.

Obviously nothing is changed in the theory if $x(t)$ and $y(t)$ are not physical variables of the same kind. For instance, if $x(t)$ is a voltage between the end points of an electrical circuit, and $y(t)$ the corresponding intensity, the relation between $x(t)$ and $y(t)$ can also be written

$$y = \mathscr{H}x$$

or, by the same arguments as above,

$$| Y\rangle = \tilde{H} \,| X\rangle$$

Then $\tilde{H}$ is an *admittance matrix*, and $\tilde{H}^{-1}$ is an *impedance matrix*.

As an example, the impedance matrix of the rectifier whose characteristic is shown in Fig. 5 can be readily determined. If one limits the series expansions to the first two terms

$$x(t) = a_0 + a_1 \cos \omega t$$
$$y(t) = b_0 + b_1 \cos \omega t + \cdots$$

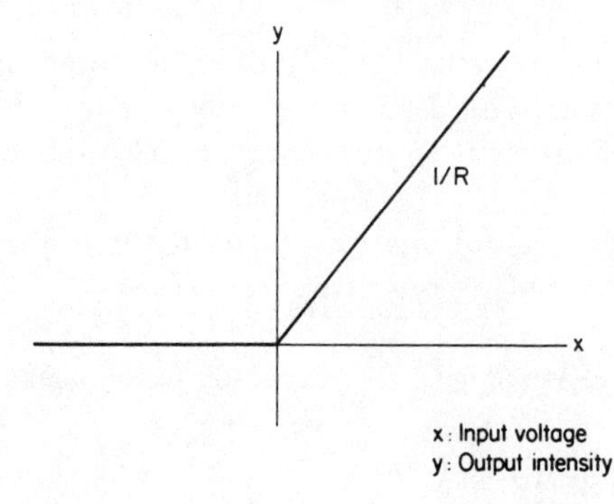

FIG. 5. Characteristic curve of a rectifier.

† See [7].

namely,

$$\hat{x}(t) = a_0 + a_1 e^{j\omega t}$$
$$\hat{y}(t) = b_0 + b_1 e^{j\omega t}$$

one finds, when a_0 is assumed to be small, the impedance matrix,

$$\begin{pmatrix} \frac{R}{2} & \frac{R}{\pi} \\ \frac{2R}{\pi} & \frac{R}{2} \end{pmatrix}$$

Note that, in this example, in contrast with the earlier ones, the direct current b_0 plays an essential role. Accordingly, in the definition of the impedance matrix, we have been led to introduce terms associated with the frequency $\omega = 0$. Also note that in this example the impedance matrix neither depends upon frequency ω nor upon the amplitudes of the currents. Such a linear-like example is unique.

If instead of dealing with voltages and currents in electricity one deals with forces and displacements, or with forces and velocities, in mechanics one is led to defining *stiffness matrices* or *force-velocity matrices* in the same way.

REFERENCES

1. K. F. Theodorchik, "Auto-Oscillatory Systems." Moscow, 1948.
2. R. J. Kochenburger, A Frequency Response Method for Analyzing and Synthesizing Contactor Servomechanisms. *Trans. AIEE*, **69**, 270–283 (1950).
3. A. Blaquière, Extension de la théorie de Nyquist au cas de caractéristiques non-linéaires. *Compt. Rend.* **233**, 345 (1951).
4. J. Loeb, Un Criterium général de stabilité des servomécanismes sièges de phénomènes héréditaires. *Compt. Rend.* **233**, 344 (1951).
5. A. Blaquière, Adaptation générale de la méthode du diagramme de Nyquist dans le domaine non linéaire. *J. Phys. Radium* **13**, 527–540, 636–644 (1952).
6. R. C. Booton, Jr., The Measurement and Representation of Nonlinear Systems. *IRE Trans. Circuit Theory* **1**, 32–34 (1954).
7. F. H. Clauser, The Behaviour of Nonlinear Systems. *J. Aerospace Sci.* **23**, 409–32 (1956).
8. A. Blaquière, Mécanique non-linéaire, les oscillateurs à régimes quasi-sinusoidaux. Thèse, Paris, 1957. (Edited in "Mémorial des Sciences Mathématiques," Fasc. 141, Gauthier-Villars, Paris, 1960.)
9. K. Klotter, How to Obtain describing Functions for Nonlinear Feedback Systems. *Trans. ASME* (April 1957).
10. F. Bertein, Sur quelques aspects du calcul matriciel des circuits non linéaires à oscillation locale. *J. Phys. Radium* **21** (1960).

11. R. H. Lyon, Equivalent Linearization of the Hard Spring Oscillator. *J. Acoust. Soc. Am.* **32**, 1161–1162 (1960).
12. J. C. West, "Analytical Techniques for Non-Linear Control Systems." English Universities Press, London, 1960.
13. S. H. Crandall, Random Vibrations of Systems with Non-Linear Restoring Forces. *Rept. AFOSR 708, MIT*, Cambridge, Mass., 1961.
14. A. Blaquière, Une nouvelle méthode de linéarization locale des opérateurs non-linéaires; approximation optimale. *Second Conf. Nonlinear Vibrations, Warsaw*, 1962.

CHAPTER VI

The Describing Function Method

The describing function method is a useful tool in the analysis of feedback systems. It is a straightforward generalization of the Nyquist [1] and Mikaïlov [2] diagrams. Since these diagrams are probably sufficiently well known, we shall only summarize the main results which apply to linear feedback loops, with a view to extending them to nonlinear systems.

1. EQUATION OF FEEDBACK LOOPS

Consider first a feedback loop as shown in Fig. 1. The components of this diagram are

(a) The feedback element (F), whose feedback coefficient is Φ, which transforms the signal x_s into Φx_s.

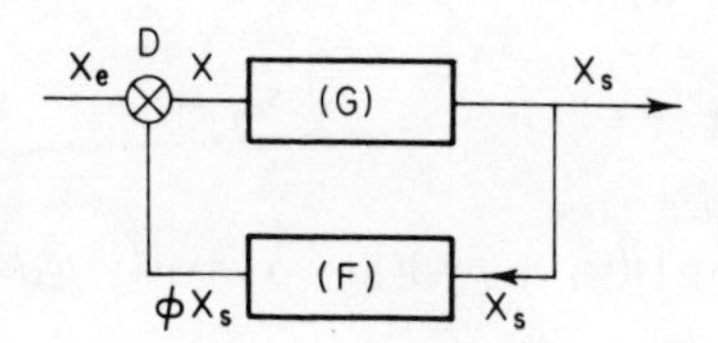

FIG. 1. Feedback loop.

(b) The discriminator (D), which adds (algebraically) the feedback signal Φx_s to the input signal x_e , to form the control signal x:

$$x = x_e + \Phi x_s$$

The sign $+$ is only a matter of convention. Should we choose to change the sign of the feedback coefficient, then the feedback signal would be subtracted from the input x_e according to the formula $x = x_e - \Phi x_s$.

(c) The amplifier (G), whose gain without feedback is Γ, which applies at the input of (F) the signal $\Gamma x = x_s$.

From these relations we deduce the basic equation of feedback loops:

$$x_s = \frac{\Gamma}{1 - \Gamma\Phi} x_e \tag{1}$$

or

$$(1 - \Gamma\Phi)x_s = \Gamma x_e \tag{2}$$

Next we shall assume that $x_e \equiv 0$. As was pointed out earlier, Γ and Φ can be thought of either as *operators* or as *functions of the variable* s, which is usually complex. In this chapter we shall not differentiate between Γ and Φ used as operators or functions.

If Γ and Φ are operators involving time derivatives d/dt, d^2/dt^2, etc., (2) is a differential equation, whose solution can be stable or unstable. Indeed the question of its stability will be the major business of this chapter. It involves determination of the roots of the equation

$$1 - \Gamma(s)\Phi(s) = 0 \tag{3}$$

To each root is associated a "transient,"

$$x_s = X_0 e^{st}$$

which can be oscillating or nonoscillating according to whether s is real or complex.

In general we shall write

$$s = j\omega \qquad \omega = \omega_1 + j\omega_2$$

and

$$x_s = X_0 \exp[j(\omega_1 + j\omega_2)t] = X_0 \exp(-\omega_2 t) \exp(j\omega_1 t)$$

ω_1 is the real angular frequency of the transient and ω_2 is an amplitude damping (positive or negative) coefficient.

The different cases are shown in Fig. 2.

Case 1. If Re $(s) < 0$, say $\omega_2 > 0$, the damping of the transient is positive; i.e., it is *stable* (Fig. 2a and c).

Case 2. If Re $(s) > 0$, say $\omega_2 < 0$, the damping is negative; i.e., the transient is *unstable* (Fig. 2b and d).

Case 3. If Re $(s) = 0$, which can only be considered in theory, since it does not occur in practice in linear systems, $x_s(t)$ is either a constant term or an oscillation with constant amplitude.

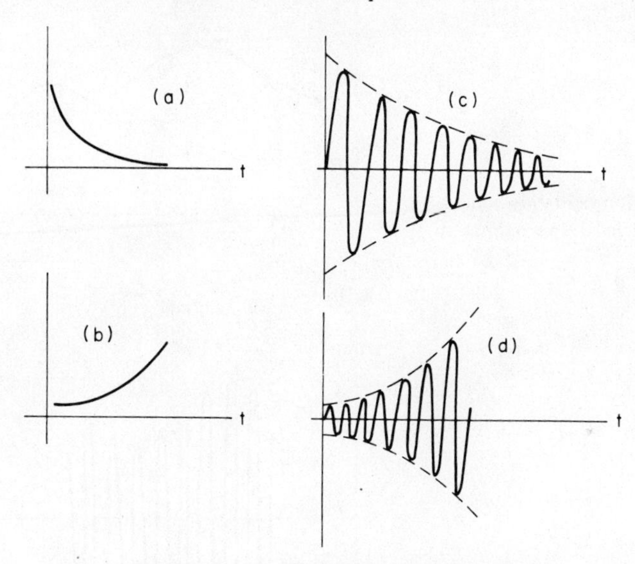

FIG. 2. Transients: (a, c) stable; (b, d) unstable.

In linear systems the solution of (2) is a linear combination of the transients, each of which is associated with a root of (3). A number of constants are involved, whose values depend on the initial conditions.

It is probably worthwhile at this point to speak of the nonlinear systems, since the superposition principle does not apply to them. Again two cases can be considered: The motion $x_s(t)$ is either *oscillating* or *nonoscillating*, during a time interval $[t_1, t_2]$ (Figs. 3 and 4).

Case 1. If the function $x_s(t)$ is nonoscillating, it can be approached, in a sufficiently small neighborhood of any time t_0, $t_1 < t_0 < t_2$, by the function

$$x_s^* = X_0 \exp(s^0 t)$$

where

$$x_s^*(t_0) = x_s(t_0) = X_0 \exp(s^0 t_0) \qquad s^0 \text{ real}$$

$$s^0 = \frac{\dot{x}_s(t_0)}{x_s(t_0)}$$

Case 2. If $x_s(t)$ is oscillating, and if it is a *quasi-sinusoidal function*, a function of the form

$$x_s = a \exp(j\omega_1 t)$$

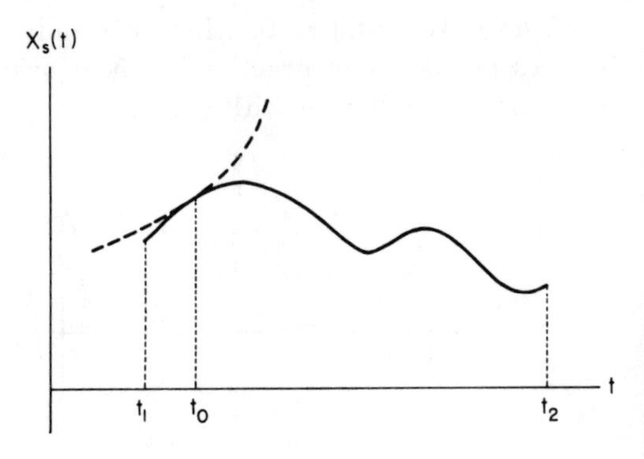

FIG. 3. Nonoscillating motion, exponential local approximation.

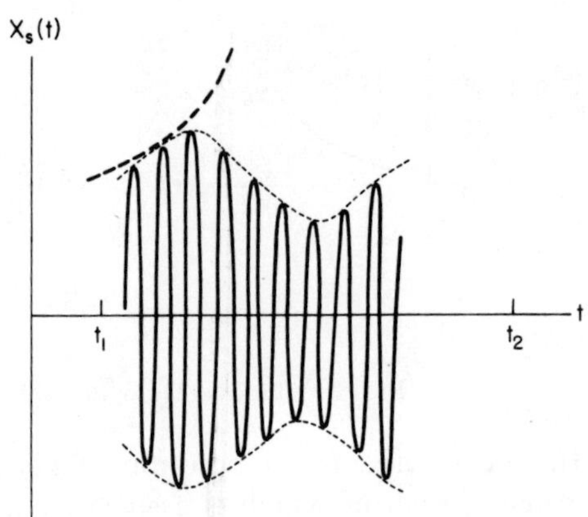

FIG. 4. Oscillating motion, local approximation for the variation of amplitude $a(t)$.

where $a = a(t)$, $\omega_1 = \omega_1(t)$, are real slowly varying functions of t, then again in a sufficiently small neighborhood of any time t_0, $t_1 < t_0 < t_2$, we can approach $a(t)$ by

$$a^*(t) = A_0 \exp(-\omega_2{}^0 t)$$

with

$$a^*(t_0) = a(t_0) = A_0 \exp(-\omega_2{}^0 t_0) \qquad \omega_2{}^0 \text{ real}$$

$$-\omega_2{}^0 \doteq \frac{\dot{a}(t_0)}{a(t_0)}$$

and $\omega_1(t)$ by

$$\omega_1{}^0 = \omega_1(t_0)$$

This means that we can approach $x_s(t)$, in the neighborhood of any time t_0, $t_1 < t_0 < t_2$, by

$$x_s{}^* = A_0 \exp(s^0 t)$$

with

$$s^0 = j\omega^0 \qquad \omega^0 = \omega_1{}^0 + j\omega_2{}^0$$

It can be considered a "local" transient, whose shape slowly varies with time. The actual nonlinear motion may thus be considered a kind of envelope of this moving transient.

These introductory remarks represent another way of taking note of a fundamental difference between linear and nonlinear mechanics. Indeed, it is commonplace to say that the superposition principle of linear mechanics is no longer valid in nonlinear mechanics. However, all the linear concepts do not necessarily fail. As a matter of fact, it may be that we again meet these concepts in nonlinear mechanics, but we do not deal with them in the same way.

For instance, instead of superposing a number of basic states to get the over-all picture of the behavior of the system, we need consider these states separately and get the over-all picture by determining the envelope of *the state* which has been selected, say its history during the interval of time under consideration.

This conclusion should be compared with the one which led Rosenberg to introduce the concept of normal modes (which again is a linear concept) in nonlinear mechanics (Chapter IV, Section 5).

2. LINEAR AND NONLINEAR FEEDBACK LOOPS†

The coefficients of the differential equation which governs the motion of a *linear* feedback system are determined by its plan. In contrast with this simple bearing, the coefficients of the differential equation of a *nonlinear* feedback loop depend on the plan and *also* on the motion—an unpleasant situation, since one wishes to determine the motion by solving the differential equation.

In quasi-linear systems the situation is not so hopeless, since the dependency between the motion and the form of the differential equation which governs the motion gets in by means of a number of parameters which are slowly varying functions of time. These parameters depend on some characteristics of the motion which are slowly varying functions of time. For instance, in a quasi-sinusoidal oscillator these parameters may be the amplitude of oscillation, its phase angle with respect to a given sinusoidal signal, etc.

Then in a conveniently chosen neighborhood of any time t_0,

† See [7, 10].

$t_1 < t_0 < t_2$, these parameters can be "frozen"; they can be considered constant at first approximation, which means that the differential equation of motion is linearized in each such interval. The "local" solution thus obtained depends on these slowly varying parameters and, again, it may be considered as enveloping the actual motion.

Once the problem has been stated in this way, it becomes obvious that the describing function method will provide one of the most convenient tools for studying quasi-linear feedback loops from an engineering viewpoint.

The function which has been introduced in Chapter 5,

$$H(j\omega, \lambda_1, \lambda_2, ..., \lambda_n) \tag{4}$$

where $\lambda_1, \lambda_2, ..., \lambda_n$ are the parameters which depend on the actual state of motion will serve as the starting point of the analysis. This idea will be made more precise in the following sections.

3. NYQUIST'S DIAGRAM†

In the linear case, according to Section 1, the feedback loop is *stable*; i.e., all the transients tend to zero as $t \to +\infty$, if and only if the equation

$$1 - \Gamma(j\omega)\Phi(j\omega) = 0 \tag{5}$$

has *all* roots with positive imaginary parts. This condition can be analyzed by means of a graphical solution due to Nyquist.

Nyquist's theory is based on the representation of the steady-state frequency-response locus of the open loop in the complex plane:—the locus (N) of the images of complex numbers

$$\mu^*(\omega) \triangleq \Gamma(j\omega)\Phi(j\omega) = R(\omega) + jI(\omega)$$

when ω is given *real* values from $-\infty$ to $+\infty$.

The coordinates of the image, for each given *real* value of ω, are $R(\omega)$ and $I(\omega)$, the real and the imaginary parts respectively, of $\mu^*(\omega)$. The locus (N), which is called the *Nyquist diagram* for the feedback loop, is the *transform* of the real axis of the complex plane (ω):

$$(\omega): \qquad \omega_1 + j\omega_2 \qquad \omega_1, \omega_2 \text{ real}$$

by the mapping defined by the *analytical function*

$$\mu^*(\omega) \triangleq \Gamma(j\omega)\Phi(j\omega)$$

† See [1].

That is, $\mu^*(\omega)$ establishes a correspondance between:

(a) The points $p(\omega_1, \omega_2)$ of plane (ω) and the points $P[R(\omega_1, \omega_2), I(\omega_1, \omega_2)]$ of plane (Ω):

$$(\Omega): \qquad R(\omega_1, \omega_2) + jI(\omega_1, \omega_2) \qquad R, I \text{ real}$$

(b) The points of the real axis $(\Delta_r) \in (\omega)$:

$$(\Delta_r): \qquad \omega_2 = 0 \qquad -\infty < \omega_1 < +\infty$$

and the points of the Nyquist diagram $(N) \in (\Omega)$:

$$(N): \qquad R = R(\omega_1, 0) \qquad I = I(\omega_1, 0)$$

Now if we denote by $\omega^{(k)} = \omega_1^{(k)} + j\omega_2^{(k)}$ the roots of (5) ($k = 1, 2, \ldots$) and by $p^{(k)} \in (\omega)$ their representative points in plane (ω), the above mapping transforms these points into the *same point* $P^{(k)} = P_1$ of plane (Ω), since from (5)

$$1 - \mu^*[\omega^{(k)}] = 0$$

from which follows

$$R[\omega_1^{(k)}, \omega_2^{(k)}] + jI[\omega_1^{(k)}, \omega_2^{(k)}] = 1$$

The coordinates of P_1 are

$$R[\omega_1^{(k)}, \omega_2^{(k)}] = 1 \qquad I[\omega_1^{(k)}, \omega_2^{(k)}] = 0$$

Furthermore, the analyticity of $\mu^*(\omega)$ implies that the situation of points $p^{(k)}$ with respect to (Δ_r) is the same as the situation of point P_1 with respect to (N), since

$$\{p^{(k)}\} \leftrightarrow P_1 \qquad (\Delta_r) \leftrightarrow (N)$$

From these remarks a stability criterion is readily obtained, owing to the fact that all the roots of (5) should have positive imaginary parts—all the points $p^{(k)}$ should be on the left side of (Δ_r) as (Δ_r) is traversed in the direction of increasing frequency.

Then:

THEOREM 1. The feedback loop is stable if and only if the point $P_1(1, 0)$ is on the left side of the frequency response locus (N) as the locus is traversed in the direction of increasing frequency.

The Nyquist diagram can be obtained experimentally by applying a sinusoidal signal of varying frequency to the input of the open loop; then by measuring gain and phase angle, one gets the complex number

$$\mu^*(\omega) = \mu(\omega)e^{j\psi(\omega)} \tag{6}$$

as a function of the *real frequency* ω.

The new notations are related to the earlier ones by

$$\mu = (R^2 + I^2)^{1/2} \qquad \tan\psi = \frac{I}{R}$$

$\mu(\omega)$ is the amplification factor, or gain, of the open loop and $\psi(\omega)$ is the phase angle.

It may be proved that the Nyquist diagram is a closed curve which can be constructed, as pointed out above, by letting ω vary from $-\infty$ to $+\infty$, and which passes through the origin of the complex plane (Ω). As a matter of fact, the Nyquist diagram can be constructed by letting ω vary from 0 to $+\infty$, since it passes through the origin of (Ω) for $\omega = 0$ and $\omega \to +\infty$; e.g.,

$$\mu^*(0) = \mu^*(\infty) = 0$$

This closed curve is the boundary of a domain $(D) \in (\Omega)$.

Furthermore, if one takes account of the direction in which the closed curve is traversed as ω is increased, which corresponds to a clockwise rotation, the above criterion can be replaced by the following one:

THEOREM 2. The feedback loop is stable if and only if

$$P_1(1, 0) \notin (D)$$

Theorem 2 is the Nyquist criterion of stability; Theorem 1 is due to Dzung [8].

4. MIKAÏLOV'S HODOGRAPH[†]

Let us refer to a simple example; consider the vacuum-tube oscillator shown in Fig. 5, whose equation is written, in the linear approximation,

$$LC\ddot{x} + rC\dot{x} + x = M\frac{di_a}{dt} \tag{7}$$

with

$$i_a = S_0 x \tag{8}$$

† See [2].

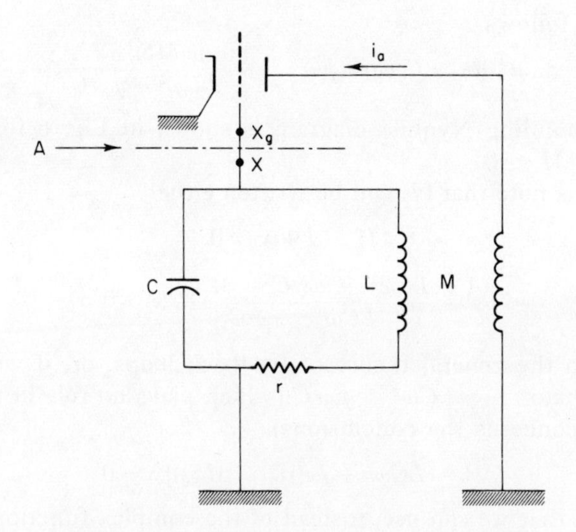

FIG. 5. Vacuum-tube oscillator; the loop is opened at point A.

If one opens the loop at point A and determines separately $\Gamma(j\omega)$ and $\Phi(j\omega)$, x_g and x being, respectively, the input and the output signal with respect to the open loop, one easily finds

$$\Gamma(j\omega) = j\omega M S_0$$

$$\Phi(j\omega) = \frac{1}{1 - LC\omega^2 + j\omega rC}$$

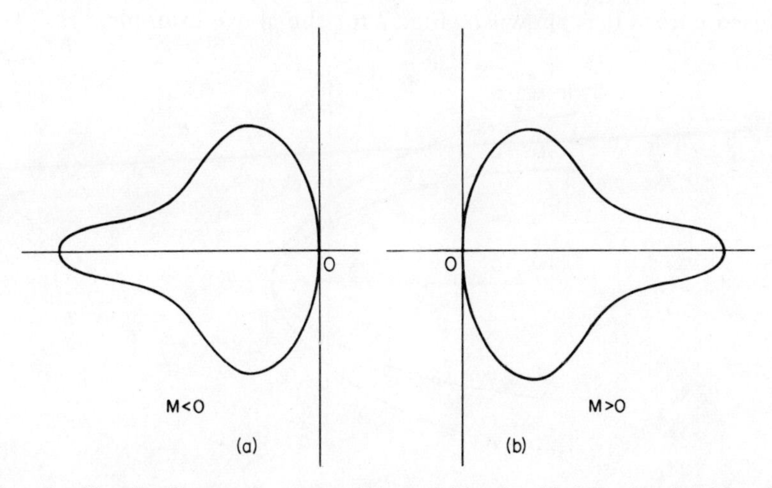

FIG. 6. Nyquist diagram for the oscillator of Fig. 5. (a) $M < 0$; (b) $M > 0$.

from which follows

$$\mu^*(\omega) = \Gamma(j\omega)\Phi(j\omega) = \frac{j\omega MS_0}{1 - LC\omega^2 + j\omega rC}$$

The corresponding Nyquist diagram is shown in Fig. 6 for the cases $M > 0$ and $M < 0$.

Now let us note that (7) can be written either

$$(1 - \Gamma\Phi)x = 0$$

say

$$\frac{1 - LC\omega^2 + j\omega(rC - MS_0)}{1 - LC\omega^2 + j\omega rC}\, x = 0$$

according to the general theory of feedback loops, or, if we disregard the denominator $1 - LC\omega^2 + j\omega rC$ (which plays no role in the present analysis as concerns the conclusions),

$$[1 - LC\omega^2 + j\omega(rC - MS_0)]x = 0$$

This means that we can use, instead of the complex function

$$\mu^*(\omega) \triangleq \Gamma(j\omega)\Phi(j\omega)$$

the complex function

$$H(j\omega) \triangleq [1 - LC\omega^2 + j\omega(rC - MS_0)]$$

Then the arguments of Section 3 hold, if $\mu^*(\omega)$ is replaced by $H(j\omega)$, except for the fact that the point $P_1(1, 0)$ must be replaced by the origin 0 of the complex plane (Ω).

The diagram which is associated with the complex function $H(j\omega)$ is Mikaïlov's hodograph. In contrast with Nyquist's diagram, it is not a closed curve. It is shown in Fig. 7 for the above example.

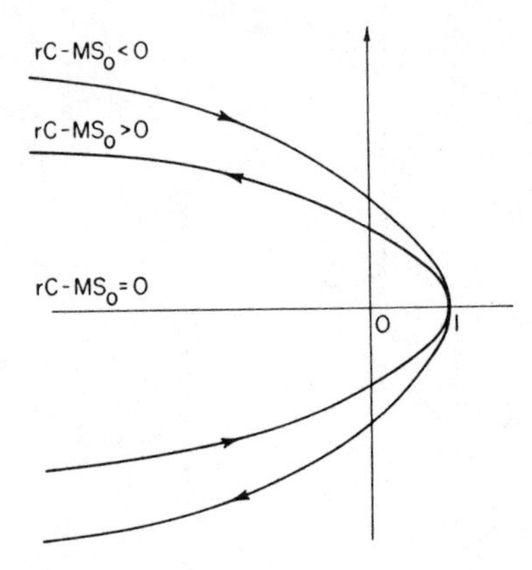

FIG. 7. Mikaïlov hodograph for the oscillator of Fig. 5.

It is possible to associate a Mikaïlov hodograph to every linear differential equation with constant coefficients, and we shall extend the procedure to quasi-linear differential equations. Examples will be given in the following.

We have the following criterion of stability:

THEOREM 3. The feedback loop is stable if and only if the origin 0 of the complex plane (Ω) is on the left side of Mikaïlov's hodograph, when this hodograph is traversed in the direction of increasing frequency.

This theorem can be proved by the same arguments as those used above.

5. GENERALIZATION OF MIKAÏLOV'S HODOGRAPH FOR NONLINEAR SYSTEMS†

5.1. Equiamplitude and Equifrequency Curves

In the case of quasi-linear systems, we shall follow the method developed in Chapter 5 and apply to the differential equation of motion the optimal linearization procedure, which will lead us to replace the actual equation by a differential equation whose coefficients depend on a number of slowly varying parameters, $\lambda_1, \lambda_2, ..., \lambda_n$.

Then the describing function associated with this equation has the form

$$H(j\omega, \lambda_1, ..., \lambda_n)$$

so that, *for each given set* $\lambda_1, ..., \lambda_n$, we can represent a Mikaïlov hodograph, by giving ω real values from $-\infty$ to $+\infty$.

In contrast with the linear case, we see that to each nonlinear equation we are led to associate a *family* of Mikaïlov's hodographs. Or, from another viewpoint, we can say that since $\lambda_1, ..., \lambda_n$ are parameters which slowly vary with time following the history of the system, the Mikaïlov hodograph slowly gets out of shape with time.

It may be convenient to consider the case where the hodograph depends upon one parameter only. For instance, if the system is a quasi-sinusoidal oscillator, this parameter may be the amplitude of oscillation, which we shall denote a_t [or $a(t)$].

Figure 8 shows a family of such hodographs, for a given oscillator, each of which corresponds to a value of amplitude a_t.

† See [10, 19, 25].

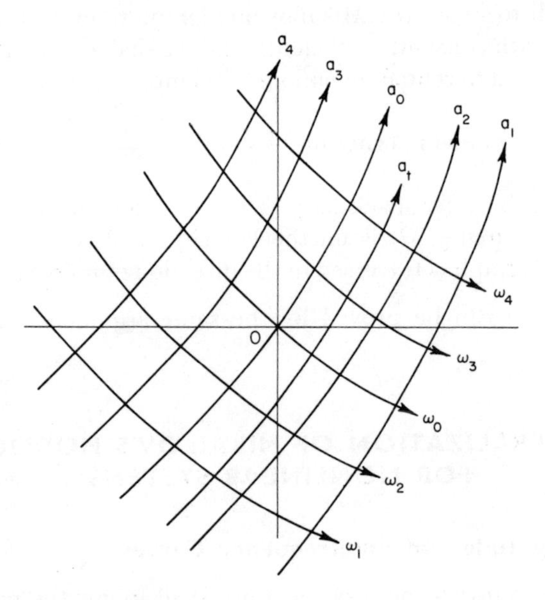

FIG. 8. Generalization of Mikaïlov's hodograph for nonlinear systems, equiamplitude and equifrequency curves.

A hodograph which is defined by the value a_t of the parameter is called an *equiamplitude curve* (a_t), since all its points correspond to the same value of the amplitude.

One of these hodographs passes through the origin 0. Assume that it corresponds to amplitude a_0. It separates the family into two sets:

(a) The hodographs associated with values of the parameter $a_t > a_0$.

(b) The hodographs for which $a_t < a_0$.

The arrows, in Fig. 8 indicate increasing angular frequency along the curves of the family.

Furthermore, to each given ω, *real*, there corresponds a point on each hodograph. The locus of such points is a curve which is called the *equifrequency curve* (ω): To each real value of ω there corresponds an equifrequency curve (ω).

One of these equifrequency curves passes through the origin 0. Assume that it corresponds to angular frequency ω_0. It separates the family of equifrequency curves into two sets:

(a) The equifrequency curves associated with the parameters $\omega > \omega_0$.

(b) The equifrequency curves for which $\omega < \omega_0$.

Finally, the complex plane (Ω) is ruled in squares by the two families: the family of equifrequency curves and the family of equiamplitude curves. They will enable us to discuss the various aspects of the motion of the nonlinear oscillator under study.

As a matter of fact, parameters a_t and ω(real) play an almost symmetrical role in this representation. Indeed each equiamplitude curve is generated from the function $H(j\omega, a_t)$ by giving a_t a constant value and letting ω, *real*, vary from $-\infty$ to $+\infty$. On the other hand, each equifrequency curve is obtained from the same function $H(j\omega, a_t)$ by giving ω a constant real value and letting a_t vary from 0 to $+\infty$. The symmetry is complete if one considers only positive frequencies, in which case ω, like a_t, will vary from 0 to $+\infty$.

As was pointed out above, the arrows in Fig. 8 along the equiamplitude curves indicate increasing angular frequency; the arrows along the equifrequency curves indicate increasing amplitude. For instance, on Fig. 8:

$$a_4 < a_3 < a_0 < a_2 < a_1 < \cdots$$

$$\omega_1 < \omega_2 < \omega_0 < \omega_3 < \omega_4 < \cdots$$

Next the left and right sides of curves (a_t) and (ω) will be defined with respect to these arrows.

5.2. Discussion

It may easily be verified that, for each given a_t, the actual frequency ω of the oscillation verifies the equation

$$H(j\omega, a_t) = 0 \tag{9}$$

That is to say, the nonlinear system is replaced for each a_t by the associated tangent linear system.

In general ω is complex. It is a function of the real parameter a_t:

$$\omega(a_t) = \omega_1(a_t) + j\omega_2(a_t) \qquad \omega_1, \omega_2 \text{ real}$$

which means that the tangent oscillation has the form

$$x = X_0 \exp[-\omega_2(a_t)t] \exp[j\omega_1(a_t)t]$$

Hence, in general, a_t is not the amplitude of the steady motion.

Once the oscillator has reached steady motion, we have

$$a_t = a_0$$
$$\omega_2(a_0) = 0$$
$$\omega_1(a_0) = \omega_0$$
$$\left.\begin{aligned} x &= X_0 \exp(j\omega_0 t) \\ H(j\omega_0, a_0) &= 0 \end{aligned}\right\} \quad \text{where } \omega_0 \text{ is real}$$

Also note that Theorem 3 remains valid for each hodograph of the family which does not pass through 0, according to which:

Case 1. If a_t is such that 0 is on the left side of equiamplitude curve (a_t) (with respect to increasing values of real parameter ω), then the oscillation will tend to fade, i.e., a_t will tend to decrease, which is the expression of "local" stability.† In Fig. 8 this condition is verified, for instance, for equiamplitude curves (a_1) and (a_2). Thus since $a_0 < a_2 < a_1$, the amplitude will tend to decrease to a_0 in both cases.

Case 2. If a_t is such that 0 is on the right side of equiamplitude curve (a_t), then the oscillation will tend to grow; i.e., a_t will tend to increase. This is now the expression of "local" instability. In Fig. 8 this condition is verified for equiamplitude curves (a_3) and (a_4). Since $a_4 < a_3 < a_0$, the amplitude will tend to increase to a_0 in both cases.

By comparing Cases 1 and 2 we see that a_0 is a *stable steady-state amplitude*, which means that if any perturbation tends to modify the amplitude of the oscillation from a_0, the deviation will tend to fade whatever its sign. In other words, equiamplitude (a_0) corresponds to a *stable limit cycle*.

This example shows clearly how a "local" stability and a "local" instability can conjugate their actions so as to result in a *steady-motion stability*.

Finally, note that equiamplitude curve $a_t = 0$ bounds the family of hodographs, since $a_t \geqslant 0$.

It is also interesting to plot on the same figure the equiamplitude curves which bound the domain corresponding to oscillatory motions. Outside this domain the equation

$$H(j\omega, a_t) = 0$$

† "Local" means in the near neighborhood of a_t.

has purely imaginary roots in ω. Accordingly, its boundary is determined by the conditions according to which the equation

$$H(s, a_t) = 0 \tag{10}$$

has real roots in s. One can deduce the limiting values of a_t from the curve $a_t(s)$, obtained from (10).

5.3. Stability Criterion for Steady Motion

Obviously the stability of the steady motion which we considered in the preceding section is due to the fact that the parameter a_t increases on the right side of the equiamplitude curve (a_0) and decrease on the left side. Should we reverse this assumption, the steady motion would become instable, as can be readily verified. Indeed the new assumption would imply $a_1 < a_2 < a_0 < a_3 < a_4$. Accordingly:

(a) In case 1, as in the earlier discussion, the amplitude will tend to decrease from the initial value (a_1), or (a_2); but now this will result in increasing deviation from a_0.

(b) In case 2 the amplitude will again tend to increase from (a_3), or from (a_4), and this will result in divergence from the steady state, as in case 1.

The net result is that if any perturbation modifies the amplitude of the oscillation from a_0, the deviation will tend to increase whatever its sign. That is the expression of instability for steady motion. These conclusions will be summarized in Theorem 4.

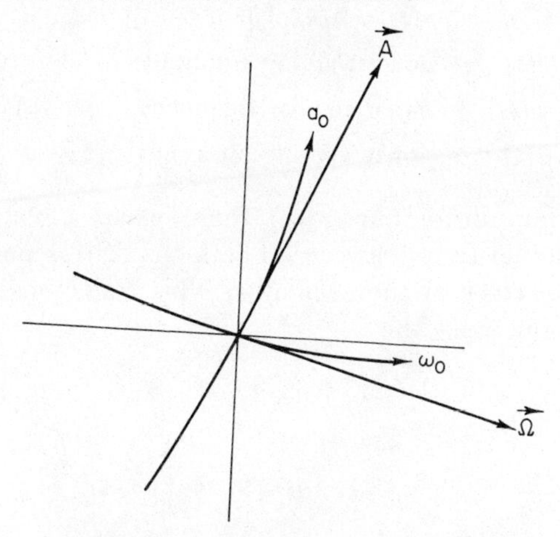

FIG. 9. Stability criterion for steady motion.

Let

$$H(j\omega, a_t) = U(\omega, a_t) + jV(\omega, a_t) \qquad \omega \text{ real} \tag{11}$$

and consider at point 0 the two vectors (Fig. 9)

$$\mathbf{A} \quad \text{with components} \quad \left(\frac{\partial U}{\partial \omega}\right)_0, \left(\frac{\partial V}{\partial \omega}\right)_0$$

$$\mathbf{\Omega} \quad \text{with components} \quad \left(\frac{\partial U}{\partial a}\right)_0, \left(\frac{\partial V}{\partial a}\right)_0$$

(partial derivatives are to be computed at point 0).

According to the definition of the equiamplitude and equifrequency curves, we see that **A** and **Ω** are tangent to the equiamplitude and equifrequency curves through 0, respectively. Furthermore, they are pointing toward increasing values of ω (real) and a_t, respectively.

Then from the above discussion the following criterion is readily deduced:

THEOREM 4. The steady motion defined by equiamplitude (a_0) and equifrequency (ω_0) through 0 is stable if an only if

$$\mathbf{\Omega} \times \mathbf{A} > 0.$$

5.4. Trajectory of a Quasi-Sinusoidal Oscillator

Assume that the system under consideration is a quasi-sinusoidal oscillator, whose law of motion is

$$x(t) = a_t \exp[j(\omega_t t - \varphi_t)]$$

$$\begin{aligned} a_t &= a(t) && \text{slowly varying amplitude} \\ \omega_t &= \omega(a_t) && \text{actual angular frequency,} \qquad \omega_t \text{ is real} \\ \varphi_t &= \varphi(t) && \text{slowly varying phase angle} \end{aligned}$$

On each equiamplitude curve (a_t) there exists a point which corresponds to the actual frequency ω_t. The locus of this point, as a_t varies, is called the *trajectory* of the oscillator. This trajectory passes through 0 at steady motion, since then

$$a_t = a_0$$

$$\omega_t = \omega_0$$

$$x(t) = a_0 \exp[j(\omega_0 t - \varphi_0)]$$

and

$$H(j\omega_0, a_0) = 0 \qquad \omega_0 \text{ real}$$

Furthermore, we shall prove the following property:

Property 1. At point 0, the trajectory of the oscillator is normal to the equiamplitude curve (a_0) through that point.

Consider an oscillatory motion in the neighborhood of the steady state, such that

$$\begin{gathered} a_t = a_0 + \delta a \\ \omega = \omega_0 + \delta\omega_1 + j\delta\omega_2 \qquad \delta\omega_1, \delta\omega_2 \text{ real} \\ H(j\omega, a_t) = 0 \end{gathered} \tag{12}$$

If δa, $\delta\omega_1$, and $\delta\omega_2$ are of first-order smallness, say if $\delta\omega_1 = k_1\delta a + o_1(\delta a)$, $\delta\omega_2 = k_2\delta a + o_2(\delta a)$ where k_1 and k_2 are nonzero numbers and

$$\frac{o_1(\delta a)}{\delta a} \to 0 \qquad \frac{o_2(\delta a)}{\delta a} \to 0 \qquad \text{uniformly as} \quad \delta a \to 0$$

we get from (12)

$$\begin{aligned} H[j(\omega_0 + \delta\omega_1 + j\delta\omega_2), a_0 + \delta a] &= U(a_0, \omega_0) + jV(a_0, \omega_0) \\ &+ \delta a \left(\frac{\partial U}{\partial a} + j\frac{\partial V}{\partial a}\right)_0 + (\delta\omega_1 + j\delta\omega_2)\left(\frac{\partial U}{\partial \omega} + j\frac{\partial V}{\partial \omega}\right)_0 + o(\delta a) = 0 \end{aligned}$$

where partial derivatives are to be computed at point 0 and $[o(\delta a)/\delta a] \to 0$ uniformly as $\delta a \to 0$.

Since $U(a_0, \omega_0) = V(a_0, \omega_0) = 0$ we have

$$\delta a \left(\frac{\partial U}{\partial a} + j\frac{\partial V}{\partial a}\right)_0 + (\delta\omega_1 + j\delta\omega_2)\left(\frac{\partial U}{\partial \omega} + j\frac{\partial V}{\partial \omega}\right)_0 + o(\delta a) = 0$$

from which follows, when dividing by δa and taking $\delta a \to 0$

$$\begin{aligned} \left(\frac{\partial U}{\partial a}\right)_0 + \left(\frac{\partial U}{\partial \omega}\right)_0 k_1 &= \left(\frac{\partial V}{\partial \omega}\right)_0 k_2 \\ \left(\frac{\partial V}{\partial a}\right)_0 + \left(\frac{\partial V}{\partial \omega}\right)_0 k_1 &= -\left(\frac{\partial U}{\partial \omega}\right)_0 k_2 \end{aligned} \tag{13}$$

Now, in the neighborhood of 0, the trajectory is determined by letting

$$\begin{aligned} a_t &= a_0 + \delta a \\ \omega_t &= \omega_0 + \delta\omega_1 \qquad \delta\omega_1 \text{ real} \end{aligned}$$

and the coordinates of its moving point, are

$$U(\omega_0 + \delta\omega_1\,, a_0 + \delta a) = \left(\frac{\partial U}{\partial a}\right)_0 \delta a + \left(\frac{\partial U}{\partial \omega}\right)_0 \delta\omega_1 + o(\delta a)$$

$$V(\omega_0 + \delta\omega_1\,, a_0 + \delta a) = \left(\frac{\partial V}{\partial a}\right)_0 \delta a + \left(\frac{\partial V}{\partial \omega}\right)_0 \delta\omega_1 + o(\delta a)$$

Then

$$\lim_{\delta a \to 0} \frac{1}{\delta a} U(\omega_0 + \delta\omega_1\,, a_0 + \delta a) = \left(\frac{\partial U}{\partial a}\right)_0 + k_1 \left(\frac{\partial U}{\partial \omega}\right)_0$$

$$\lim_{\delta a \to 0} \frac{1}{\delta a} V(\omega_0 + \delta\omega_1\,, a_0 + \delta a) = \left(\frac{\partial V}{\partial a}\right)_0 + k_1 \left(\frac{\partial V}{\partial \omega}\right)_0 \tag{14}$$

Finally, since $(\partial U/\partial \omega)_0$ and $(\partial V/\partial \omega)_0$, in the right side of (13), are the components of vector $\mathbf{A}$, which is tangent to (a_0) at the origin, (13) with (14) establish Property 1.

5.5. Frequency-Amplitude Relation

From equations (13) we deduce†

$$k_1 = -\left\{\left(\frac{\partial U}{\partial \omega}\frac{\partial U}{\partial a} + \frac{\partial V}{\partial \omega}\frac{\partial V}{\partial a}\right)\Big/\left[\left(\frac{\partial U}{\partial \omega}\right)^2 + \left(\frac{\partial V}{\partial \omega}\right)^2\right]\right\} \tag{15}$$

$$k_2 = \left\{\left(\frac{\partial U}{\partial a}\frac{\partial V}{\partial \omega} - \frac{\partial U}{\partial \omega}\frac{\partial V}{\partial a}\right)\Big/\left[\left(\frac{\partial U}{\partial \omega}\right)^2 + \left(\frac{\partial V}{\partial \omega}\right)^2\right]\right\} \tag{16}$$

where all the partial derivatives are to be computed at point 0. Equation (15) expresses the change in the real angular frequency of the oscillator caused by a small amplitude deviation δa, from the steady-state amplitude a_0, at the first-order approximation. Note that the numerator of (15) is the dot product $\mathbf{\Omega} \cdot \mathbf{A}$. Accordingly, we have:

Property 2. A small amplitude deviation δa from the steady-state amplitude a_0 will not affect the real frequency of the oscillation, at the first-order approximation, provided that the angle between the equifrequency curve (ω_0) and the equiamplitude curve (a_0) at point 0 be $\pi/2$.

This property is also an obvious consequence of the fact that the trajectory of the oscillator is normal to the equiamplitude curve (a_0) at point 0. If the equifrequency curve (ω_0) is also normal to (a_0) at point 0,

† Next we shall drop the index zero.

then a small piece of trajectory is in coincidence with a small piece of equifrequency curve in the neighborhood of 0, at the first-order approximation. Since by definition the real frequency is constant all along an equifrequency curve, the frequency of the oscillation will not be modified, at the first-order approximation, when the amplitude will fluctuate in the neighborhood of the steady state. Oscillators which have this property are especially interesting for practical applications.

On the other hand, $\delta\omega_2$ is the amplitude damping coefficient in the neighborhood of steady motion: The steady-motion amplitude being a_0, in the neighborhood of $a_0 + \delta a$, the law of variation with time of the amplitude is

$$(a_0 + \delta a)\exp(-\delta\omega_2 t) \tag{17}$$

Accordingly, for the steady oscillation to be stable, it is necessary and sufficient that, for sufficiently small δa,

$$\delta a > 0 \Rightarrow \delta\omega_2 > 0$$

$$\delta a < 0 \Rightarrow \delta\omega_2 < 0$$

in which case any change in amplitude, in the neighborhood of steady oscillation, will tend to fade, whatever its sign.

In other words, the condition for steady-motion stability is

$$\frac{\partial U}{\partial a}\frac{\partial V}{\partial \omega} - \frac{\partial U}{\partial \omega}\frac{\partial V}{\partial a} > 0$$

Since the left side of this inequality is the cross product $\mathbf{\Omega} \times \mathbf{A}$, the condition is $\mathbf{\Omega} \times \mathbf{A} > 0$. We again find Theorem 4.

5.6. Time Constants of a Quasi-Sinusoidal Oscillator

In the near neighborhood of steady motion, the law of variation with time of the amplitude of oscillation is given by (17). Next we shall put

$$\tau_1 = \frac{1}{\delta\omega_2} \tag{18}$$

which is the definition of the "*local*" *time constant* of the oscillation. By "local" we mean that it is the time constant of the linear approximation which is tangent to the actual oscillation at the amplitude $a_0 + \delta a$.

Another time constant, τ_2, will play an important role in the following —the one which characterizes the law according to which the oscillator returns back to steady motion after the oscillation has been displaced from it by a small perturbation. Indeed, here we assume that the steady motion is stable (Fig. 10).

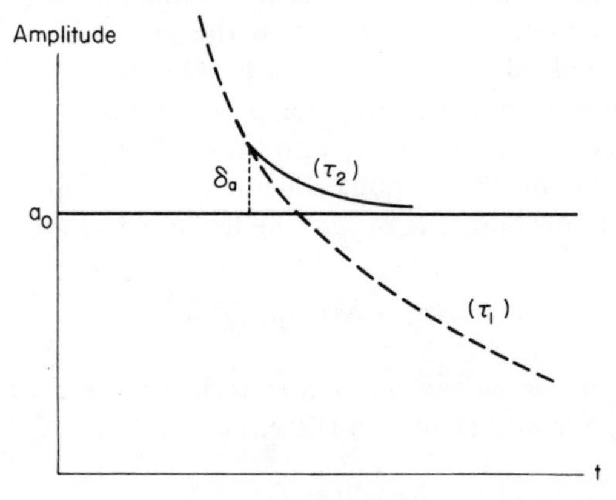

FIG. 10. Time constants of a quasi-sinusoidal oscillator.

To get a convenient expression for τ_2, note that the "local" time constant occurs in the formula

$$\frac{1}{a_t}\frac{da_t}{dt} = -\delta\omega_2 \tag{19}$$

where $a_t = a(t)$ is the amplitude of oscillation at time t. Then from (16) we have

$$\frac{1}{a_t}\frac{da_t}{dt} = -k_2\,\delta a + o(\delta a) \tag{20}$$

with

$$k_2 = \left(\frac{\partial U}{\partial a}\frac{\partial V}{\partial \omega} - \frac{\partial U}{\partial \omega}\frac{\partial V}{\partial a}\right)\Big/\left[\left(\frac{\partial U}{\partial \omega}\right)^2 + \left(\frac{\partial V}{\partial \omega}\right)^2\right]$$

Since we are interested in the law of variation of δa, let us put

$$\delta a = \epsilon \qquad \text{with} \quad \epsilon = a_t - a_0$$

and rewrite (20)

$$\frac{d\epsilon}{dt} + k_2 a_0 \epsilon + o'(\epsilon) = 0$$

where $o'(\epsilon)/\epsilon \to 0$ uniformly as $\epsilon \to 0$. At last we get

$$\lim_{\epsilon\to 0}\frac{1}{\epsilon}\frac{d\epsilon}{dt} = -k_2 a_0$$

Accordingly,

$$\tau_2 \triangleq \frac{1}{k_2 a_0} = \frac{1}{a_0}\left(\frac{\partial U}{\partial a}\frac{\partial V}{\partial \omega} - \frac{\partial U}{\partial \omega}\frac{\partial V}{\partial a}\right)\Big/\left[\left(\frac{\partial U}{\partial \omega}\right)^2 + \left(\frac{\partial V}{\partial \omega}\right)^2\right] \tag{21}$$

6. APPLICATIONS TO AUTONOMOUS SYSTEMS†

6.1. Second-Order Systems

To illustrate theory just discussed, we shall consider a few applications to autonomous and nonautonomous systems. Let us begin with autonomous systems governed by second-order differential equations. For instance, look at a Van der Pol oscillator governed by

$$\ddot{x} + \mu(x^2 - 1)\dot{x} + x = 0 \qquad \mu > 0 \tag{22}$$

As stated earlier, the tangent linear equation is

$$\ddot{x} + \mu\left(\frac{a^2}{4} - 1\right)\dot{x} + x = 0 \tag{23}$$

From this follows the describing function:

$$H(j\omega, a) \equiv 1 - \omega^2 + j\omega\mu\left(\frac{a^2}{4} - 1\right) \tag{24}$$

Three equiamplitude curves are shown in Fig. 11. They correspond

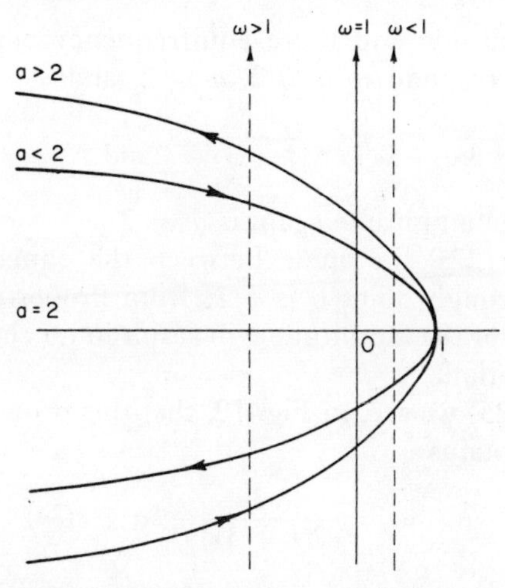

FIG. 11. Generalized Mikaïlov's hodograph for a second-order system; the Van der Pol oscillator.

† See [4, 10, 25].

to $a > 2$, $a = 2$, and $a < 2$. For $a = 2$ the equiamplitude reduces to a piece of the real axis. Accordingly, it passes through 0 and thus it determines the steady motion. The other equiamplitude curves are parabolas which are described following the arrows (Fig. 11) as the real parameter ω varies from $-\infty$ to $+\infty$. Theorem 4 can be readily applied to them.

The equifrequency curves are straight lines which are parallel to the imaginary axis. Three of them are shown in Fig. 11. They correspond to $\omega > 1$, $\omega = 1$, and $0 < \omega < 1$. The arrows indicate the increase of parameter a from 0 to $+\infty$.

If a nonlinear restoring force is taken into account in (22), one obtains, for instance,

$$\ddot{x} + \mu(x^2 - 1)\dot{x} + x + \alpha x^3 = 0 \qquad \text{with} \quad \mu > 0, \alpha > 0 \tag{25}$$

The tangent linear equation again is

$$\ddot{x} + \mu\left(\frac{a^2}{4} - 1\right)\dot{x} + \left(1 + \frac{3\alpha a^2}{4}\right)x = 0 \tag{26}$$

and the describing function is

$$H(j\omega, a) \equiv 1 + \frac{3\alpha a^2}{4} - \omega^2 + j\omega\mu\left(\frac{a^2}{4} - 1\right) \tag{27}$$

Three equiamplitude and three equifrequency curves are shown in Fig. 12. They correspond to $a > 2$, $a = 2$, and $a < 2$, and to $\omega > 0$:

$$\omega > \sqrt{1 + 3\alpha} \qquad \omega = \sqrt{1 + 3\alpha} \qquad \text{and} \qquad \omega < \sqrt{1 + 3\alpha}$$

The steady-motion amplitude again is $a = 2$.

In the case of (22), the angle between the equiamplitude and the equifrequency through point 0 is $\pi/2$; from Property 2 it follows that small fluctuations of the amplitude will result in no change in frequency, at first approximation.

In the case of (25) we see, in Fig. 12, that this property does not hold. From (15) one obtains

$$\delta\omega_1 = \frac{3\alpha}{2(1 + 3\alpha)^{1/2}}\,\delta a + o(\delta a)$$

In both cases we have

$$\delta\omega_2 = \frac{\mu}{2}\,\delta a + o(\delta a) \qquad \tau_2 = \frac{\mu}{4}$$

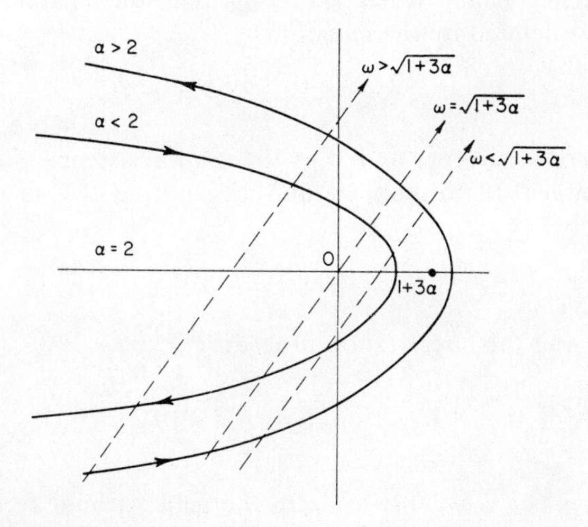

FIG. 12. Generalized Mikaïlov's hodograph; Van der Pol oscillator with a cubic restoring force.

6.2. Third-Order Systems

An example of a third-order system is the phase-shift oscillator shown in Fig. 13. Note that this kind of oscillator contains no selective tank circuit. A linear network, made of three equal capacitors C and three equal resistances R, is connected between the anode and the grid of the vacuum tube. Let V_1 be the input voltage and V_2 the output voltage; the feedback coefficient of this network is readily computed. It is

$$\Phi(j\omega) = \frac{-j\omega^3 R^3 C^3}{-j\omega^3 R^3 C^3 - 6R^2C^2\omega^2 + 5j\omega RC + 1}$$

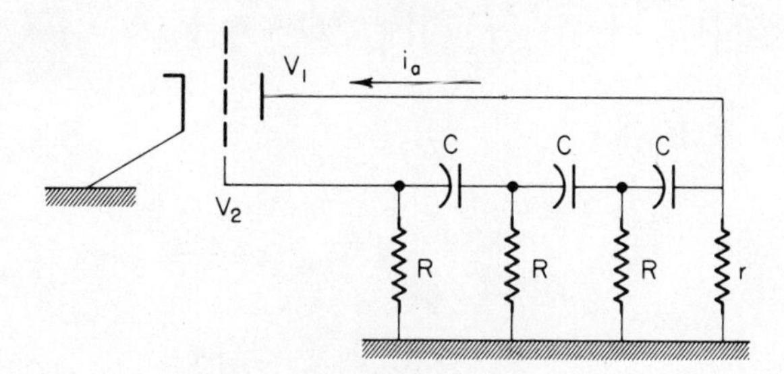

FIG. 13. Phase-shift oscillator.

On the other hand, when assuming that the characteristic curve of the tube is defined by the equation

$$i_a = S_0 V_2 + S_2 V_2^3$$

(i_a is the anode current) and that the load resistance r is small, the relation between the input V_2 and the output V_1 of the tube is approximately

$$V_1 = -r(S_0 V_2 + S_2 V_2^3)$$

say, by applying the linearization procedure,

$$V_1 = -r\left(S_0 + \frac{3S_2}{4} a^2\right) V_2$$

with $V_2 \simeq a \cos \omega t$. This leads to the gain without feedback of the nonlinear tube

$$\Gamma = -r\left(S_0 + \frac{3S_2}{4} a^2\right)$$

Note that it does not depend on ω. Finally, we get

$$1 - \Gamma\Phi = \frac{-j\omega^3 R^3 C^3\{1 + r[S_0 + (3S_2/4)a^2]\} - 6R^2C^2\omega^2 + 5j\omega RC + 1}{-j\omega^3 R^3 C^3 - 6R^2C^2\omega^2 + 5j\omega RC + 1}$$

and the describing function,

$$-j\omega^3 R^3 C^3 \left[1 + r\left(S_0 + \frac{3S_2}{4} a^2\right)\right] - 6R^2C^2\omega^2 + 5j\omega RC + 1$$

The corresponding differential equation of motion of the oscillator is

$$R^3C^3 \left[1 + r\left(S_0 + \frac{3S_2}{4} a^2\right)\right] \dddot{x} + 6R^2C^2\ddot{x} + 5RC\dot{x} + x = 0$$

by putting $x = V_2$, or

$$[2\alpha + \beta S(a)]\dddot{x} + \ddot{x} + 2\delta\dot{x} + \omega_0^2 x = 0 \qquad (28)$$

with

$$\omega_0^2 = \frac{1}{6R^2C^2} \qquad 2\alpha = \frac{RC}{6} \qquad \beta = \frac{RCr}{6} \qquad 2\delta = \frac{5}{6RC} \qquad S(a) = S_0 + \frac{3S_2}{4} a^2$$

When r is no longer negligible with respect to R, the form of (28) is the same, but the coefficients are to be replaced by

$$\omega_0^2 = \frac{1}{(RC)^2}\frac{R}{6R+4r} \qquad 2\alpha = RC\frac{R+3r}{6R+4r}$$

$$\beta = RC\frac{Rr}{6R+4r} \qquad 2\delta = \frac{1}{RC}\frac{5R+r}{6R+4r}$$

The describing function will be rewritten

$$H(j\omega, a) = -[2\alpha + \beta S(a)]j\omega^3 - \omega^2 + 2j\omega\delta + \omega_0^2$$

and the parametric equations of the Mikaïlov hodographs are

$$U = \omega_0^2 - \omega^2$$
$$V = \{2\delta - [2\alpha + \beta S(a)]\omega^2\}\omega$$

Their shapes are shown in Fig. 14 in the three cases $a < a_0$, $a = a_0$, and $a > a_0$, with

$$a_0 = 2\left[\frac{2\delta - (2\alpha + \beta S_0)\omega_0^2}{3S_2\beta\omega_0^2}\right]^{1/2} = 2\left[\frac{29 - rS_0 + 23(r/R) + 4(r/R)^2}{3S_2 r}\right]^{1/2}$$

a_0 is the steady-oscillation amplitude and

$$\omega = \omega_0 = \frac{1}{RC}\left(\frac{R}{6R+4r}\right)^{1/2}$$

is the corresponding angular frequency.

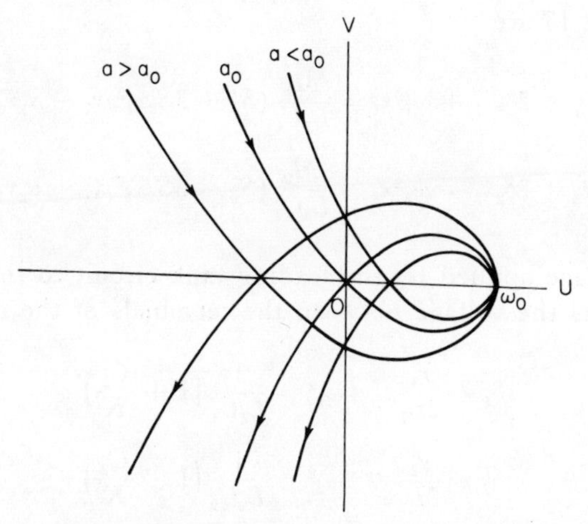

FIG. 14. Generalized Mikaïlov's hodograph for a third-order system, the phase-shift oscillator.

Note that the oscillation can start and tend to the steady state only if $rS_0 > 29$, in which case the origin 0 has the position shown in Fig. 15 with respect to the limiting hodograph $a = 0$. In this case the hodograph (a_0) through the origin exists. If $rS_0 < 29$, we get the opposite conclusion, as shown in Fig. 16.

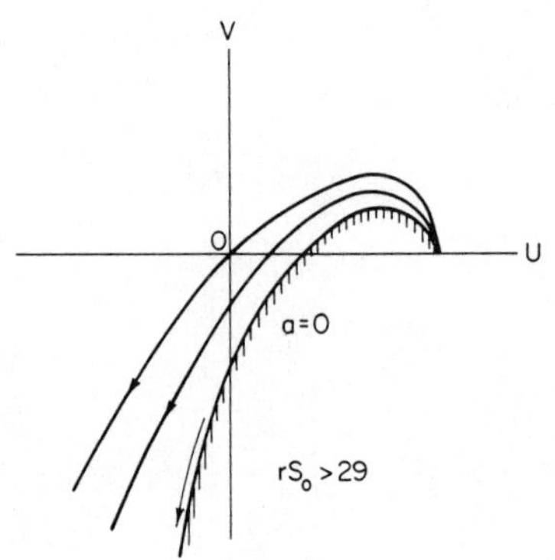

FIG. 15. Self-sustained oscillation tending to steady state, $rS_0 > 29$.

FIG. 16. Oscillation not self-sustained, $rS_0 < 29$.

6.3. Fourth-Order Systems

Let us return to the problem we considered in Chapter IV, Section 4—the mutual synchronization of two coupled self-sustained oscillators. With the same notation as in that section, the equations of the circuit shown in Fig. 17 are

$$\ddot{x}_1 + 2\delta_1\dot{x}_1 + \omega_1{}^2x_1 = \frac{M_1}{L_1C_1}(S_0 + 3S_2x_1{}^2)\dot{x}_1 + \alpha_1x_2 \tag{29}$$

$$\ddot{x}_2 + 2\delta_2\dot{x}_2 + \omega_2{}^2x_2 = \frac{M_2}{L_2C_2}(S_0 + 3S_2x_1{}^2)\dot{x}_1 + \alpha_2x_1 \tag{30}$$

x_1 is the voltage applied by the leading tank circuit to the grid of the tube and x_2 is the voltage between the terminals of the capacitor C_2.

$$\delta_1 = \frac{r_1}{2L_1} \qquad \omega_1{}^2 = \frac{1}{L_1C_1}\left(1 + \frac{C_1}{K}\right)$$

$$\delta_2 = \frac{r_2}{2L_2} \qquad \omega_2{}^2 = \frac{1}{L_2C_2}\left(1 + \frac{C_2}{K}\right)$$

The meaning of r_1, r_2, L_1, L_2, C_1, C_2, K, M_1, and M_2 is pointed out in Fig. 17. If we substitute in (30) the expression for x_2 given by (29), we get a fourth-order differential equation with respect to the variable x_1.

As a matter of fact, the way in which we have studied this problem in Chapter IV, Section 4, can be considered as an introduction to the describing function method; accordingly we shall not go through the whole derivation again, but shall stress the graphical discussion.

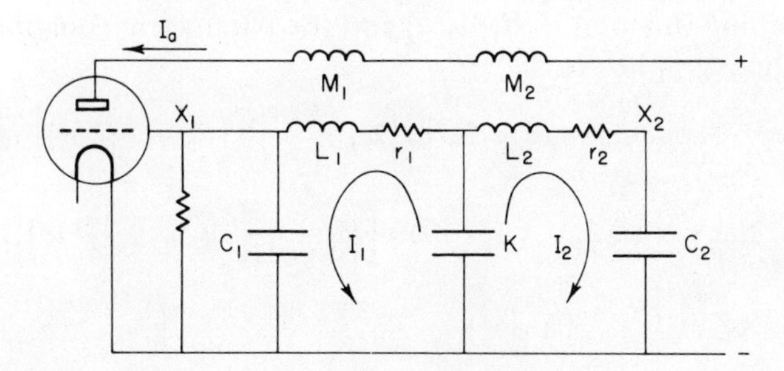

FIG. 17. Fourth-order system, self-sustained radioelectric oscillator.

First of all, let us state the main steps of the solution. By means of the linearization procedure, (29) and (30) are replaced by

$$\ddot{x}_1 + 2\delta_1\dot{x}_1 + \omega_1^2 x_1 = \frac{M_1}{L_1C_1}\left(S_0 + \frac{3S_2}{4}a_1^2\right)\dot{x}_1 + \alpha_1 x_2 \tag{31}$$

$$\ddot{x}_2 + 2\delta_2\dot{x}_2 + \omega_2^2 x_2 = \frac{M_2}{L_2C_2}\left(S_0 + \frac{3S_2}{4}a_1^2\right)\dot{x}_1 + \alpha_2 x_1 \tag{32}$$

where a_1 is the amplitude of the function x_1 which is supposed to be quasi-sinusoidal. Then going to complex exponential functions,

$$x_1 = a_1 e^{j\omega t} \qquad x_2 = a_2 e^{j(\omega t - \varphi)}$$

and introducing the coupling coefficient $\Lambda = \lambda e^{j\varphi}$, with $\lambda = a_1/a_2$, one gets

$$\left\{(\omega_1^2 - \omega^2) + j\omega\left[2\delta_1 - \frac{M_1}{L_1C_1}\left(S_0 + \frac{3S_2}{4}a_1^2\right)\right]\right\}\Lambda = \alpha_1$$

$$[(\omega_2^2 - \omega^2) + 2j\omega\delta_2] = \left[j\omega\frac{M_2}{L_2C_2}\left(S_0 + \frac{3S_2}{4}a_1^2\right) + \alpha_2\right]\Lambda$$

from which, by eliminating Λ,†

$$H(j\omega, a) \equiv (\omega_1^2 - \omega^2)(\omega_2^2 - \omega^2) - 2\omega^2\delta_2 \left[2\delta_1 - \frac{M_1}{L_1C_1}\left(S_0 + \frac{3S_2}{4}a^2\right)\right]$$
$$+ j\omega \left\{2(\omega_1^2 - \omega^2)\,\delta_2 + (\omega_2^2 - \omega^2)\left[2\delta_1 - \frac{M_1}{L_1C_1}\left(S_0 + \frac{3S_2}{4}a^2\right)\right]\right\}$$
$$- \alpha_1\alpha_2 - j\omega\alpha_1 \frac{M_2}{L_2C_2}\left(S_0 + \frac{3S_2}{4}a^2\right) = 0$$

The describing function is $H(j\omega, a)$ and the parametric equations of the Mikaïlov hodographs are

$$U = (\omega_1^2 - \omega^2)(\omega_2^2 - \omega^2) - 2\omega^2\delta_2 \left[2\delta_1 - \frac{M_1}{L_1C_1}\left(S_0 + \frac{3S_2}{4}a^2\right)\right] - \alpha_1\alpha_2 \tag{33}$$

$$V = \omega \left\{2(\omega_1^2 - \omega^2)\,\delta_2 + (\omega_2^2 - \omega^2)\left[2\delta_1 - \frac{M_1}{L_1C_1}\left(S_0 + \frac{3S_2}{4}a^2\right)\right] - \alpha_1 \frac{M_2}{L_2C_2}\left(S_0 + \frac{3S_2}{4}a^2\right)\right\} \tag{34}$$

One of the equiamplitude curves is readily obtained (Fig. 18), namely the one which corresponds to amplitude

$$a = 2\left(-\frac{S_0}{3S_2}\right)^{1/2} \qquad (S_0 > 0,\ S_2 < 0)$$

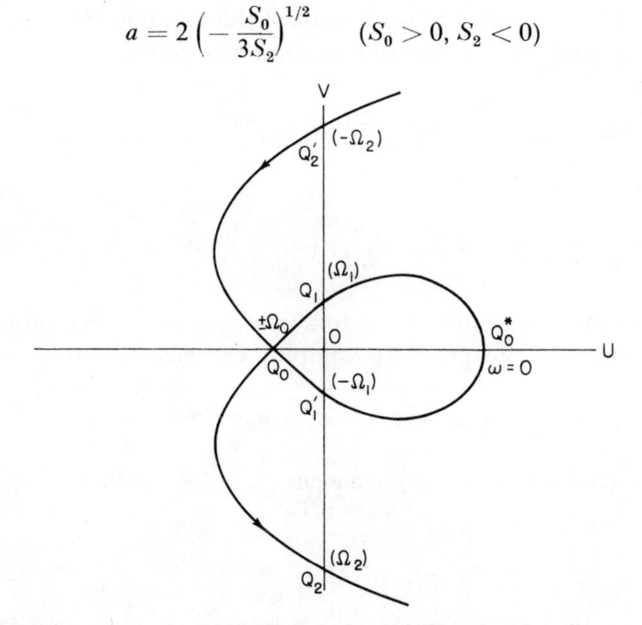

Fig. 18. Generalized Mikaïlov's hodograph for a fourth-order system.

† Next we shall drop the index 1; we shall write a instead of a_1.

Then (33) and (34) reduce to

$$U = (\omega_1^2 - \omega^2)(\omega_2^2 - \omega^2) - 4\omega^2\delta_1\delta_2 - \alpha_1\alpha_2 \tag{35}$$

$$V = \omega[2(\omega_1^2 - \omega^2)\,\delta_2 + 2(\omega_2^2 - \omega^2)\,\delta_1] \tag{36}$$

Its intersection points Q_1, Q_2, Q_1', and Q_2', with axis $U = 0$, correspond to the frequencies

$$\omega = \pm\Omega_1 \qquad \text{as concerns} \quad Q_1, Q_1'$$

$$\omega = \pm\Omega_2 \qquad \text{as concerns} \quad Q_2, Q_2'$$

(assume $\Omega_1 < \Omega_2$) and the intersection points Q_0^* and Q_0, with axis $V = 0$, correspond to the frequencies

$$\omega = 0 \qquad\qquad \text{point } Q_0^*$$

$$\omega = \pm\Omega_0 \quad \text{with} \quad \Omega_0 = \left(\frac{\delta_2\omega_1^2 + \delta_1\omega_2^2}{\delta_1 + \delta_2}\right)^{1/2} \qquad \text{point } Q_0$$

It may easily be seen that

$$\Omega_1 < \Omega_0 < \Omega_2$$

Accordingly, this equiamplitude has the shape shown in Fig. 18. This curve, as well as all the other curves of the family when a varies, passes through the point

$$Q_0^*: \qquad \omega = 0 \quad \begin{cases} V = 0 \\ U = \omega_1^2\omega_2^2 - \alpha_1\alpha_2 \end{cases}$$

which is independent of a.

Now let us illustrate this problem more completely by discussing from the graphical viewpoint the case $M_1 = 0$, $M_2 \neq 0$, in which the parametric equations of the Mikaïlov hodographs are

$$U = (\omega_1^2 - \omega^2)(\omega_2^2 - \omega^2) - 4\omega^2\delta_1\delta_2 - \alpha_1\alpha_2 \tag{37}$$

$$V = \omega\left[2(\omega_1^2 - \omega^2)\,\delta_2 + 2(\omega_2^2 - \omega^2)\,\delta_1 - \alpha_1\frac{M_2}{L_2C_2}\left(S_0 + \frac{3S_2}{4}a^2\right)\right] \tag{38}$$

We see that the intersection points $Q_1^{(a)}$ and $Q_2^{(a)}$, with axis $U = 0$, for $0 < \omega < +\infty$, correspond to the frequencies $+\Omega_1$ and $+\Omega_2$ defined above.

The deformation of the equiamplitude curve can be easily described, starting from the above case, $a = 2[-(S_0/3S_2)]^{1/2}$, when a is given increasing or decreasing values.

Note that the equiamplitude $a = 2[-(S_0/3S_2)]^{1/2}$ corresponds to an oscillation which is "locally stable," that is, its amplitude will tend spontaneously to decrease. Accordingly we have to investigate whether the oscillation will tend to steady oscillation and, if so, to determine the characteristics of this steady motion, say its amplitude and frequency. As a matter of fact, the result will depend on the sign of M_2. In Chapter IV, Section 4 we had $M_2 > 0$; now we shall relax this assumption.

If we assume $M_2 > 0$, the decreasing of amplitude a will result in the decreasing of

$$-\alpha_1 \frac{M_2}{L_2 C_2}\left(S_0 + \frac{3S_2}{4} a^2\right)$$

in (38) (since $S_2 < 0$, and the other coefficients are positive).

Accordingly, the intersection points $Q_1^{(a)}$ and $Q_2^{(a)}$, with axis $U = 0$, which correspond to positive frequencies $+\Omega_1$ and $+\Omega_2$, respectively, will move downward from the starting positions Q_1 and Q_2. This results in the deformation shown in Fig. 19.

At the limit, points $Q_1^{(a)}$ and 0 will coalesce, say the oscillation will tend to steady oscillation whose frequency is $\omega_0 = \Omega_1$. Then the steady-

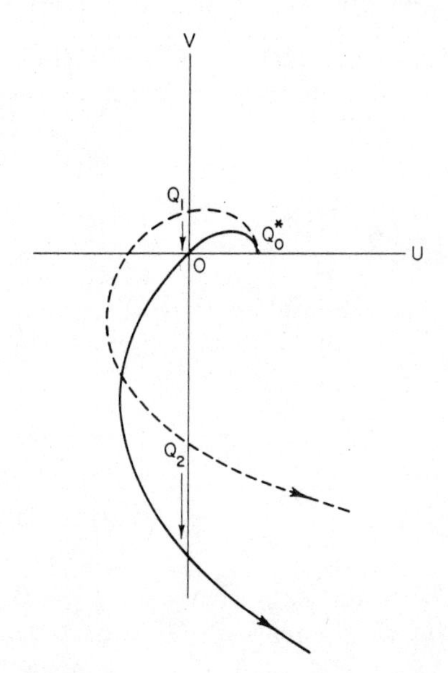

FIG. 19. Fourth-order system shown in Fig. 17, $M_1 = 0$, $M_2 > 0$. Deformation of equiamplitude curves when the system tends to steady motion.

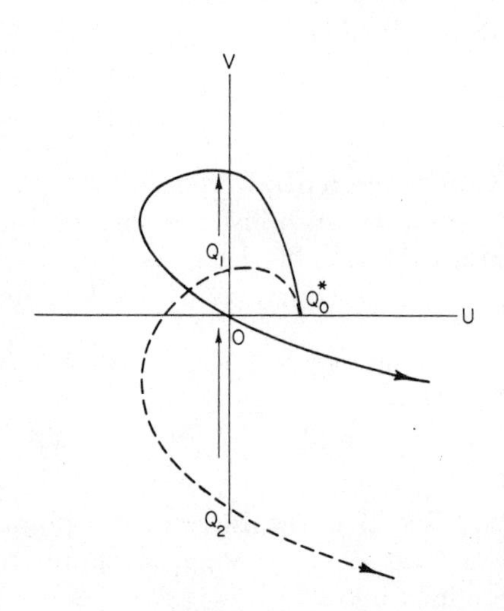

FIG. 20. Fourth-order system shown in Fig. 17, $M_1 = 0$, $M_2 < 0$. Deformation of equiamplitude curves when the system tends to steady motion.

state amplitude a_0 is deduced from (38) by letting $V = 0$, $\omega = \Omega_1$. One gets

$$a_0 = 2\left[\left(\frac{M_2 S_0}{L_2 C_2} - 2\delta_2 \frac{\omega_1^2 - \Omega_1^2}{\alpha_1} - 2\delta_1 \frac{\omega_2^2 - \Omega_1^2}{\alpha_1}\right) \Big/ -3 \frac{M_2}{L_2 C_2} S_2\right]^{1/2}$$

On the other hand, if $M_2 < 0$, the decreasing of a will result in the increase of

$$-\alpha_1 \frac{M_2}{L_2 C_2}\left(S_0 + \frac{3S_2}{4} a^2\right)$$

Accordingly, $Q_1^{(a)}$ and $Q_2^{(a)}$, will move upward from the initial positions Q_1 and Q_2. This results in the deformation shown in Fig. 20.

At the limit, $Q_2^{(a)}$ and 0 will coalesce, say the oscillation will tend to a steady state whose frequency is $\omega_0 = \Omega_2$. The corresponding steady-state amplitude a_0 as given by (38) is

$$a_0 = 2\left[\left(\frac{|M_2|}{L_2 C_2} S_0 - 2\delta_2 \frac{\Omega_2^2 - \omega_1^2}{\alpha_1} - 2\delta_1 \frac{\Omega_2^2 - \omega_2^2}{\alpha_1}\right) \Big/ -3 \frac{|M_2|}{L_2 C_2} S_2\right]^{1/2}$$

The case $M_2 = 0$, $M_1 \neq 0$, would lead to a similar discussion, starting with the same initial state as above, namely the equiamplitude curve defined by $a = 2[-(S_0/3S_2)]^{1/2}$.

The conclusions that can be obtained by looking at the way in which the hodographs get out of shape are the ones which have been obtained in Chapter IV, Section 4.

6.4. Higher-Order Systems

We should not close this section without adding a few comments which will possibly help in extending the method to higher-order systems (Fig. 22, 23).

The describing function associated with a nth-order differential equation of motion, according to our definition, assumes the form

$$H = f(\omega^2) + j\omega g(\omega^2)$$

say the parametric equations of the equiamplitude curves are

$$U = f(\omega^2) \qquad V = \omega g(\omega^2)$$

where f and g are polynomials, with respect to the variable ω^2, which in addition depend on a number of slowly varying parameters. On each equiamplitude, points which are associated with opposite values of frequency ω are symmetric with respect to the real axis, since

$$U(-\omega) = U(\omega) \qquad \text{and} \qquad V(-\omega) = -V(\omega)$$

Accordingly, it is sufficient to plot equiamplitude curves for ω varying from 0 to $+\infty$.

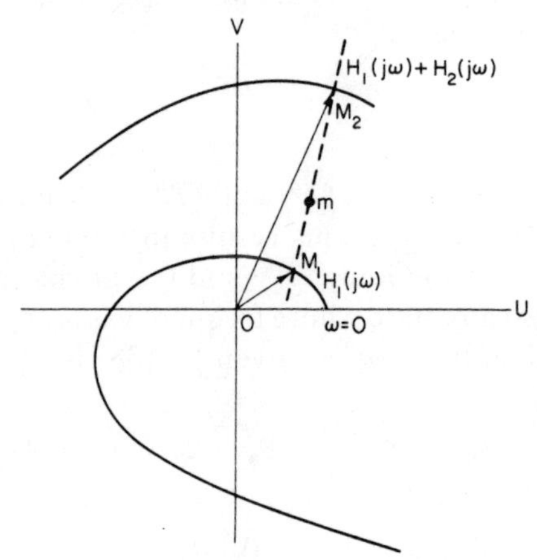

FIG. 21. Geometric construction of Mikaïlov's hodograph.

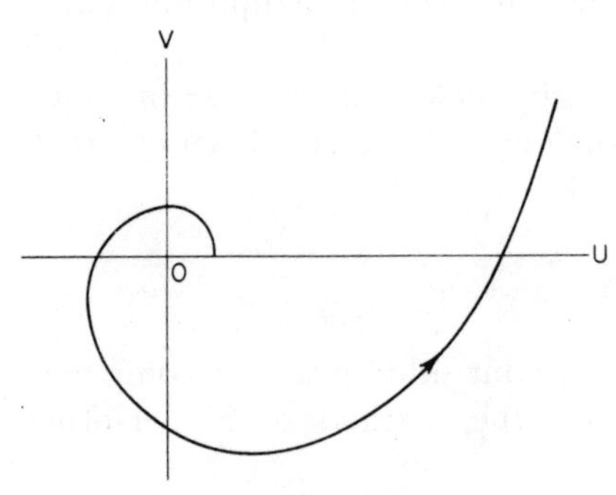

FIG. 22. Generalized Mikaïlov's hodograph for a fifth-order system.

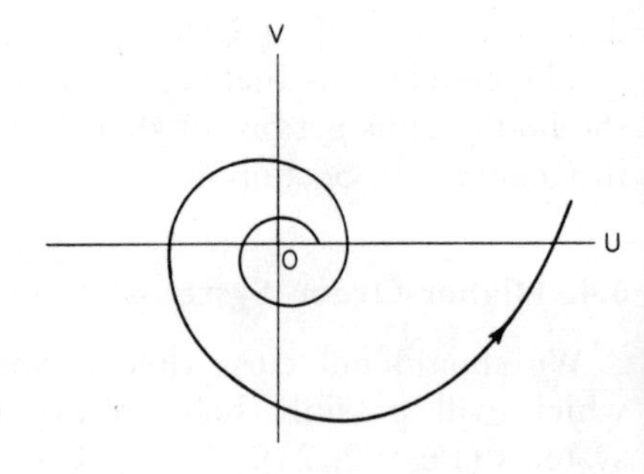

FIG. 23. Generalized Mikaïlov's hodograph for higher-order systems.

In many problems there is only one slowly varying parameter involved, a, and this parameter occurs in such a way that the describing function can be written

$$H(j\omega, a) = H_1(j\omega) + h(a)H_2(j\omega)$$

where $h(a)$ is a real function of parameter a, which does not otherwise occur in the components $H_1(j\omega)$ and $H_2(j\omega)$.

Then the discussion is much simplified, since the equiamplitude curves, for each value of parameter a, can be easily deduced from two basic hodographs which do not depend on a and, therefore, can be plotted at the outset. These basic hodographs are the ones associated with the describing functions $H_1(j\omega)$ and $H_1(j\omega) + H_2(j\omega)$.

Indeed, given any real value $\omega = \omega_r$, point M_1 is associated with $H_1(j\omega_r)$, point M_2 is associated with $H_1(j\omega_r) + H_2(j\omega_r)$, and point m is associated with $H_1(j\omega_r) + h(a)H_2(j\omega_r)$.
Then we have (Fig. 21)

$$\mathbf{Om} = \mathbf{OM}_1 + h(a)\mathbf{M}_1\mathbf{M}_2$$

Accordingly, m can be easily obtained for each value of parameter a; furthermore, when a varies, for each given $\omega = \omega_r$, m moves along the straight line M_1M_2, which is the equifrequency curve associated with $\omega = \omega_r$ (Fig. 21).

7. APPLICATIONS TO NONAUTONOMOUS SYSTEMS†

7.1. Another Definition of Tangent Linearization

In this section we shall use another linearization procedure, which will sometimes prove more convenient in the case of nonautonomous systems.

With this purpose in mind, let us consider the following example:

$$\ddot{x} + \omega_0{}^2 x + \mu f(x, \dot{x}, t) = 0 \tag{39}$$

a second-order nonlinear differential equation, where μ is a *small* parameter and $f(x, \dot{x}, t)$ a nonlinear function of x and $\dot{x}$, in which the time occurs explicitly.

Let us start with the first approximation

$$x = a \cos(\omega_0 t - \varphi)$$

which is an exact solution when $\mu = 0$, and, instead of going to a better approximation,

$$x = a(t) \cos[\omega(t)t - \varphi]$$

where amplitude $a(t)$ and *frequency* $\omega(t)$ are slowly varying functions of time, φ being a constant phase angle, let us search for a better approximation:

$$x = a(t) \cos[\omega_0 t - \varphi(t)] \tag{40}$$

† See [25].

in which frequency ω_0 is fixed, while amplitude $a(t)$ and *phase angle* $\varphi(t)$ are slowly varying functions of time. Indeed, ω_0 is determined by the first approximation.

This new idea will rely on physical arguments which play an important role in time-measurement techniques, in which a sinusoidal signal issued from an actual clock is compared with the pure sinusoidal signal of a standard of time.

It is impossible to determine whether the erratic deviations between the clock and the standard of time are due to frequency fluctuations or to random phase shifts, so much the more so because the sinusoidal signal which is issued from the clock is almost perfect, when considered during a sufficiently short interval of time. The discrepancy between the running of the two oscillators is revealed by the local phase shift, and can be explained either in terms of a slow and random phase shifting or of slow (and random) changes in frequency.

These remarks needs to be complemented by computing the phase shift $\delta\varphi$, which is equivalent to the change in frequency $\delta\omega$ *over one cycle*. The principle of this computation is shown in Fig. 24. One obtains readily

$$\frac{\delta\varphi}{\omega} = \delta T \qquad \text{or} \qquad \delta\varphi = -2\pi\frac{\delta\omega}{\omega} \tag{41}$$

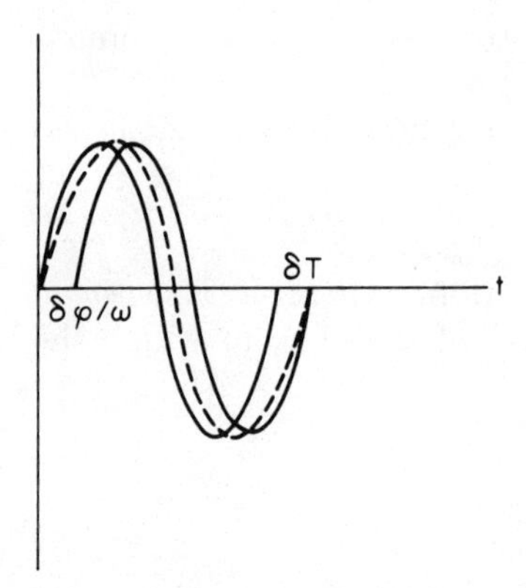

FIG. 24. Phase-shift $\delta\varphi$ equivalent to the change in frequency $\delta\omega$ over one cycle: $\omega = 2\pi/T$.

Now let us come back to the problem of tangent linearization. By substituting (40) in (39) and using the first-harmonic technique, one arrives at the new equation

$$\ddot{x} + \eta(a, \varphi)\dot{x} + [\omega_0{}^2 + \xi(a, \varphi)]x = 0 \tag{42}$$

where ξ and η are correcting functions which depend on the slowly varying parameters a and φ (instead of a and ω, as considered previously).

Equation (42) is the new *tangent linear equation*,[†] from which the describing function

$$H(a, \varphi) \triangleq \xi(a, \varphi) + j\omega_0\eta(a, \varphi) \tag{43}$$

is defined by replacing x in (42) by $a(t) \exp\{j[\omega_0 t - \varphi(t)]\}$, and assuming that $a(t)$ and $\varphi(t)$ are slowly varying functions of time.

On the other hand, it follows from (42) that

$$\frac{1}{a}\frac{da}{dt} = -\frac{1}{2}\eta(a, \varphi) \tag{44}$$

say *over one cycle* ($\delta t = 2\pi/\omega_0$):

$$\frac{\delta a}{a} = -\frac{\pi}{\omega_0}\eta(a, \varphi) \tag{45}$$

As for the change in "local" frequency, one gets

$$\omega^2 = \omega_0{}^2 + \xi(a, \varphi) \tag{46}$$

$$\frac{\delta\omega}{\omega} = \frac{\omega - \omega_0}{\omega_0} \simeq \frac{\xi}{2\omega_0{}^2} \tag{47}$$

$$\delta\varphi = -2\pi\frac{\delta\omega}{\omega_0} = -\frac{\pi}{\omega_0{}^2}\xi(a, \varphi) \tag{48}$$

After these preliminary manipulations, the steady oscillations, if they exist, will be determined as previously, by writing the conditions according to which the hodograph associated to $H(a, \varphi)$ passes through the origin 0. When these conditions are fulfilled, one gets $\delta a = \delta\varphi = 0$. The amplitude and the phase angle of the steady state are determined by

$$\xi(a, \varphi) = \eta(a, \varphi) = 0$$

7.2. Application to Mathieu Oscillators

We shall apply the above version of the describing function method to oscillators governed by the Mathieu differential equations:

$$\ddot{x} + (1 + \alpha \cos 2t)x = 0 \tag{49}$$

and

$$\ddot{x} + \beta\dot{x} + (1 + \alpha \cos 2t)x + \gamma x^3 = 0 \tag{50}$$

the first one being linear and the second nonlinear. α, β, and γ are assumed to be small positive quantities.

[†] Linear for each given a and φ, but nonlinear when a and φ are released.

Such oscillators play an important role in physics. They have been chiefly introduced by the works of Mandelstam and Papalexi on parametric excitation.

According to the above remarks, let us start with the first approximation $x = a\cos(t - \varphi)$, then try to fit the more general solution

$$x = a(t)\cos[t - \varphi(t)] \tag{51}$$

where a and φ are slowly varying functions.

Substituting in (50) and following the linearization procedure, one gets

$$\ddot{x} + \left(\beta + \frac{\alpha}{2}\sin 2\varphi\right)\dot{x} + \left(1 + \frac{3\gamma}{4}a^2 + \frac{\alpha}{2}\cos 2\varphi\right)x = 0 \tag{52}$$

That is the *tangent linear equation.*† The associated describing function is

$$H(a, \varphi) \triangleq \frac{3\gamma}{4}a^2 + \frac{\alpha}{2}\cos 2\varphi + j\left(\beta + \frac{\alpha}{2}\sin 2\varphi\right) \tag{53}$$

A family of equiamplitude and equiphase curves is shown in Fig. 25. The equiphase curves are straight lines which are parallel to the real axis, and the equiamplitude curves are circles whose centers move along a parallel to the real axis. The distance between this locus and the real axis is β, and the radii of all the circles are $\alpha/2$. The arrows in Fig. 25 show the directions in which the parameters a and φ are increasing.

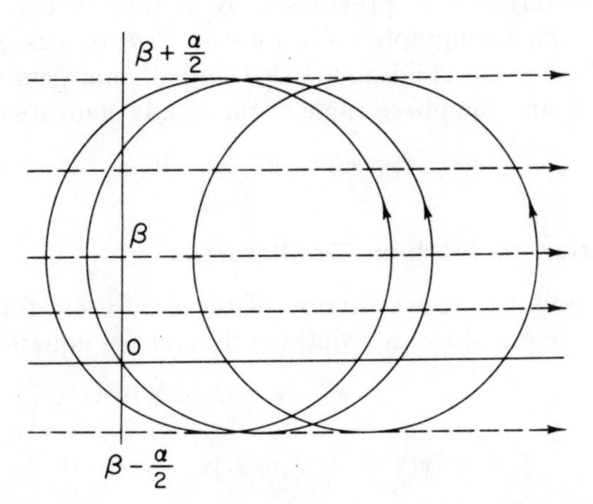

FIG. 25. Equiamplitude and equiphase curves of a nonlinear Mathieu oscillator.

† It was also obtained by the optimal linearization method in Chapter V, Section 5.3 and Eq. (32).

The amplitude and the phase angle for steady oscillation are determined by the following conditions:

$$\beta + \frac{\alpha}{2} \sin 2\varphi = 0 \tag{54}$$

$$\frac{3\gamma}{4} a^2 + \frac{\alpha}{2} \cos 2\varphi = 0 \tag{55}$$

We see that a necessary condition for the existence of an equiamplitude through 0, namely of steady oscillation, is

$$\beta - \frac{\alpha}{2} < 0 < \beta + \frac{\alpha}{2}$$

In the case of (49), in which $\beta = \gamma = 0$, Fig. 25 is degenerated into a circle, whose radius is $\alpha/2$ and whose center is the origin 0 (Fig. 26). This is a consequence of the linearity of this equation.

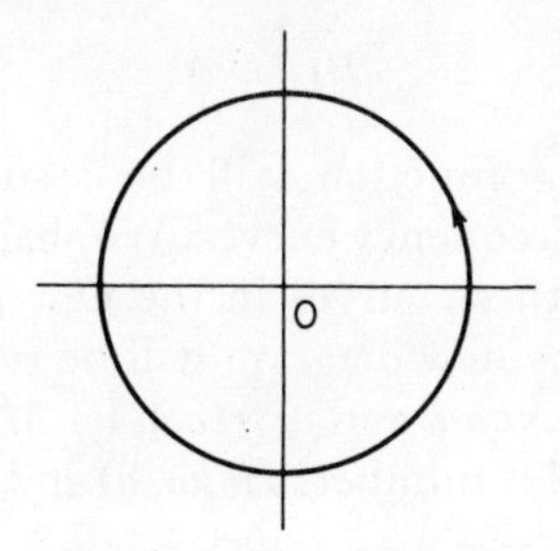

FIG. 26. Diagram of a linear Mathieu oscillator.

7.3. Connection between the Describing Function and Stroboscopic Method†

The above procedure shows clearly the close relationship between our version of the describing function method and the stroboscopic method of Minorsky. As a matter of fact, they appeared at about the same time, and they both derive from the principle of equivalent linearization of Krylov and Bogoliubov.

The stroboscopic method leads to the construction of the integral curves of the system

$$\frac{\delta a}{a} = -\frac{\pi}{\omega_0} \eta(a, \varphi) \qquad \delta\varphi = -\frac{\pi}{\omega_0^{\,2}} \xi(a, \varphi)$$

in the usual phase plane, whereas our method makes use of the complex plane.

† See [26].

In both cases a slowly varying parameter is introduced; it characterizes the state of evolution of the physical system under consideration. The most significant parameters are amplitude a, angular frequency ω, and phase angle φ. One finally gets the relations $\omega(a)$ or $\varphi(a)$.

Note that to each point of the stroboscopic locus an "equiamplitude" curve, is associated by a kind of "duality," whereas the stroboscopic locus itself corresponds to the trajectory of the oscillator which we have defined above.

7.4. Synchronization of a Nonlinear Oscillator†

7.4.1. Extension of the Representation in the Complex Plane

To study the synchronization of a nonlinear self-oscillator, we are led now to using an improved describing function, which we shall define at the outset as

$$aH(j\omega, a)$$

To this new describing function will be associated new families of equiamplitude and equifrequency curves. We shall again use the notations (a) and (ω) to indicate these curves in the new representation.

As a matter of fact, the new diagram will be readily deduced from the earlier one as follows: Given a and ω (*real*), let M and m be the representative points of the complex numbers $H(j\omega, a)$ and $aH(j\omega, a)$, respectively, We have

$$\frac{\mathbf{Om}}{\mathbf{OM}} = a$$

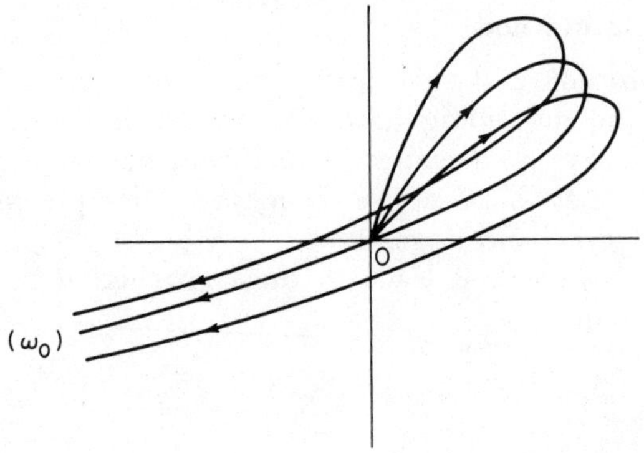

FIG. 27. Family of equifrequency curves in the new representation: $aH(j\omega, a)$.

† See [10, 11].

Since along each equifrequency curve a varies with the position of point M, the shape of the equifrequency curves is notably altered by the transformation (Fig. 27).

On the other hand, since a is constant along each equiamplitude curve, each new equiamplitude is similar to the corresponding earlier one; that is, they are deduced from one another by an homothety of constant ratio.

The general shape of the new equifrequency curves can be specified as follows:

(a) $a = 0$ implies $aH(j\omega, a) = 0$ whatever ω. Hence, in the new representation, every equifrequency curve is issuing from the origin 0.

(b) When $\omega = \omega_0$ (where ω_0 is the steady-oscillation frequency which we shall assume to exist), we know that there exists a value for a, say $a = a_0$, such that

$$H(j\omega_0, a_0) = 0$$

a_0 is the steady-oscillation amplitude.

Hence $a_0 H(j\omega_0, a_0) = 0$, which implies that the new equifrequency curve (ω_0) passes through 0 for $a = a_0$. Since it also passes through 0 for $a = 0$, it is in general a loop which, under certain conditions, can collapse into a piece of straight line.

Equifrequency curves which correspond to values of ω next to ω_0 can be deduced from the loop (ω_0) by continuity, as shown in Fig. 27.

7.4.2. Graphical Resolution for Synchronized Oscillations

Let us return to the differential equation which governs the oscillator under study, a Van der Pol oscillator, for instance, and introduce in its right side the sinusoidal forcing function

$$F \sin(\omega t + \varphi)$$

where F and ω are given. The equation of motion will again be replaced by its tangent linear approximation, the forcing function by $Fe^{j\varphi}e^{j\omega t}$, and the expected synchronized oscillation by $a_s \exp(j\psi_s) \exp(j\omega t)$, according to which the equation will be rewritten

$$H(j\omega, a_s)a_s \exp(j\psi_s) = F \exp(j\varphi) \tag{56}$$

or

$$a_s H(j\omega, a_s) = F \exp[j(\varphi - \psi_s)] \tag{57}$$

The convenience of the new representation is visible in (57). As a matter of fact, it will enable us to determine by graphical construction the amplitude of the synchronized oscillation and its phase lag $\psi_s - \varphi$ with respect to the forcing function. Here we shall only outline the principle of this graphical construction, which can be easily ascertained. The construction will be based on the assumption that a stable synchronized solution exists, which assumption will be discussed next.

Let us represent this solution by point M, at the intersection of equifrequency curve (ω) and of equiamplitude curve (a_s); i.e., we consider the families of equifrequency and equiamplitude curves (in the new representation) as a system of curvilinear coordinates, in which we select the two curves that determine the expected solution.

On the other hand, let M_0 be the representative point of the complex number $Fe^{j\varphi}$.

From (57) it appears that M is also on the circumference (F) whose center is at the origin and whose radius is $OM_0 = F$. Hence M is at the intersection of the equifrequency curve (ω), of the equiamplitude curve (a_s), and of the above circumference. This intersection can be determined, since (ω) can be constructed and F is known (Fig. 28).

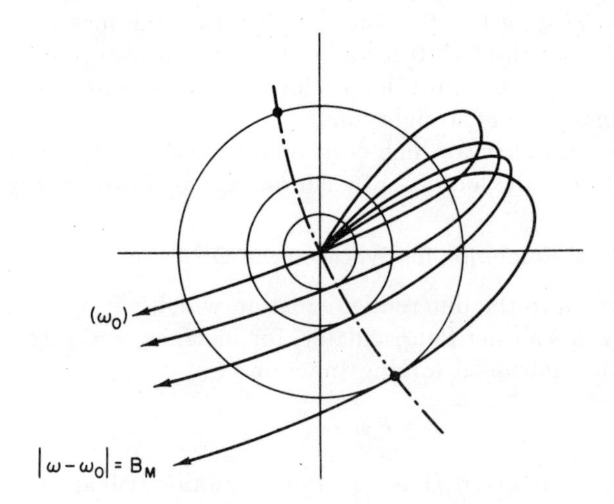

FIG. 28. Threshold interval B_M for synchronized oscillations.

Then:

(a) Angle $(\mathbf{OM}_0, \mathbf{OM}) = \psi_s$.

(b) The equiamplitude curve through M determines the amplitude of the synchronized solution.

Next we shall discuss this graphical resolution.

7.4.3. Discussion of the Graphical Resolution

In Figs. 28 and 29 we see that, given the forcing function $F \sin(\omega t+\varphi)$, and provided that

(a) F is sufficiently small (but not too small)

(b) ω is sufficiently close to ω_0 (ω_0 is the characteristic frequency of the nonlinear oscillator—its angular frequency without forcing function, at the steady state)

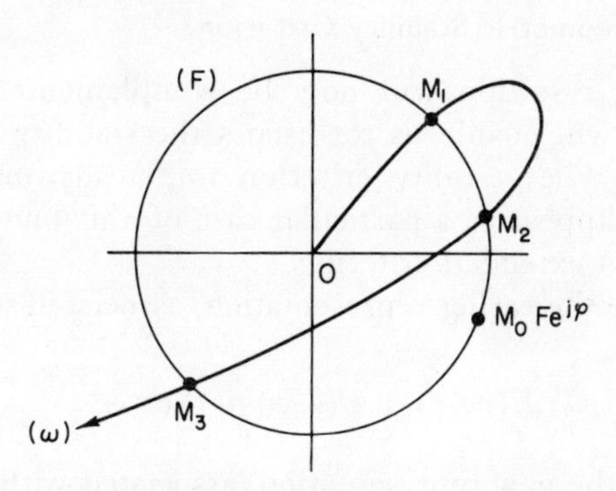

FIG. 29. Graphical resolution for synchronized oscillations.

then in general the equifrequency curve (ω) and the circumference (F) have three intersection points. Accordingly, it seems that there exist three synchronized solutions. However, this number will be reduced by stability considerations.

If (ω) is kept constant (close to ω_0) and F is *increased*, two of the above real solutions will coalesce, then become complex. There will remain only one solution, whose *stability* will be ascertained in the next section.

On the other hand, if (ω) is again kept constant but F is *decreased*, another threshold becomes visible, corresponding to another tangency condition between (ω) and the circumference (F). Below this threshold, there remains again only one intersection point, which is *unstable*, as will be verified later.

The conclusion of this first part of the discussion is the one we arrived at in Chapter 2: ω being given, the synchronized solution (stable) disappears when the amplitude of the forcing function becomes less than a threshold value F_m.

Now if F is kept constant and ω is varied, we see that, as $|\omega - \omega_0|$ is increased, the equifrequency curves deviate more and more from the loop (ω_0), and, again, there will exist a threshold corresponding to a tangency condition between (ω) and the circumference (F). Beyond this threshold there will exist only one intersection point, which is *unstable*. Accordingly, F being given, the stable synchronized solution disappears when $| \omega - \omega_0 |$ exceeds a threshold interval B_M.

The locus of all the tangency points between the equifrequency curves (ω) and the circumferences (F), which, as pointed out above, plays an important role as concerns the existence of a stable synchronized solution, is shown in Fig. 28.

7.4.4. Extended Geometric Stability Criterion†

The graphical resultion will now be complemented by a stability criterion, which will enable us to discuss the stability of the solutions obtained above. The stability criterion for steady motion, stated in Section 5.3, will appear as a particular case of this one which is, from this viewpoint, an extended criterion.

Let (H) denote the earlier representation, associated with the describing function

$$H(j\omega, a) = U(a, \omega) + jV(a, \omega)$$

Let (H_1) denote the new representation, associated with the function

$$aH(j\omega, a) = aU(a, \omega) + jaV(a, \omega)$$

We shall also have to consider the representation (H_2) associated with the function

$$a^2H(j\omega, a) = a^2U(a, \omega) + ja^2V(a, \omega)$$

This function will be generated from (H_1) in the same way as (H_1) was generated from (H).

Now let us come back to the determination of a synchronized oscillation, to the equation

$$H(a_s, \omega)a_s \exp(j\psi_s) = F \tag{58}$$

in which we have put, for the sake of simplification, $\varphi = 0$ and

$$H(a_s, \omega) \triangleq H(j\omega, a_s).$$

A synchronized solution is a stationary motion, in the neighborhood

† See [25].

of which the amplitude and phase angle of the oscillation, a and ψ, are slowly varying functions of time. Accordingly, we shall put

$$\frac{1}{a}\frac{da}{dt} = -\omega_2 \qquad \frac{d\psi}{dt} = q \tag{59}$$

where ω_2 and q are small real parameters.

This amounts to approaching the laws following which the oscillator goes back to the steady state, for any given deviation:

(a) By an exponential law as concerns the amplitude.

(b) By a linear law as concerns the phase angle.

Analytically this amounts to representing the transient motion, in the neighborhood of the stationary state $a_s \exp(j\psi_s) \exp(j\omega t)$ by

$$a \exp(j\psi) \exp[j(\omega + j\omega_2 + q)t] = a \exp(-\omega_2 t) \exp[j(\psi + qt)] \exp(j\omega t)$$

where a and ψ are the amplitude and the phase angle at time $t = 0$.

Then (58) is replaced by

$$H(a, \omega + j\omega_2 + q)ae^{j\psi} = F \tag{60}$$

which gives, at first-order approximation with respect to ω_2 and q,

$$aH(a, \omega) + (j\omega_2 + q)a\left(\frac{\partial H}{\partial \omega}\right)_{a,\omega} = Fe^{-j\psi} \tag{61}$$

One obtains

$$(j\omega_2 + q)a = \frac{1}{(\partial H/\partial \omega)_{a,\omega}}[Fe^{-j\psi} - aH(a, \omega)] \tag{62}$$

and, according to (59),

$$-j\frac{da}{dt} + a\frac{d\psi}{dt} = \frac{1}{(\partial H/\partial \omega)_{a,\omega}}[Fe^{-j\psi} - aH(a, \omega)] \tag{63}$$

Now let us introduce the deviations from the stationary state, δa and $\delta\psi$, by putting

$$\psi = \psi_s + \delta\psi \qquad a = a_s + \delta a$$

By substituting in (63) one gets the variational equation

$$-j\frac{d(\delta a)}{dt} + a_s\frac{d(\delta\psi)}{dt} = \frac{1}{(\partial H/\partial \omega)_{a,\omega}}\left\{-jFe^{-j\psi_s}\delta\psi - \frac{\partial}{\partial a}[aH(a, \omega)]_{a_s}\delta a\right\} \tag{64}$$

from which follows, by taking $F \exp(-j\psi_s) = a_s H(a_s, \omega)$ into account,

$$-j\frac{d(\delta a)}{dt} + a_s \frac{d(\delta\psi)}{dt} = \frac{1}{(\partial H/\partial\omega)_{a,\omega}} \left\{ -ja_s H(a_s, \omega)\, \delta\psi - \frac{\partial}{\partial a}[aH(a, \omega)]_{a_s}\, \delta a \right\} \tag{65}$$

This equation can be split into two equations,

$$\begin{aligned} a_s \frac{\partial U}{\partial\omega}\frac{d(\delta\psi)}{dt} + \frac{\partial V}{\partial\omega}\frac{d(\delta a)}{dt} &= a_s V\, \delta\psi - \frac{\partial}{\partial a}(aU)\, \delta a \\ a_s \frac{\partial V}{\partial\omega}\frac{d(\delta\psi)}{dt} - \frac{\partial U}{\partial\omega}\frac{d(\delta a)}{dt} &= -a_s U\, \delta\psi - \frac{\partial}{\partial a}(aV)\, \delta a \end{aligned} \tag{66}$$

where U, V, $\partial U/\partial\omega$, $\partial V/\partial\omega$, $\partial(aU)/\partial a$, and $\partial(aV)/\partial a$ are to be computed at the point (a_s, ω), which represents the stationary state. Finally, one gets the characteristic equation associated with this variational equation:

$$a_s \left[\left(\frac{\partial U}{\partial\omega}\right)^2 + \left(\frac{\partial V}{\partial\omega}\right)^2\right]\lambda^2 + \frac{1}{a_s^2}\left[\frac{\partial(a_s^2 U)}{\partial a_s}\frac{\partial(a_s^2 V)}{\partial\omega} - \frac{\partial(a_s^2 V)}{\partial a_s}\frac{\partial(a_s^2 U)}{\partial\omega}\right]\lambda$$
$$+ \frac{1}{2}\frac{\partial}{\partial a_s}[(a_s U)^2 + (a_s V)^2] = 0 \tag{67}$$

The stability of the solutions we have obtained by the above graphical construction will now be discussed, starting with the characteristic equation (67). We shall determine the roots of (67) and base the discussion on the sign of their real part. The discussion will rely on some geometric properties of the representations (H), (H_1), and (H_2), which we have introduced at the begining of the section.

Let M_0, M_1, and M_2 denote the moving points in the representations (H), (H_1), and (H_2), respectively, and, first of all, note that the expression

$$(a_s U)^2 + (a_s V)^2$$

which appears in the last term of the left side of (67) has a simple geometric meaning:

$$\mathbf{OM}_1^2 = |\, aH(a, \omega)|^2_{a_s, \omega}$$

Note that this expression needs to be computed at the stationary point a_s and ω (ω is the frequency of the forcing function), whose stability is now being discussed.

Accordingly, the sign of the last coefficient in (67) will depend on the position of the stationary point under consideration with respect to the *extrema* of $\mathbf{OM}_1^2$, *along the equifrequency curve* (ω_1), *in the representation* (H_1). As a matter of fact, the last coefficient will vanish if

$$\frac{\partial}{\partial a}\mathbf{OM}_1^2 = 0$$

at the stationary point.

The locus of these extrema, when ω varies, is a line which plays an important role, since the sign of the last term in (67) depends on the position of the stationary point with respect to it in the representation (H_1). This line will be called a *separatrix of the first kind* and will be denoted (S_1).

We shall call *switching points* the points at which $(\partial/\partial a)\,\mathbf{OM}_1^2$ changes sign along an equifrequency curve, in the representation (H_1). The general shape of (S_1) is shown in Fig. 30. In Fig. 31, where an equi-

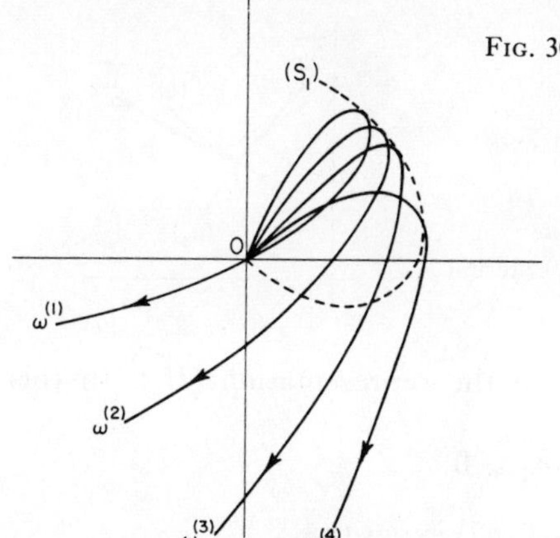

FIG. 30. Separatrix of the first kind (S_1).

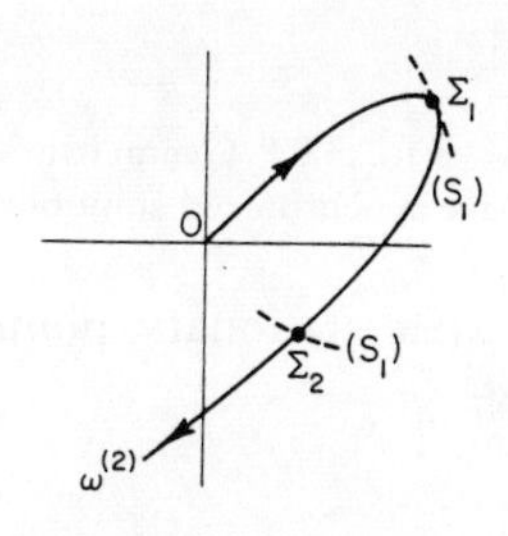

FIG. 31. Switching points.

frequency curve [in the representation (H_1)] has been drawn apart, the switching points are Σ_1 and Σ_2.

The second coefficient in (67) also has a simple geometric meaning. Indeed,

$$\frac{\partial(a_s^2U)}{\partial a_s}\frac{\partial(a_s^2V)}{\partial\omega} - \frac{\partial(a_s^2V)}{\partial a_s}\frac{\partial(a_s^2U)}{\partial\omega}$$

is the cross product

$$\frac{\partial\mathbf{OM}_2}{\partial a} \times \frac{\partial\mathbf{OM}_2}{\partial\omega}$$

This expression needs to be computed for the stationary state a_s, ω.

Note that the vectors

$$\mathbf{\Omega}_2 = \frac{\partial \mathbf{OM}_2}{\partial a} \qquad \text{and} \qquad \mathbf{A}_2 = \frac{\partial \mathbf{OM}_2}{\partial \omega}$$

are tangent to the equifrequency curve and the equiamplitude curve, respectively, at the stationary point, *in the representation* (H_2). They are pointing, respectively, in the directions of increasing amplitude and increasing frequencie. Accordingly, the sign of the second coefficient in (67) is determined by the sign of

$$\mathbf{\Omega}_2 \times \mathbf{A}_2$$

The locus of points at which $\mathbf{\Omega}_2 \times \mathbf{A}_2 = 0$ *in the representation* (H_2) *will be called a separatrix of the second kind* (S_2). At such points the equifrequency curve and the equiamplitude curve are tangent to one another in the (H_2) representation. In Fig. 32 vectors $\mathbf{A}_2$ and $\mathbf{\Omega}_2$ are

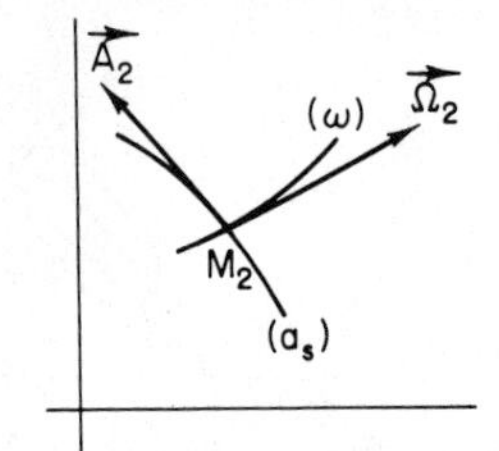

FIG. 32. Geometric stability criterion for synchronized solutions.

shown, at a stationary point M_2, in the representation (H_2). In this example,

$$\mathbf{\Omega}_2 \times \mathbf{A}_2 > 0$$

Accordingly, the second coefficient of (67) is positive.

It will be possibly useful to refer to a theoretical example, such as the one shown in Fig. 33. Here we have drawn the equifrequency curves which correspond to the frequency ω of the forcing function in the representations (H_1) and (H_2). Let us denote these equifrequency curves by $[\omega^{(1)}]$ and $[\omega^{(2)}]$.

On $[\omega^{(1)}]$ the stationary points are M_1, M_1', and M_1'', following the construction discussed in section 7.4.2. On this equifrequency curve the switching points, on the separatrix (S_1), are Σ_1 and Σ_2.

Let us consider, for example, the stationary point M_1 on $[\omega^{(1)}]$, and the point M_2 which is associated with M_1 on $[\omega^{(2)}]$. Points M_1 and M_2 correspond to the same values of a_s and ω in both representations, say

$$\mathbf{OM}_2 = a\mathbf{OM}_1$$

On the other hand, let Q be the point where $[\omega^{(2)}]$ reaches the separatrix (S_2). That is, at point Q we have $\mathbf{\Omega}_2 \times \mathbf{A}_2 = 0$. Otherwise, the sign of $\mathbf{\Omega}_2 \times \mathbf{A}_2$ is as shown in Fig. 32 at all the other points of $[\omega^{(2)}]$.

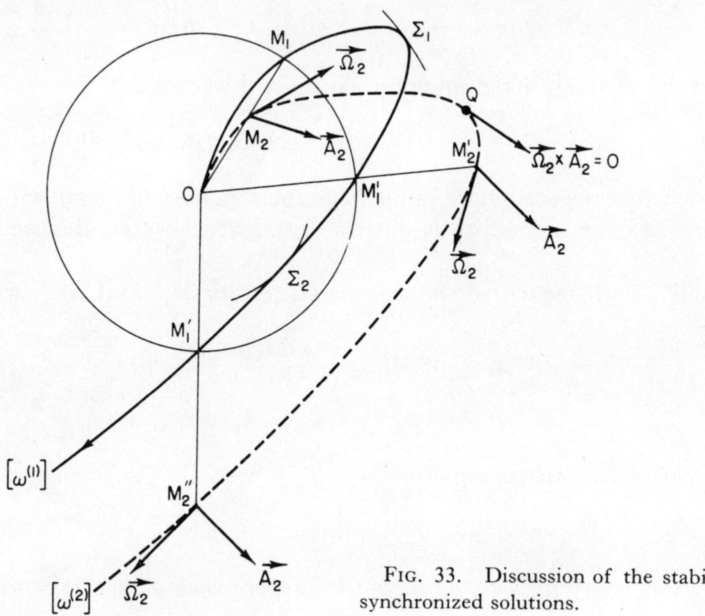

FIG. 33. Discussion of the stability of synchronized solutions.

Then:

(a) Along the branch $0\Sigma_1$ of $[\omega^{(1)}]$ we have

$$\frac{\partial \mathbf{OM}_1^2}{\partial a} > 0 \qquad \text{say} \quad \frac{\partial}{\partial a}[(aU)^2 + (aV)^2] > 0$$

Then the third coefficient in (67) is positive.

(b) At point M_2 on $[\omega^{(2)}]$ we have

$$\mathbf{\Omega}_2 \times \mathbf{A}_2 < 0$$

Hence the second coefficient in (67) is negative.

(c) The first coefficient in (67) always being positive, the characteristic equation assumes the form

$$\alpha\lambda^2 - \beta\lambda + \gamma = 0 \qquad \text{with} \quad \alpha > 0, \beta > 0, \gamma > 0$$

Accordingly, the stationary point M_1 is an *unstable point*. Furthermore, we see that this instability is due to an oscillatory transient whose amplitude is growing.

Likewise, as concerns the associated points M_1' and M_2' we see that

$$\text{at point } M_1': \quad \frac{\partial \mathbf{OM}_1'^2}{\partial a} < 0$$

$$\text{at point } M_2': \quad \mathbf{\Omega}_2 \times \mathbf{A}_2 > 0$$

Hence the characteristic equation assumes the form

$$\alpha\lambda^2 + \beta\lambda - \gamma = 0 \qquad \text{with} \quad \alpha > 0, \beta > 0, \gamma > 0$$

It follows that the stationary point M_1' is also an *unstable point*. However, at point M_1' the instability is due to a transient of the nonoscillatory kind.

Finally, with regard to the associated points M_1'' and M_2'', we have

$$\text{at point } M_1'': \quad \frac{\partial \mathbf{OM}_1''^2}{\partial a} > 0$$

$$\text{at point } M_2'': \quad \mathbf{\Omega}_2 \times \mathbf{A}_2 > 0$$

Then the characteristic equation is

$$\alpha\lambda^2 + \beta\lambda + \gamma = 0 \qquad \text{with} \quad \alpha > 0, \beta > 0, \gamma > 0$$

Hence this solution is *stable*. It is the *proper synchronized solution.*

7.4.5. Application to Van der Pol's Equation

When the equation of motion of the oscillator with forcing excitation is

$$\ddot{x} + 2\mu(x^2 - 1)\dot{x} + x = F \sin \omega t$$

the describing function is, as pointed out earlier,

$$H(j\omega, \alpha) \equiv 1 - \omega^2 + 2j\omega\mu\left(\frac{a^2}{4} - 1\right)$$

Then

$$U = 1 - \omega^2 \qquad V = 2\mu\omega\left(\frac{a^2}{4} - 1\right)$$

Accordingly, the characteristic equation (67) assumes the form

$$\lambda^2 - 2\mu\left(1 - \frac{a_s^2}{2}\right)\lambda + \mu^2\left(1 - \frac{a_s^2}{4}\right)\left(1 - \frac{3a_s^2}{4}\right) + (\Delta\omega)^2 = 0$$

$$\text{with} \quad \Delta\omega = \frac{1 - \omega^2}{2\omega}$$

In the (H) representation the equiamplitude curves are parabolas, and the equifrequency curves are straight lines, which are shown on Fig. 34 for $a = 0, 1, 2, 3$, and $\omega = 1, \sqrt{2}, \sqrt{2.5}, \sqrt{3}, 2$ respectively. The equifrequency curves in the (H_1) representation and the separatrix (S_1) are deduced from the latter ones.

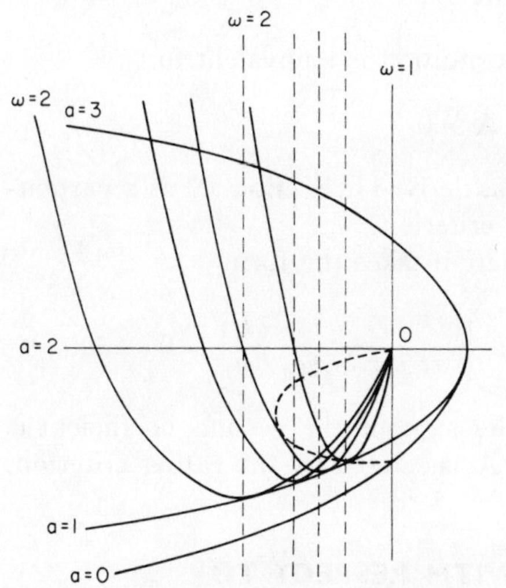

FIG. 34. Equiamplitude and equifrequency curves of Van der Pol's oscillator in the (H) representation. Equifrequency curves in the (H_1) representation. Dotted curve, separatrix (S_1).

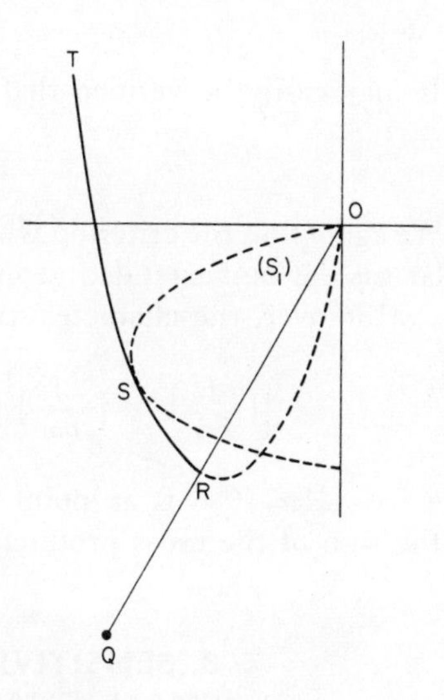

FIG. 35. Limiting case: Equifrequency curve in the (H_1) representation, tangent to the separatrix (S_1).

The limiting case, where the equifrequency curve in the (H_1) representation is tangent to the separatrix (S_1), has been drawn apart in Fig. 35. Point Q determines, in the (H_2) representation, the switching of the sign of the second coefficient of the characteristic equation. The branch $0R$ is an unstable region, since the third coefficient of the characteristic equation is positive and the second is negative. At point R there is a switching of the sign of the second coefficient, whereas the sign of the third coefficient remains unchanged; therefore, RST is a stable region.

7.4.6. Remark about the Criterion of Stability without Forcing Function

All the above arguments remain valid if $F = 0$. The discussion is thus simplified, since then the equifrequency curve which corresponds to the stationary state in the (H_1) representation passes through 0 for the stationary nonzero value of the amplitude [this property also holds in the (H) and (H_2) representations].

Point 0 corresponds to the stationary state, and at the same time is on the separatrix (S_1). Accordingly, there remains only the stability condition

$$\mathbf{\Omega}_2 \times \mathbf{A} > 0$$

It may easily be verified that this condition is equivalent to

$$\mathbf{\Omega} \times \mathbf{A} > 0$$

We again find the criterion which was derived in Section 5.3 as a particular case of our extended geometric criterion.

Moreover, the characteristic equation takes the form

$$\left[\left(\frac{\partial U}{\partial \omega}\right)^2 + \left(\frac{\partial V}{\partial \omega}\right)^2\right] \lambda + a_0 \left[\frac{\partial U}{\partial a_0}\frac{\partial V}{\partial \omega} - \frac{\partial V}{\partial a_0}\frac{\partial U}{\partial \omega}\right] = 0$$

since $U = V = 0$ at point 0. The sign of the second coefficient is the sign of the cross product $\mathbf{\Omega} \times \mathbf{A}$, as stated by the earlier criterion.

8. SENSITIVITY WITH RESPECT TO SMALL CHANGES IN PARAMETERS†

8.1. Amplitude and Frequency Sensitivity

The steady-state amplitude and frequency of a vacuum-tube oscillator depend on a number of parameters, such as the temperature of some resistive element, the voltages applied to the tube, etc. Any change in these parameters will result in a change in the amplitude and frequency of the steady oscillation, which we shall now compute.

In engineering practice, if the change resulting from these variables is great, the stability is said to be low, and conversely. However, in this situation the term "stability" may be somewhat misleading when considered from a more general viewpoint. Indeed, although there exist many definitions of the stability of a system, the concept of stability is usually defined starting with the following question: Does a small deviation from the steady state, *for (given) constant values of the parameters*, decrease with time?

Thus in the present problem the initial state is assumed to be a stable steady state, and the final state is also a stable steady state. We have in

† See [4].

mind to determine the magnitude of the transition between the initial and final states. The term "sensitivity," with respect to small changes in the parameters, is possibly more consistent with common sense.

Now let ϵ be the parameter under consideration, and rewrite the steady-state conditions:

$$U(a_0, \omega_0, \epsilon) = 0 \qquad V(a_0, \omega_0, \epsilon) = 0$$

Then by differentiating we get

$$\frac{\partial U}{\partial a_0} da_0 + \frac{\partial U}{\partial \omega_0} d\omega_0 + \frac{\partial U}{\partial \epsilon} d\epsilon = 0$$

$$\frac{\partial V}{\partial a_0} da_0 + \frac{\partial V}{\partial \omega_0} d\omega_0 + \frac{\partial V}{\partial \epsilon} d\epsilon = 0$$

from which follows

$$da_0 = -\left[\left(\frac{\partial U}{\partial \epsilon}\frac{\partial V}{\partial \omega_0} - \frac{\partial V}{\partial \epsilon}\frac{\partial U}{\partial \omega_0}\right) \Big/ \left(\frac{\partial U}{\partial a_0}\frac{\partial V}{\partial \omega_0} - \frac{\partial U}{\partial \omega_0}\frac{\partial V}{\partial a_0}\right)\right] d\epsilon \tag{68}$$

$$d\omega_0 = -\left[\left(\frac{\partial V}{\partial \epsilon}\frac{\partial U}{\partial a_0} - \frac{\partial U}{\partial \epsilon}\frac{\partial V}{\partial a_0}\right) \Big/ \left(\frac{\partial U}{\partial a_0}\frac{\partial V}{\partial \omega_0} - \frac{\partial U}{\partial \omega_0}\frac{\partial V}{\partial a_0}\right)\right] d\epsilon \tag{69}$$

In many practical applications

$$\frac{\partial U}{\partial \omega_0} \simeq 0 \qquad \text{and} \qquad \frac{\partial V}{\partial a_0} \simeq 0$$

Hence formulas (68) and (69) result in

$$| da_0 | = \frac{| \partial U/\partial \epsilon |}{| \partial U/\partial a_0 |} | d\epsilon | \tag{70}$$

$$| d\omega_0 | = \frac{| \partial V/\partial \epsilon |}{| \partial V/\partial \omega_0 |} | d\epsilon | \tag{71}$$

Finally, as a matter of convenience, the following coefficients are introduced:

$$\sigma_a = \left| \frac{\partial U}{\partial a_0} \right| = | \mathbf{\Omega} | \qquad \sigma_\omega = \left| \frac{\partial V}{\partial \omega_0} \right| = | \mathbf{A} |$$

or, according to more general conventions, since the *amplitude sensitivity* and *frequency sensitivity* are defined by $(1/a_0) | da_0 |$ and $(1/\omega_0) | d\omega_0 |$,

$$S_a = a_0 \left| \frac{\partial U}{\partial a_0} \right| \qquad S_\omega = \omega_0 \left| \frac{\partial V}{\partial \omega_0} \right|$$

When S_a is large, the amplitude sensitivity is small, and conversely. When S_ω is large, the frequency sensitivity is small, and conversely. From the engineering viewpoint, as discussed above, the values of the parameters S_a and S_ω can be thought of as measuring the stability of the amplitude and the frequency of the oscillator, respectively.

8.2. Sensitivity of Bridge Oscillators

Let us apply the above considerations to a bridge oscillator, shown in Fig. 36. The bridge consists of three resistances R', R'', and R_T and the impedance

$$Z(\omega) = X(\omega) + jY(\omega)$$

which is, in general, the impedance of a resonant circuit.

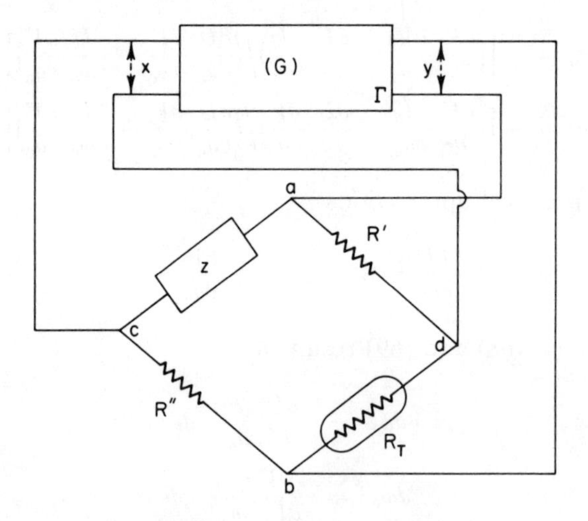

FIG. 36. Bridge oscillator.

R' and R'' are constant, whereas R_T is a *temperature-dependent resistance*, which acts as a stabilizing element with respect to fluctuations of different kinds. R_T may be the resistance of a tungsten incandescent lamp.

Now the well-known relation between the input y and the output x of the bridge is

$$x = \left(\frac{R''}{R'' + Z(\omega)} - \frac{R_T}{R' + R_T}\right) y$$

and, through the amplifier, we have

$$y = \Gamma x$$

Hence the equation of the feedback loop is

$$\left[1 - \Gamma\left(\frac{R''}{R'' + Z(\omega)} - \frac{R_T}{R' + R_T}\right)\right] x = 0$$

or

$$\{[R'' + Z(\omega)](R_T + R') - \Gamma[R'R'' - R_T Z(\omega)]\}x = 0$$

Let us assume, to simplify, that Γ is real and independent of frequency ω, and let

$$U = [R'' + X(\omega)][R_T + R'] - \Gamma[R'R'' - R_T X(\omega)] \tag{72}$$

$$V = [R' + (1 + \Gamma)R_T]Y(\omega) \tag{73}$$

The conditions for the stationary state are $U = 0$, $V = 0$. From (72) we get

$$R_T + R' = \frac{\Gamma[R'R'' - R_T X(\omega)]}{X(\omega) + R''} \tag{74}$$

Then substituting in (73),

$$\frac{(R_T + R')R''}{X(\omega) + R''}\,\Gamma Y(\omega) = 0$$

from which, since R', R'', R_T, and $X(\omega)$ are positive,

$$Y(\omega) = 0 \tag{75}$$

This condition determines the frequency of oscillation. Substituting in (72) we get the steady-state amplitude, which occurs only in the expression of the temperature-dependent resistance R_T, if we assume that the amplifier is *linear* (say Γ is a constant).

On the other hand, we see from (74) that

$$R'R'' - R_T X(\omega) > 0 \qquad \text{requires} \quad \Gamma > 0$$

and

$$R'R'' - R_T X(\omega) < 0 \qquad \text{requires} \quad \Gamma < 0$$

The sensitivity of the oscillator is obtained by differentiating (72) and

(73) with respect to ω and a. By differentiating (73) with respect to ω, and by taking (74) into account, we get

$$S_\omega = \omega_0 \left| \frac{\partial V}{\partial \omega_0} \right| = \frac{R' + R_T}{R'' + X(\omega_0)} R'' \omega_0 \mid \Gamma \mid \left| \frac{\partial Y}{\partial \omega_0} \right| .$$

Accordingly, from the engineering viewpoint, the greater the amplification factor $\mid \Gamma \mid$, the higher the frequency stability of the oscillator. Then by differentiating (72) with respect to a,

$$S_a = a_0 \left| \frac{\partial U}{\partial a_0} \right| = a_0 \mid R'' + (1 + \Gamma) X(\omega_0) \mid \left| \frac{\partial R_T}{\partial a_0} \right|$$

The amplitude stability is proportional to the temperature coefficient $\mid \partial R_T / \partial a_0 \mid$ of the resistance R_T.

9. RETARDED ACTIONS†

Theodorchik pointed out that the describing function method can also be applied to the analysis of the so-called retarded actions, which are described by such equations as

$$\ddot{x} + 2(\delta + \delta_2 x^2)\dot{x} + \nu_0^2 x + \nu^2 x_\tau = 0 \tag{76}$$

in which

$$x_\tau \triangleq x(t - \tau)$$

Again assuming that there exists a quasi-sinusoidal solution, we shall write

$$x \simeq a \sin \omega t$$

and substitute in (76). However, first let us remark that x_τ can be rewritten

$$x_\tau = a \sin \omega(t - \tau) = x \cos \omega\tau - \dot{x} \frac{\sin \omega\tau}{\omega}$$

Then substituting in (76) and making use of the linearization procedure, we get

$$\ddot{x} + 2\left(\delta + \frac{\delta_2 a^2}{4} - \frac{\nu^2}{2\omega} \sin \omega\tau\right) \dot{x} + \omega_0^2 \left(1 + \frac{\nu^2}{\nu_0^2} \cos \omega\tau\right) x = 0 \tag{77}$$

† See [4].

The describing function associated with (77) is

$$H(\omega, a, \tau) \equiv -\omega^2 + \omega_0^2 \left(1 + \frac{\nu^2}{\nu_0^2} \cos \omega\tau\right) + 2j\omega \left(\delta + \frac{\delta_2 a^2}{4} - \frac{\nu^2}{2\omega} \sin \omega\tau\right)$$

Hence the stationary conditions are

$$-\omega^2 + \nu_0^2 + \nu^2 \cos \omega\tau = 0 \tag{78}$$

$$2\delta\omega + \tfrac{1}{2}\delta_2 \omega a^2 - \nu^2 \sin \omega\tau = 0 \tag{79}$$

Now if we assume that the system is unstable at zero amplitude, i.e., that the coefficient of the damping term in (77) is negative when the amplitude is very small, say

$$\delta - \frac{\nu^2}{2\omega} \sin \omega\tau < 0 \qquad a \simeq 0$$

the amplitude will start increasing and tend to the stationary amplitude given by (79):

$$a_0 = \left(\frac{2\nu^2 \sin \omega_0\tau - 4\delta\omega_0}{\delta_2\omega_0}\right)^{1/2} \tag{80}$$

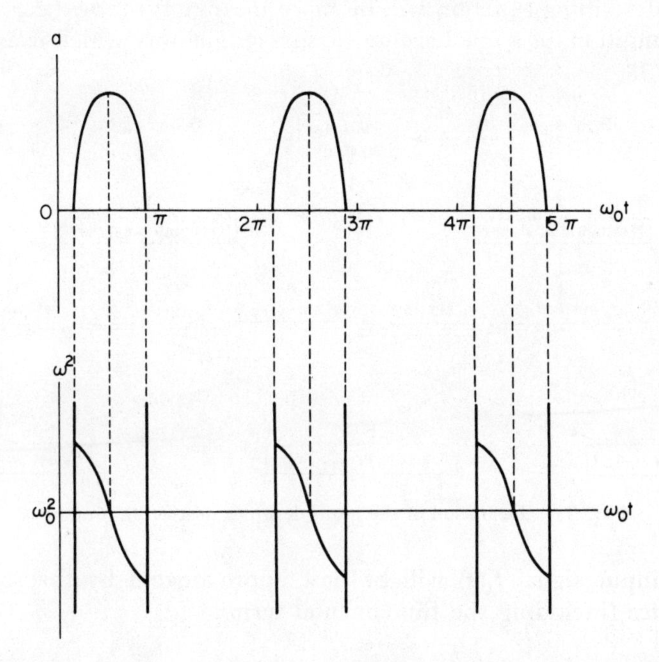

FIG. 37. Retarded action. Amplitude a and frequency ω of the quasi-sinusoidal oscillation as a function of $\omega_0\tau$: ω_0, angular frequency of the steady oscillation; τ, delay of the retarded action.

(ω_0 is the angular frequency of the steady oscillation), *provided that*

$$\nu^2 > \frac{2\delta\omega_0}{\sin \omega_0 \tau} \tag{81}$$

On the other hand, the frequency of the steady state, which occurs in (80), is given by (78). Variations of a and ω^2 are plotted in Fig. 37 as functions of $\omega_0 \tau$.

10. MULTIPLE-INPUT DESCRIBING FUNCTION†

The describing function method needs to be improved to explain other phenomena, such as, for example, subharmonic resonance. The reason the earlier formulation was inadequate for dealing with such situations lies in the fact that we have resigned ourselves to approximating periodical signals by sinusoidal functions, which leads to simple computations, but whose counterpart is a loss of information.

The describing function will be more meaningful if we take account, in its definition, of a few harmonics, in a simple way which is explained by Fig. 38.

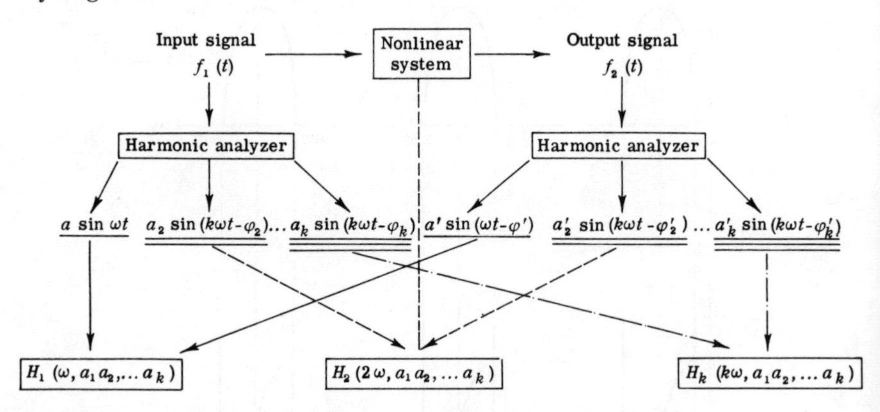

FIG. 38. Definition of the multiple-input describing function.

The input signal $f_1(t)$ will be now approximated by the sum of k harmonics (including the fundamental term):

$$f_1(t) = a \sin \omega t + a_2 \sin(2\omega t - \varphi_2) + \cdots + a_k \sin(k\omega t - \varphi_k)$$

† See [24, 25].

and the output signal $f_2(t)$ will be introduced into a harmonic analyzer which provides its first k harmonics:

$$f_2(t) = a' \sin(\omega t - \varphi) + a_2' \sin(2\omega t - \varphi_2') + \cdots + a_k' \sin(k\omega t - \varphi_k')$$

Then the first harmonic of the output signal will be compared to the first harmonic of the input signal, which gives a describing function

$$H_1(\omega, a, a_2, a_3, ..., a_k)$$

which depends on a and on the other amplitudes $a_2, a_3, ..., a_k$.

Other describing functions can be defined by comparing the second harmonic of the output with the second harmonic of the input, the third harmonic of the output with the third harmonic of the input, and so on, up to the kth harmonic. In this way one obtains

$$\begin{gathered} H_2(2\omega, a, a_2, a_3, ..., a_k) \\ \vdots \\ H_k(k\omega, a, a_2, a_3, ..., a_k) \end{gathered}$$

As a matter of fact we again meet the method which was introduced in Chapter V, Section 9, and which opened the way to the matrix representation of nonlinear systems. Accordingly, we shall not return to the general theory, but rather will lay stress on the applications.

However, first let us remark that when the nonlinear system is not a quadrupole, the method can also be applied by considering its differential equation of motion and the nonlinear transformation which is associated with it, as pointed out in Chapter 5.

The method leads to a set of equations; i.e., when no forcing function is acting on the system:

$$\begin{aligned} H_1(\omega, a, a_2, a_3, ..., a_k) &= 0 \\ H_2(2\omega, a, a_2, a_3, ..., a_k) &= 0 \\ &\vdots \\ H_k(kw, a, a_2, a_3, ..., a_k) &= 0 \end{aligned}$$

The meaning of $H_1, H_2, ..., H_k$ for this type of problem was discussed in Chapter 5.

On the other hand, when a sinusoidal forcing function is introduced, it appears in the right side of the above equation, which corresponds to the proper frequency, as will be shown more precisely by the following examples. In any case, the frequency and the steady-state amplitudes $a, a_2, ..., a_k$ can be obtained. In some problems their computation can be simplified by heuristic remarks.

10.1. Subharmonic Resonance of a Van der Pol Oscillator

Consider a Van der Pol oscillator, with a forcing function $F \sin 2t$, whose equation of motion is

$$\ddot{x} + \mu(\beta x^2 - \alpha)\dot{x} + x = F \sin 2t \tag{82}$$

μ is a small positive parameter and α and β are positive constants.

In this problem the nonlinear system is not a quadrupole; however, as pointed out earlier, we can consider $x(t)$ as an input function and define the corresponding output function $y(t)$ by the law of transformation

$$y(t) \equiv \ddot{x} + \mu(\beta x^2 - \alpha)\dot{x} + x \tag{83}$$

The first approximation deduced from (82) in the case $\mu = 0$ is

$$x = A \sin t + B \cos t - \frac{F}{3} \sin 2t$$

More generally, when $\mu \neq 0$, let us substitute in (83) an expression such as

$$x = x_1 + m \sin 2\omega t$$

where m is a parameter and $x_1 = A \sin \omega t + B \cos \omega t$.

By the linearization procedure, which takes account of the first harmonic of the transform only, we get

$$\ddot{x}_1 + \mu \left[\beta \left(\frac{A^2 + B^2}{4} + \frac{m^2}{2}\right) - \alpha\right] \dot{x}_1 + x_1$$

from which we deduce the describing function

$$H_1(\omega, A, B, m) \equiv 1 - \omega^2 + j\omega\mu \left[\beta \left(\frac{A^2 + B^2}{4} + \frac{m^2}{2}\right) - \alpha\right] \tag{84}$$

Since we have considered the two input frequencies ω and 2ω, this describing function is called a *dual-input* describing function, which is a particular case of the *multiple-input* describing function.

Now the steady-state conditions are determined by

$$1 - \omega^2 = 0$$

$$\beta \left(\frac{A^2 + B^2}{4} + \frac{m^2}{2}\right) = \alpha \tag{85}$$

However, we need an additional condition to determine parameter m.

We shall simplify the derivation by taking the value which is provided by the first approximation, say $m = -F/3$. Then we get the steady amplitude of the subharmonic response

$$(A^2 + B^2)^{1/2} = 2\left(\frac{\alpha}{\beta} - \frac{m^2}{2}\right)^{1/2} = 2\left(\frac{\alpha}{\beta} - \frac{F^2}{18}\right)^{1/2}$$

with the proviso that $\alpha/\beta - F^2/18 > 0$, say $F < \sqrt{18\alpha/\beta}$.

Note that the above heuristic argument amounts to considering the describing function *relative to the second harmonic*, say

$$H_2 \equiv 1 - (2\omega)^2 \tag{86}$$

when μ is neglected.

Then, since we are interested in the subharmonic resonance of order $\frac{1}{2}$, the forcing function needs to be introduced into the equation relative to the second harmonic, *not* into the equation relative to the first one. This explains more precisely why the equation which gives the first harmonic of the steady oscillation, namely

$$H_1(\omega, A, B, m) = 0$$

[from which equations (85) are deduced] does not contain the forcing function explicitly. On the other hand, the forcing function will occur explicitly in the equation relative to the second harmonic.

The form of the latter equation, when the describing function relative to the second harmonic is approximated by (86), is readily obtained. It is

$$H_2 \equiv 1 - (2\omega)^2 = \frac{F}{m}$$

say, by taking account of the first condition (85), $\omega = 1$:

$$m = -\frac{F}{3}$$

A later example will illustrate the use of the dual-input describing function.

10.2. Superharmonic Resonance of a Duffing Oscillator†

Consider now a nonlinear oscillator excited by the forcing function $F \cos \omega t$, whose law of motion is determined by Duffing's equation

$$\ddot{x} + \mu\dot{x} + x^3 = F \cos \omega t \tag{87}$$

† See [21].

Then let us consider generation of the harmonic of order 3, which is the so-called phenomenon of *superharmonic resonance*.

Here we shall try to fit a solution of the form

$$x = a\cos(\omega t - \psi_1) + a_3\cos(3\omega t - \psi_3) \tag{88}$$

To do this we shall substitute (88) in the left side of (87).

The expansion of x^3 into a sum of sinusoidal terms has been carried out in Chapter V, Section 9.2, to which we shall refer. According to this earlier computation, we have

$$\begin{aligned} y &= \ddot{x} + \mu\dot{x} + x^3 \\ &= -a\omega^2\cos(\omega t - \psi_1) - 9a_3\omega^2\cos(3\omega t - \psi_3) \\ &\quad - \mu a\omega\sin(\omega t - \psi_1) - 3\mu a_3\omega\sin(3\omega t - \psi_3) \\ &\quad + \frac{3a}{4}(a^2 + 2a_3^2)\cos(\omega t - \psi_1) + \frac{3a^2a_3}{4}\cos(\omega t - \psi_3 + 2\psi_1) \\ &\quad + \frac{a^3}{4}\cos(3\omega t - 3\psi_1) + \frac{3a_3}{4}(a_3^2 + 2a^2)\cos(3\omega t - \psi_3) + \cdots \end{aligned}$$

Then following the discussion of Chapter V, Section 9, we shall replace $x(t)$ by

$$\hat{x}(t) = \hat{x}_1(t) + \hat{x}_3(t)$$

with

$$\hat{x}_1(t) = a\exp(-j\psi_1)\exp(j\omega t) \qquad \hat{x}_3(t) = a_3\exp(-j\psi_3)\exp(3j\omega t)$$

and $y(t)$ by

$$\begin{aligned} \hat{y}(t) &= \ddot{\hat{x}} + \mu\dot{\hat{x}} + \left\{\frac{3a}{4}(a^2 + 2a_3^2)\exp(-j\psi_1) \right. \\ &\quad \left. + \frac{3a^2a_3}{4}\exp[-j(\psi_3 - 2\psi_1)]\right\}\exp(j\omega t) \\ &\quad + \left[\frac{a^3}{4}\exp(-3j\psi_1) + \frac{3a_3}{4}(a_3^2 + 2a^2)\exp(-j\psi_3)\right]\exp(3j\omega t) + \cdots \end{aligned}$$

say

$$\begin{aligned} \hat{y}(t) &= \ddot{\hat{x}}_1 + \mu\dot{\hat{x}}_1 + \left\{\tfrac{3}{4}(a^2 + 2a_3^2) + \frac{3aa_3}{4}\exp[-j(\psi_3 - 3\psi_1)]\right\}\hat{x}_1 \\ &\quad + \ddot{\hat{x}}_3 + \mu\dot{\hat{x}}_3 + \left\{\tfrac{3}{4}(a_3^2 + 2a^2) + \frac{a^3}{4a_3}\exp[-j(3\psi_1 - \psi_3)]\right\}\hat{x}_3 \\ &\triangleq \hat{y}_1(t) + \hat{y}_3(t) \end{aligned} \tag{89}$$

where the indices 1 and 3 refer to frequencies ω and 3ω, respectively.

Then by splitting (89) into

$$\hat{y}_1 = \ddot{\hat{x}}_1 + \mu\dot{\hat{x}}_1 + \left\{\tfrac{3}{4}(a^2 + 2a_3^2) + \frac{3aa_3}{4}\exp[-j(\psi_3 - 3\psi_1)]\right\}\hat{x}_1$$

$$\hat{y}_3 = \ddot{\hat{x}}_3 + \mu\dot{\hat{x}}_3 + \left\{\tfrac{3}{4}(a_3^2 + 2a^2) + \frac{a^3}{4a_3}\exp[-j(3\psi_1 - \psi_3)]\right\}\hat{x}_3$$

we return either to the matrix representation of Chapter V, Section 9 or to the dual-input describing function, whose two components are

$$H_1(\omega, a, a_3) \equiv -\omega^2 + \left\{\tfrac{3}{4}(a^2 + 2a_3^2) + \frac{3aa_3}{4}\exp[-j(\psi_3 - 3\psi_1)]\right\} + j\omega\mu$$

$$H_3(3\omega, a, a_3) \equiv -(3\omega)^2 + \left\{\tfrac{3}{4}(a_3^2 + 2a^2) + \frac{a^3}{4a_3}\exp[-j(3\psi_1 - \psi_3)]\right\} + 3j\omega\mu$$

Finally, when replacing the forcing function in (87) by

$$Fe^{j\omega t}$$

we get the two equations

$$\begin{aligned} H_1(\omega, a, a_3) &= \frac{F}{a\exp(-j\psi_1)} \\ H_3(3\omega, a, a_3) &= 0 \end{aligned} \tag{90}$$

If we assume that ω is a dominant frequency, i.e., that

$$a \gg a_3 \tag{91}$$

then by neglecting a_3^2, (91) reduces to

$$\begin{aligned} -\omega^2 a + j\omega\mu a + \frac{3}{4}a^3 + \frac{3}{4}a^2 a_3\exp[-j(\psi_3 - 3\psi_1)] &= F\exp(i\psi_1) \\ -9\omega^2 a_3 + 3j\omega\mu a_3 + \frac{a^3}{4}\exp[-j(3\psi_1 - \psi_3)] + \frac{3}{2}a^2 a_3 &= 0 \end{aligned} \tag{92}$$

From these equations, a and a_3 can be readily computed and plotted as functions of ω, as shown in Figs. 39 and 40. These curves were obtained by Clauser by using the matrix representation. A more complete discussion, in which the simplifying assumption (91) is not taken into account, is given in [21].

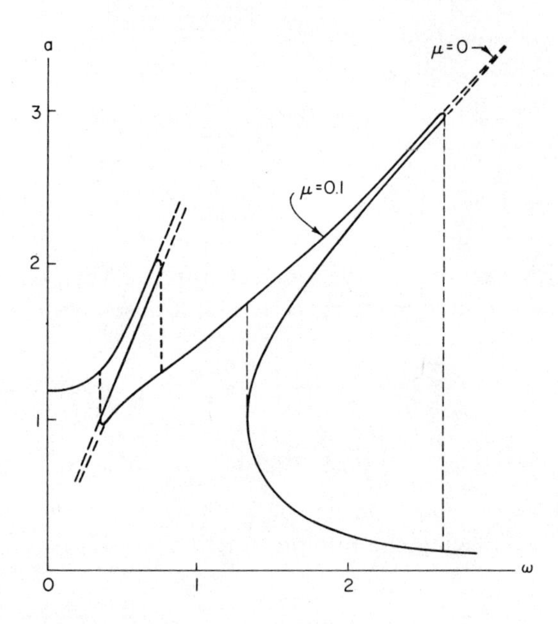

FIG. 39. Superharmonic resonance of a Duffing's oscillator; amplitude of the first harmonic a vs. frequency ω (from F. H. Clauser).

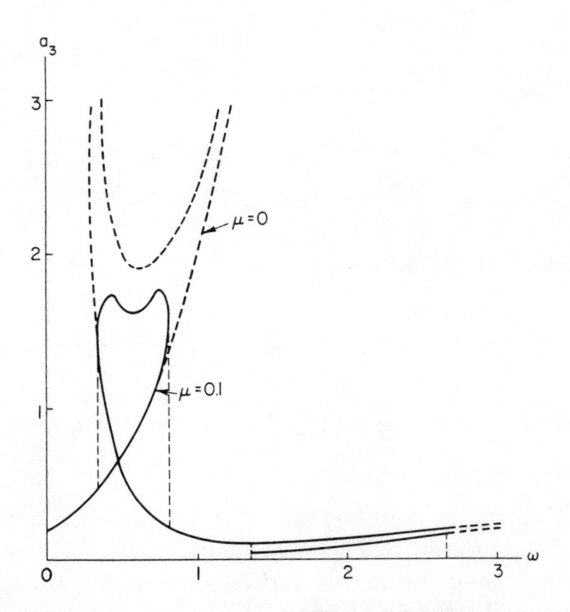

FIG. 40. Superharmonic resonance of a Duffing's oscillator; amplitude of the third harmonic a_3 vs. frequency ω (from F. H. Clauser).

REFERENCES

1. H. Nyquist, Regeneration Theory. *Bell System Tech. J.* **11**, 176 (1932).
2. A. Mikaïlov, Method garmoniceskogo analiza v teorii regulirovanija. *Avtomatika i Telemekhanika* **3**, 27 (1938).
3. L. Goldfarb, "Frequency Response." Ed. Oldenburger, 1947; English translation O nekotorykh nelinejnostjakh v sistemakh avtomaticeskogo regulirovanija. *Avtomatika i Telemekhanika* **8**, 349–383 (1947); K. voprosu o teorii vibracionnykh reguljatorov. *Avtomatika i Telemekhanika* **9**, 413–431 (1947).
4. K. F. Theodorchik, "Auto-Oscillatory Systems." Moscow, 1948.
5. J. R. Dutilh, Théorie des servo-mécanismes non linéaires. *Onde Elec.* **30**, 438–445 (1950).
6. R. J. Kochenburger, A Frequency Response Method for Analyzing and Synthesizing Contactor Servomechanisms. *Trans. AIEE* **69**, 270–283 (1950).
7. A. Blaquière, Extension de la théorie de Nyquist au cas de catactéristiques non linéaires. *Compt. Rend.* **233**, 345 (1951).
8. M. Dzung, *Symp. Cranfield Conf. Automatic Control*, 1951.
9. J. Loeb, Un Criterium général de stabilité des servomécanismes sièges de phénomènes héréditaires. *Compt. Rend.* **233**, 344 (1951).
10. A. Blaquière, Adaptation générale de la méthode du diagramme de Nyquist dans le domaine non linéaire. *J. Phys. Radium* **13**, 527–540, 636–644 (1952).
11. M. A. Ayzerman, O postroenii rezonancnikh grafikov dlja sistem c nelinejnoj obratnoj svjazjou. *Inst. Mekhan. Akad. Nauk SSR* (1952); "Teorija avtomaticeskogo regulirovanija dvigatelej: uravnenija dvizenija i ustojcivost'." Gostekhizdat, Moscow, 1952.
12. E. C. Johnson, Sinusoidal Analysis of Feedback Control Systems Containing Nonlinear Elements. *AIEE Trans.* **71**, 169–181 (1952).
13. R. J. Kochenburger, Limiting in Feedback Control Systems. *AIEE Trans.* **72**, 180–194 (1953).
14. H. Chestnut, Approximate Frequency-Response Methods for Representing Saturation and Dead Band. *Trans. ASME* **76**, 1345–1364 (1954).
15. J. Loeb, Recent Advances in Nonlinear Servo Theory. *Trans. ASME* **76**, 1281–1290 (1954).
16. C. H. Thomas, Stability Characteristics of Closed-Loop Systems with Dead Band. *Trans. ASME* **76**, 1365–1382 (1954).
17. J. G. Truxal, "Automatic Feedback Control System Synthesis." McGraw-Hill, New York, 1955.
18. L. Goldfarb, Metod issledovanija nelinejnykh sistem regulirovanija, osnovannyj na principe garmoniceskogo balansa. *Proc. Second Conf. Automatic Control, Moscow*, 177–192, 1955.
19. Y. Tsypkin, "The Theory of Relay Control Systems." Moscow, 1955.
20. M. A. Ayzerman, Sur les méthodes approchées de définition des régimes périodiques dans les systèmes possédant la contre-réaction non linéaire. *Proc. First Intern. Symp. Cybernetics, Namur*, 1956.
21. F. H. Clauser, The Behaviour of Nonlinear Systems. *J. Aeron. Sci.* **23**, 409–432 (1956).
22. K. Klotter, How to Obtain Describing Functions for Nonlinear Feedback Systems. *Trans. ASME* (April 1957).
23. A. Blaquière, Extension de la méthode de Nyquist aux systèmes bouclés non linéaires. (Réunions d'Études et de mises au point tenues sous la présidence de Louis de Broglie 1959.) *Rev. d'Optique Théorique et Instrumentale* (1962).

24. J. C. West, "Analytical Techniques for Non-Linear Control Systems." English Universities Press, London, 1960.
25. A. Blaquière, La méthode du diagramme de Nyquist généralisé dans le domaine non linéaire. *Intern. Symp. Nonlinear Oscillations, Kiev,* 1961.
26. N. Minorsky, "Nonlinear Oscillations." Van Nostrand, Princeton, N.J., 1962.

CHAPTER VII

Nonlinear Equations with Periodic Coefficients

INTRODUCTION

In the analysis of betatron oscillations in an alternating-gradient synchrotron, we were faced in Chapter 4 with such equations as

$$
\begin{aligned}
x_1'' + [1 - n(\theta)]x_1 &= \frac{\alpha(\theta)}{2}(x_1^2 - x_2^2) \\
x_2'' + n(\theta)x_2 &= -\alpha(\theta)x_1x_2
\end{aligned} \tag{1}
$$

where $n(\theta)$ and $\alpha(\theta)$ are periodic functions of the azimuthal coordinate θ. Coupled variables x_1 and x_2 are the radial and vertical deviations from the reference orbit, respectively, and the double-prime represents $d^2/d\theta^2$ (see Chapter IV, Section 3 ff.).

If vertical deviations are disregarded, (1) reduces to

$$
x_1'' + [1 - n(\theta)]x_1 - \frac{\alpha(\theta)}{2}x_1^2 = 0 \tag{2}
$$

Such equations as (2) are known as *Hill equations*, and play a central role in many practical applications.

Simple examples can be considered. For instance, a pendulum whose length is subject to periodical change, or a radioelectric resonant circuit (L, r, C) whose capacitance is governed by the law

$$
C = C_0[1 + \varphi(t)]
$$

where $\varphi(t)$ is a periodic function of time, obey Hill equations.

Such examples are met in condensator microphones, parametric amplifiers, etc.

When $n(\theta)$ assumes the form

$$
n(\theta) \equiv \gamma \cos 2\theta \qquad \text{and} \qquad \alpha(\theta) \equiv 0
$$

(2) is a Mathieu equation. When considering the Mathieu equation we shall refer to the standard form:

$$x'' + (\mu - \gamma \cos 2\theta)x = 0 \tag{3}$$

where μ and γ are constant parameters, with respect to which values one can discuss the stability of the solutions of (3).

Equation (3) is a linear equation, whose properties as inferred from Floquet's theory [1] are probably sufficiently well known to the reader to need no further comment, except for a brief summary of its stability characteristics as given by Fig. 1. This illustration is plotted with respect

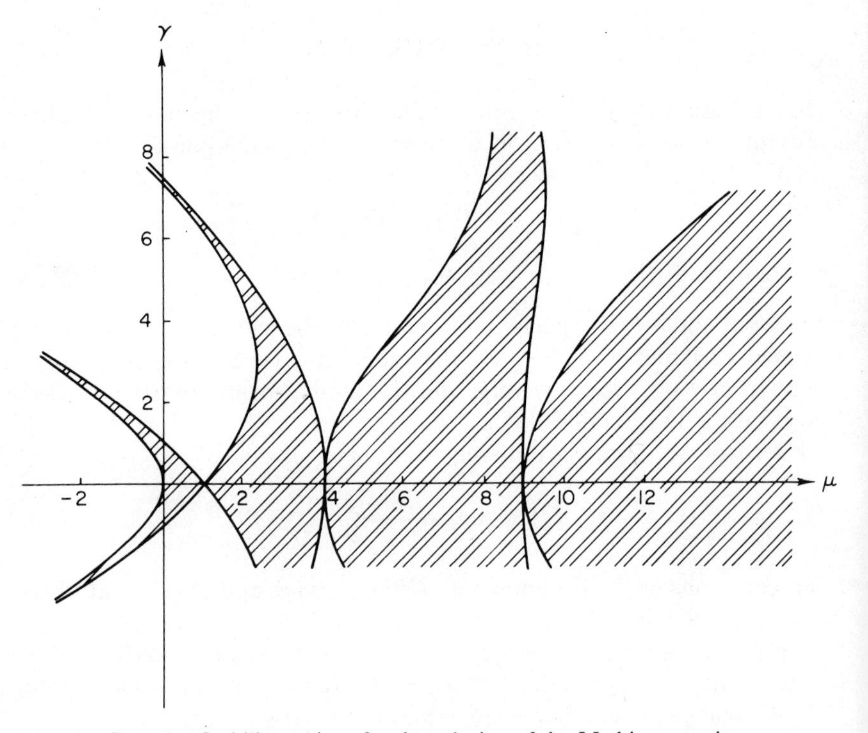

FIG. 1. Stability regions for the solution of the Mathieu equation.

to the coordinates μ and γ; the shaded areas correspond to the stability regions; i.e., to each couple of values μ and γ, say to each point in the shaded areas, there corresponds a solution of (3) which is bounded whatever the value of θ. It is an almost-periodic solution, an example of which is shown in Fig. 2a; other examples [16] are shown in Fig. 2d and e.

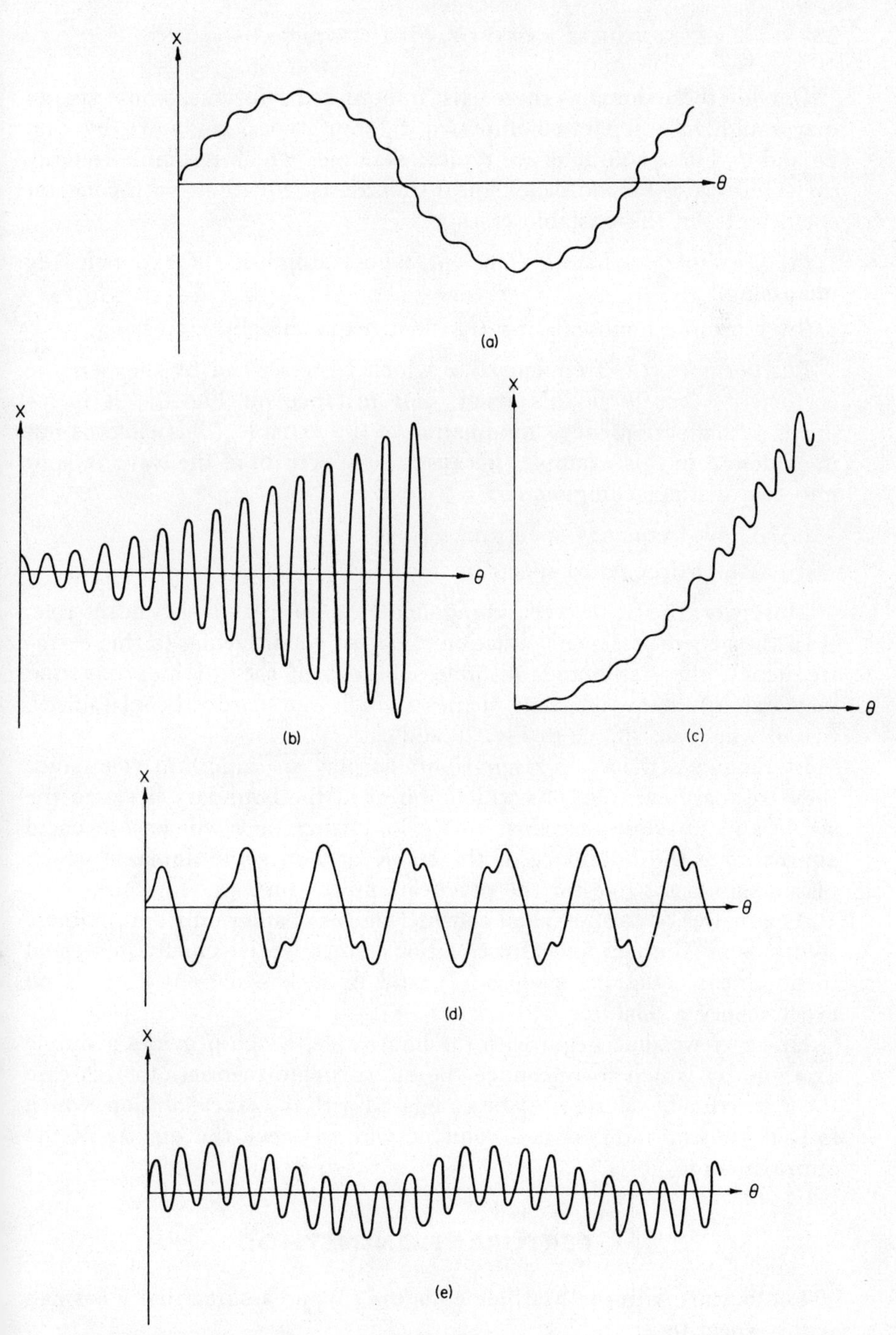

FIG. 2. Typical solutions of the Mathieu equation: (a, d, e) stable solutions; (b, c) unstable solutions.

Outside these domains there exist unbounded solutions, whose shapes may roughly be separated into two different types, as shown by Fig. 2b and c. These solutions are typical examples which resemble roughly the solutions of second-order linear differential equations with constant coefficients, in the unstable cases:

(a) Unstable oscillating transient, whose amplitude is exponentially increasing.

(b) Unstable nonoscillating transient, exponentially increasing.

Furthermore, the frequency $1/\pi$ which is prescribed by the periodic coefficient $\gamma \cos 2\theta$ reveals itself. For instance on Fig. 2a, it introduces a "high-frequency" modulation of the variable. This effect is put in evidence, in this example, because the spectrum of the wave is split into two distinct components:

(a) A low-frequency spectrum.

(b) A high-frequency spectrum.

Moreover, the in-between components do not play a significant role.

In the general case, the situation is not so good. Owing to this extra-frequency, the stable and unstable solutions of the Mathieu equation may exhibit many different shapes (which can hardly be classified), two of which are shown in Fig. 2d and e.

In Floquet's theory, *periodic* solutions play an important role, since they are marginal solutions which appear at the boundary between the stable and unstable domains. In the following they will enable us to approximate the boundary of the stable area in some domains which play a significant role for the practical applications.

As a matter of fact, we shall consider in this chapter only approximate solutions of Mathieu and Hill equations, since we are chiefly interested in nonlinear equations such as (2), and in such situations there is no exact theory available.

However we shall begin with the linear case, which provides a simple example by which to introduce the approximate theories. In this case the approximate solution can be compared with the exact solution, which is well known, and this is a valuable way to check the quality of the approximation.

1. PERTURBATION METHOD

Let us start with the Mathieu equation (3) and assume that γ is small with respect to μ:

$$\gamma \ll \mu$$

Next we shall change the notation, putting

$$2\theta = \tau \qquad \frac{\mu}{4} = Q^2 \qquad \frac{\gamma}{4} = \epsilon$$

Equation (3) is rewritten

$$\frac{d^2x}{d\tau^2} + (Q^2 - \epsilon \cos \tau)x = 0 \tag{4}$$

Following a computational procedure which Struble and Fletcher [25] developed for a solution of the artificial-satellite problem, we shall try to fit a solution in the form of a series expansion:

$$x(\tau) = a \cos(Q\tau - \varphi) + \epsilon u_1 + \epsilon^2 u_2 + \cdots + \epsilon^k u_k \tag{5}$$

$a \cos(Q\tau - \varphi)$ is referred to as the *principal part*, in which, in general, a and φ are functions of τ.

When $\epsilon = 0$, a and φ are constant, so that it may be reasonably assumed that when ϵ is small, a and φ are *slowly varying* functions of τ. The subsequent computations will justify this assumption.

$u_1, u_2, ..., u_k$ are functions of τ, which will be determined. However, in general, the series (5) does not converge for any $\epsilon \neq 0$, as $k \to \infty$. Here, given k, $u_1, u_2, ..., u_k$ will be computed following an iterative scheme, based on the assumption that ϵ is "sufficiently" small. Indeed, we only avoid by this elliptic foreword the problems raised by the convergence of (5). As a matter of fact, it is difficult to define quantitatively what is meant by "sufficiently" small, and this can better be considered as a heuristic assumption which will be justified by subsequent computation.

This method differs from the classical approach, in which two series expansions are considered:

$$x(\tau) = x_0(\tau) + \epsilon u_1(\tau) + \epsilon^2 u_2(\tau) + \cdots + \epsilon^k u_k(\tau)$$
$$\omega(\epsilon) = Q_0(\epsilon) + \epsilon q_1(\epsilon) + \epsilon^2 q_2(\epsilon) + \cdots + \epsilon^k q_k(\epsilon)$$

Then $x_0(\tau)$ is the generating solution and $\omega(\epsilon)$ the actual angular frequency.

Note that in the present method, φ being a function of τ, the instantaneous angular frequency of the principal part is

$$\omega = Q - \frac{d\varphi}{d\tau}$$

Then as the computational procedure gives $d\varphi/d\tau$ as a function of ϵ, ω is again a function of ϵ.

1.1. First-Order Approximation

Let us substitute (5) in (4); we get

$$\begin{aligned}
&\left[2Q a \frac{d\varphi}{d\tau} + \frac{d^2 a}{d\tau^2} - a\left(\frac{d\varphi}{d\tau}\right)^2\right] \cos(Q\tau - \varphi) \\
&+ \left(-2Q \frac{da}{d\tau} + 2\frac{da}{d\tau}\frac{d\varphi}{d\tau} + a\frac{d^2\varphi}{d\tau^2}\right) \sin(Q\tau - \varphi) \\
&+ \epsilon\left(\frac{d^2 u_1}{d\tau^2} + Q^2 u_1\right) + \epsilon^2\left(\frac{d^2 u_2}{d\tau^2} + Q^2 u_2\right) + \cdots + \epsilon^k\left(\frac{d^2 u_k}{d\tau^2} + Q^2 u_k\right) \\
&= \frac{\epsilon a}{2}\cos[(Q+1)\tau - \varphi] + \frac{\epsilon a}{2}\cos[(Q-1)\tau - \varphi] \\
&+ \epsilon^2 u_1 \cos\tau + \epsilon^3 u_2 \cos\tau + \cdots + \epsilon^{k+1} u_k \cos\tau
\end{aligned} \tag{6}$$

The first-order approximation will be derived by taking into account only zero-order and first-order terms in ϵ. Then we are faced with the problem of separating (6) into a set of differential equations which will enable us to compute $a(\tau)$, $\varphi(\tau)$, and $u_1(\tau)$.

Two cases need to be considered:

Case 1. *Solution far from the resonance frequency* $\frac{1}{2}$: $Q \neq \frac{1}{2}$

When Q is "sufficiently" far from $\frac{1}{2}$, we can replace (6), at the first-order approximation, by the following set:

$$\begin{aligned}
2Q a \frac{d\varphi}{d\tau} + \frac{d^2 a}{d\tau^2} - a\left(\frac{d\varphi}{d\tau}\right)^2 &= 0 \\
-2Q\frac{da}{d\tau} + 2\frac{da}{d\tau}\frac{d\varphi}{d\tau} + a\frac{d^2\varphi}{d\tau^2} &= 0
\end{aligned} \tag{7}$$

$$\epsilon\left(\frac{d^2 u_1}{d\tau^2} + Q^2 u_1\right) = \frac{\epsilon a}{2}\cos[(Q+1)\tau - \varphi] + \frac{\epsilon a}{2}\cos[(Q-1)\tau - \varphi] \tag{8}$$

Equations (7) and (8) are called *variational* and *perturbational* equations, respectively. Equations (7) are verified by

$$a \equiv a_0 \qquad \varphi \equiv \varphi_0$$

where a_0 and φ_0 are constant, and (8) gives immediately

$$u_1 = -\frac{a}{2(2Q+1)}\cos[(Q+1)\tau - \varphi] + \frac{a}{2(2Q-1)}\cos[(Q-1)\tau - \varphi] \tag{9}$$

Finally, the first-order approximation is

$$x^*(\tau) = a\cos(Q\tau - \varphi) - \frac{\epsilon a}{2(2Q+1)}\cos[(Q+1)\tau - \varphi] + \frac{\epsilon a}{2(2Q-1)}\cos[(Q-1)\tau - \varphi] \tag{10}$$

Case 2. Solution near the resonance frequency $Q = \frac{1}{2}$

When Q is near $\frac{1}{2}$, the perturbational equation (8) exhibits a secular term which needs to be avoided—the term $(\epsilon a/2)\cos[(Q-1)\tau - \varphi]$ in the right side of (8) produces a resonance and $u_1(\tau)$ becomes very large. It tends to infinity when $Q \to \frac{1}{2}$.

Accordingly, the above decomposition would lead to a contradiction with the assumptions. To overcome this difficulty, since when $Q = \frac{1}{2}$, the term $(\epsilon a/2)\cos[(Q-1)\tau - \varphi]$ in the right side of (6) has the frequency $\frac{1}{2}$, introduce this incriminated term in the variational equations (instead of the perturbational equation).

Since $\cos[(Q-1)\tau - \varphi]$ can be rewritten

$$\cos[(Q-1)\tau - \varphi] = \cos[(1-2Q)\tau + 2\varphi]\cos(Q\tau - \varphi) - \sin[(1-2Q)\tau + 2\varphi]\sin(Q\tau - \varphi)$$

where both $\cos[(1-2Q)\tau + 2\varphi]$ and $\sin[(1-2Q)\tau + 2\varphi]$ are slowly varying amplitudes, we get

$$\begin{aligned} 2Qa\frac{d\varphi}{d\tau} + \frac{d^2a}{d\tau^2} - a\left(\frac{d\varphi}{d\tau}\right)^2 &= \frac{\epsilon a}{2}\cos[(1-2Q)\tau + 2\varphi] \\ -2Q\frac{da}{d\tau} + 2\frac{da}{d\tau}\frac{d\varphi}{d\tau} + a\frac{d^2\varphi}{d\tau^2} &= -\frac{\epsilon a}{2}\sin[(1-2Q)\tau + 2\varphi] \end{aligned} \tag{11}$$

$$\frac{d^2u_1}{d\tau^2} + Q^2u_1 = \frac{a}{2}\cos[(Q+1)\tau - \varphi] \tag{12}$$

Furthermore, since a and φ are slowly varying functions of τ, we can neglect $d^2a/d\tau^2$, $(d\varphi/d\tau)^2$, and $(da/d\tau)(d\varphi/d\tau)$ in (11), which is another way to note the fact that we have been concerned, up to now, with the first-order approximation relative to ϵ.

System (11) is therefore substantially simplified. It becomes

$$\frac{d\varphi}{d\tau} = \frac{\epsilon}{4Q}\cos[(1-2Q)\tau + 2\varphi] \tag{13a}$$

$$\frac{da}{d\tau} = \frac{\epsilon a}{4Q}\sin[(1-2Q)\tau + 2\varphi] \tag{13b}$$

These equations enable us to determine the slowly varying functions $\varphi(\tau)$ and $a(\tau)$. Then from (12) we get

$$u_1 = -\frac{a}{2(2Q+1)} \cos[(Q+1)\tau - \varphi] \tag{14}$$

Finally, the first-order approximation is

$$x_r^*(\tau) = a\cos(Q\tau - \varphi) - \frac{\epsilon a}{2(2Q+1)} \cos[(Q+1)\tau - \varphi] \tag{15}$$

where a and φ are given by (13).

Now let us put $\psi = (1-2Q)\tau + 2\varphi$ and rewrite equations (13),

$$\frac{d\psi}{d\tau} = 1 - 2Q + \frac{\epsilon}{2Q}\cos\psi \tag{16a}$$

$$\frac{da}{d\tau} = \frac{\epsilon a}{4Q}\sin\psi \tag{16b}$$

from which we deduce

$$\frac{da}{a} = \frac{\epsilon}{4Q} \frac{\sin\psi \, d\psi}{1 - 2Q + (\epsilon/2Q)\cos\psi} \tag{17}$$

From (16) we immediately obtain

$$a = \frac{A_0}{|\, 1 - 2Q + (\epsilon/2Q)\cos\psi \,|^{1/2}}$$

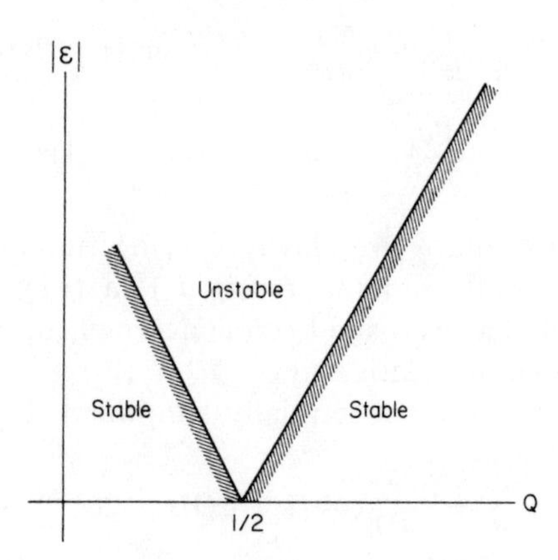

FIG. 3. Boundary of the stability domain, in the neighborhood of $Q = \frac{1}{2}$, first-order approximation.

Then we see that the solution is unstable when

$$|1 - 2Q| \leqslant \left|\frac{\epsilon}{2Q}\right|$$

say approximately (since $Q \sim \frac{1}{2}$) when

$$|1 - 2Q| \leqslant |\epsilon| \tag{18}$$

Finally, the boundary of the stability domain, *at the first approximation*, is defined by

$$|\epsilon| = |1 - 2Q| \tag{19}$$

It is shown in fig. 3.

1.2. Second-Order Approximation

Since the higher-order approximations lead to intricate (though practicable) computations, we shall only outline the method when second-order terms are taken into account.

Different cases need to be considered:

Case 1. *Solution near the resonance frequency* $Q = \frac{1}{2}$

First of all, in the neighborhood of the resonance frequency $Q = \frac{1}{2}$, we shall make use of (14), which we shall substitute in (6). Of course in (6) we shall consider only terms of zero order, first order, and second order in ϵ.

We shall again replace $\cos[(Q - 1)\tau - \varphi]$ by

$$\cos[(1 - 2Q)\tau + 2\varphi] \cos(Q\tau - \varphi) - \sin[(1 - 2Q)\tau + 2\varphi] \sin(Q\tau - \varphi)$$

Furthermore, let us note that in the right side of (6) the term $\epsilon^2 u_1 \cos \tau$ can be rewritten

$$\epsilon^2 u_1 \cos \tau = -\frac{\epsilon^2 a}{2(2Q + 1)} \cos[(Q + 1)\tau - \varphi] \cos \tau$$

$$= -\frac{\epsilon^2 a}{4(2Q + 1)} \cos[(Q + 2)\tau - \varphi] - \frac{\epsilon^2 a}{4(2Q + 1)} \cos(Q\tau - \varphi)$$

Accordingly, the term $-[\epsilon^2 a/4(2Q + 1)] \cos(Q\tau - \varphi)$ needs to be introduced in the variational equations, which become

$$2Qa \frac{d\varphi}{d\tau} + \frac{d^2 a}{d\tau^2} - a \left(\frac{d\varphi}{d\tau}\right)^2 = -\frac{\epsilon^2 a}{4(2Q + 1)} + \frac{\epsilon a}{2} \cos[(1 - 2Q)\tau + 2\varphi) \tag{20a}$$

$$-2Q \frac{da}{d\tau} + 2 \frac{da}{d\tau} \frac{d\varphi}{d\tau} + a \frac{d^2\varphi}{d\tau^2} = -\frac{\epsilon a}{2} \sin[(1 - 2Q)\tau + 2\varphi] \tag{20b}$$

On the other hand, the perturbational equation will give the expression of $u_2(\tau)$.

Here we shall limit ourselves to studying the variational equations. Again let $\psi = (1 - 2Q)\tau + 2\varphi$. From (20b) we get

$$\frac{da}{d\tau} = \frac{\epsilon a}{4Q} \sin \psi + \lambda_1 \epsilon^2 \tag{21}$$

where λ_1 is a slowly varying function of τ. Likewise, let

$$\frac{d\varphi}{d\tau} = \frac{\epsilon}{4Q} \cos \psi + \lambda_2 \epsilon^2 \tag{22}$$

and substitute (21) and (22) in (20a). At the second-order approximation we obtain

$$2Qa\lambda_2 + \frac{a}{(4Q)^2} \sin^2 \psi + \frac{2a}{(4Q)^2} \cos^2 \psi - \frac{a}{(4Q)^2} \cos^2 \psi = -\frac{a}{4(2Q+1)}$$

($d\lambda_1/d\tau$ has been neglected). From this,

$$\lambda_2 = -\frac{4Q^2 + 2Q + 1}{32Q^3(2Q+1)} \tag{23}$$

On the other hand, λ_1 is obtained from (20b). One gets

$$\lambda_1 = -\frac{1 - 2Q}{8Q^2\epsilon} a \sin \psi$$

Then (21) and (22) are rewritten

$$\frac{d\psi}{d\tau} = 1 - 2Q - \frac{4Q^2 + 2Q + 1}{16Q^3(2Q+1)} \epsilon^2 + \frac{\epsilon}{2Q} \cos \psi \tag{24a}$$

$$\frac{da}{d\tau} = \left(\frac{\epsilon}{4Q} - \frac{1 - 2Q}{8Q^2} \epsilon\right) a \sin \psi \tag{24b}$$

with $1 - 2Q \sim \epsilon$, say

$$\frac{1 - 2Q}{8Q^2} \epsilon \sim \frac{\epsilon^2}{2}$$

As for the stability condition, in the neighborhood of the resonance $Q = \frac{1}{2}$ we are only concerned with the first approximation of (24b).

By the same arguments as in section 1.1 we get the condition which determines the boundary of the stability domain:

$$\left| 1 - 2Q - \frac{4Q^2 + 2Q + 1}{16Q^3(2Q + 1)} \epsilon^2 \right| = \left| \frac{\epsilon}{2Q} \right|$$

say approximately

$$\left| 1 - 2Q - \frac{3\epsilon^2}{4} \right| = |\, \epsilon \,|$$

The new approximation of this boundary is shown in Fig. 4.

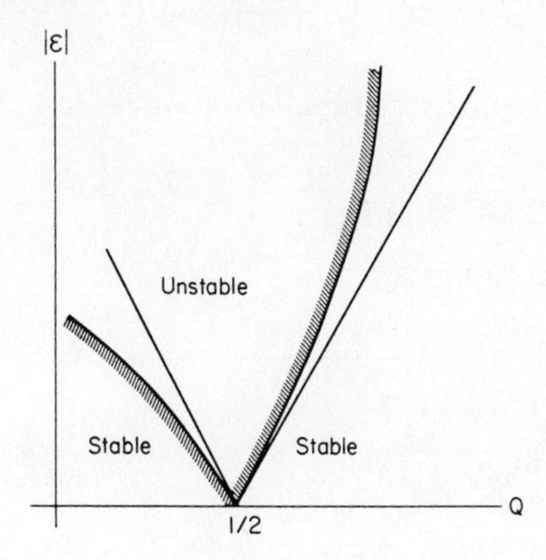

FIG. 4. Boundary of the stability domain in the neighborhood of $Q = \frac{1}{2}$, second-order approximation.

Case 2. *Solution far from the resonance frequencies* $\frac{1}{2}$ *and* 1: $Q \neq \frac{1}{2}$, $Q \neq 1$

Aside the critical region $Q = \frac{1}{2}$, and for $Q \neq 1$, the expression of u_1 is given by (9). Accordingly in the right side of (6) we get

$$\begin{aligned} u_1 \cos \tau = & - \frac{a}{2(2Q + 1)} \cos[(Q + 1)\tau - \varphi] \cos \tau \\ & + \frac{a}{2(2Q - 1)} \cos[(Q - 1)\tau - \varphi] \cos \tau \\ = & - \frac{a}{4(2Q + 1)} \{\cos[(Q + 2)\tau - \varphi] + \cos(Q\tau - \varphi)\} \\ & + \frac{a}{4(2Q - 1)} \{\cos[(Q - 2)\tau - \varphi] + \cos(Q\tau - \varphi)\} \end{aligned}$$

The variational equations become

$$2Qa\frac{d\varphi}{d\tau} + \frac{d^2a}{d\tau^2} - a\left(\frac{d\varphi}{d\tau}\right)^2 = \frac{\epsilon^2 a}{2(4Q^2-1)}$$
$$-2Q\frac{da}{d\tau} + 2\frac{da}{d\tau}\frac{d\varphi}{d\tau} + a\frac{d^2\varphi}{d\tau^2} = 0 \tag{25}$$

and the perturbational equation is

$$\frac{d^2u_2}{d\tau^2} + Q^2u_2 = -\frac{a}{4(2Q+1)}\cos[(Q+2)\tau - \varphi] + \frac{a}{4(2Q-1)}\cos[(Q-2)\tau-\varphi] \tag{26}$$

At the second-order approximation, (25) reduces to

$$\frac{d\varphi}{d\tau} = \frac{\epsilon^2}{4Q(4Q^2-1)} \qquad \frac{da}{d\tau} = 0 \tag{27}$$

and we get from (26) and (27),

$$\varphi = \varphi_0 + \frac{\epsilon^2\tau}{4Q(4Q^2-1)} \qquad a = a_0$$
$$u_2 = \frac{a}{16(2Q+1)(Q+1)}\cos[(Q+2)\tau-\varphi] + \frac{a}{16(2Q-1)(Q-1)}\cos[(Q-2)\tau-\varphi] \tag{28}$$

Case 3. *Solution near the resonance frequency* $Q = 1$

Another resonance appears when $Q = 1$, and again, in this new critical region, (28) is no longer valid.

In this situation we need to introduce the incriminated term in the variational equations instead of in the perturbational equation. This term, which was written in the right side of (26), is

$$\frac{a}{4(2Q-1)}\cos[(Q-2)\tau-\varphi]$$

Then we have

$$\frac{a}{4(2Q-1)}\cos[(Q-2)\tau-\varphi] = \frac{a}{4(2Q-1)}\{\cos[2(1-Q)\tau+2\varphi]\cos(Q\tau-\varphi) - \sin[2(1-Q)\tau+2\varphi]\sin(Q\tau-\varphi)\}$$

Accordingly, the variational equations become, after a few simplifications,

$$\frac{d\varphi}{d\tau} = \frac{\epsilon^2}{4Q(4Q^2 - 1)} + \frac{\epsilon^2}{8Q(2Q - 1)} \cos[2(1 - Q)\tau + 2\varphi]$$

$$\frac{da}{d\tau} = \frac{\epsilon^2}{8Q(2Q - 1)} \sin[2(1 - Q)\tau + 2\varphi]$$

or putting $\Phi = 2(1 - Q)\tau + 2\varphi$,

$$\begin{aligned} \frac{d\Phi}{d\tau} &= 2(1 - Q) + \frac{\epsilon^2}{2Q(4Q^2 - 1)} + \frac{\epsilon^2}{4Q(2Q - 1)} \cos \Phi \\ \frac{da}{d\tau} &= \frac{\epsilon^2}{8Q(2Q - 1)} \sin \Phi \end{aligned} \tag{29}$$

By arguments similar to those above, the boundary of the stability domain in the neighborhood of the resonance frequency $Q = 1$ is given by

$$\left| 2(1 - Q) + \frac{\epsilon^2}{2Q(4Q^2 - 1)} \right| = \frac{\epsilon^2}{4Q(2Q - 1)}$$

say approximately by

$$\left| 2(1 - Q) + \frac{\epsilon^2}{6} \right| = \frac{\epsilon^2}{4}$$

This boundary is shown in Fig. 5.

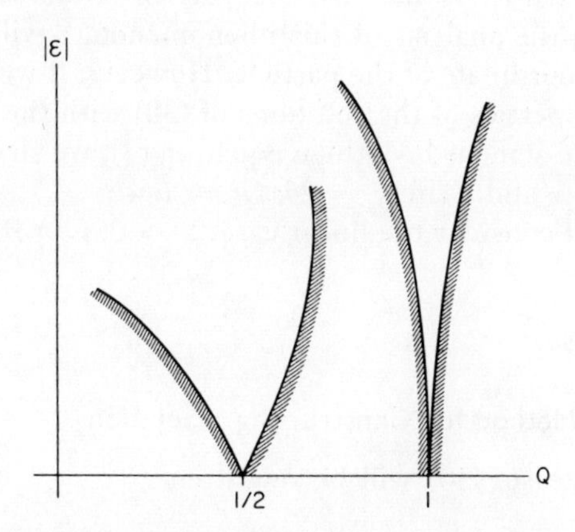

FIG. 5. Boundary of the stability domain in the neighborhood of $Q = \frac{1}{2}$ and $Q = 1$ second-order approximation.

The iteration procedure would enable us to go through with higher-order approximations, and to represent new pieces of the Mathieu stability diagram. At each step of the computation new resonances are brought out. However a more complete discussion would go beyond the scope of this book.

2. STEPWISE METHOD: APPLICATION TO THE ORBITAL STABILITY PROBLEM IN A SYNCHROTRON

Now we shall present a simple analytical approach developed by Schoch [7] to determine the orbital stability of the particles in an alternating-gradient synchrotron with nonlinear restoring forces, when these particles are subjected to periodically repeated kicks. A different method, which leads to the same results, was developed by Moser [14].

To normalize the notation with the one used in (4) of the preceding section, we shall write the equation which was analyzed by Schoch in the form

$$\frac{d^2x}{d\tau^2} + [Q^2 - \epsilon x^\nu \delta(\tau)]x - \beta x^3 = 0 \tag{30}$$

ϵ and β are constant small parameters; $\beta > 0$. $\delta(\tau)$ is a normalized periodic impulse whose period with respect to the variable τ is 2π.

In the synchrotron the kicks have periodicity 2π with respect to the azimuthal coordinate, since they are due to a local distortion of the magnetic field which is met by the particle after each full rotation. Accordingly, in the analysis of this phenomenon, τ will be though of as the azimuthal coordinate of the particle. However, if we wish to compare the stability properties of the solutions of (30) with the classical stability properties of the standard Mathieu equation (3), we shall have to change the meaning of τ and write $\tau = 2\theta$, $Q^2 = \mu/4$.

First we shall consider the linear case: $\beta = 0$, $\nu = 0$.

2.1. Linear Case

2.1.1. Stepwise Method for Constructing a Solution

In the linear case, (30) will be rewritten

$$\frac{d^2x}{d\tau^2} + Q^2 x = \epsilon\delta(\tau)x \tag{31}$$

and the successive kicks which occur at $\tau_0, \tau_1, \tau_2, \ldots, \tau_n, \tau_{n+1}, \ldots$ will be numbered 0, 1, 2, ..., n, $n+1$, ...

Now on the open interval $]\tau_n, \tau_{n+1}[$, the solution of (31) is

$$x_n = a_n \cos(Q\tau + \varphi_n) \tag{32}$$

where a_n and φ_n are integration constants which depend on the interval $]\tau_n, \tau_{n+1}[$ under consideration. As a matter of fact, these constants can be computed if the starting conditions, at $\tau = \tau_n$, are specified. This is not the case, but we can find a relation between a_n and a_{n+1}, φ_n and φ_{n+1}, which will enable us, using a stepwise procedure, to extend the solution from each interval to the next one, starting with initial conditions at $\tau = \tau_0$.

The arguments are based on the fact that, at any τ_i ($i = 0, 1, \ldots, n, n+1, \ldots$), the corresponding kick causes a sudden change in $dx/d\tau$ but no discontinuity in x. These changes are given by the elementary theory of the ballistic galvanometer. They are

$$\Delta x = 0 \qquad \Delta\left(\frac{dx}{d\tau}\right) = \int_{\Delta\tau} \epsilon\delta(\tau)x_i \, d\tau \tag{33}$$

where $\Delta\tau$ is a small neighborhood of τ_i, and $x_i = x(\tau_i)$.

Since $\delta(\tau)$ is normalized, say $\int_{\Delta\tau} \delta(\tau)\, d\tau = 1$, conditions (33) are rewritten at $\tau = \tau_{n+1}$:

$$\Delta[a_n \cos(Q\tau_{n+1} + \varphi_n)] = 0$$
$$\Delta[-Qa_n \sin(Q\tau_{n+1} + \varphi_n)] = \epsilon a_n \cos(Q\tau_{n+1} + \varphi_n)$$

from which follows

$$\Delta a_n \cos(Q\tau_{n+1} + \varphi_n) - a_n \Delta\varphi_n \sin(Q\tau_{n+1} + \varphi_n) = 0$$
$$-\Delta a_n \sin(Q\tau_{n+1} + \varphi_n) - a_n \Delta\varphi_n \cos(Q\tau_{n+1} + \varphi_n) = \frac{\epsilon a_n}{Q} \cos(Q\tau_{n+1} + \varphi_n)$$

and

$$\begin{aligned} \Delta a_n &= -\frac{\epsilon a_n}{2Q} \sin 2(Q\tau_{n+1} + \varphi_n) \\ \Delta\varphi_n &= -\frac{\epsilon}{2Q}[1 + \cos 2(Q\tau_{n+1} + \varphi_n)] \end{aligned} \tag{34}$$

Finally, the solution $x(\tau)$ can be constructed step by step by means of the relations

$$a_{n+1} = a_n + \Delta a_n \qquad \varphi_{n+1} = \varphi_n + \Delta\varphi_n$$

2.1.2. Invariant of the Motion

Let us replace the variable φ_n in (34) by the new one ψ_n, defined by

$$\psi_n = Q\tau_{n+1} + \varphi_n + k_n\pi \tag{35}$$

(k_n is an integer which depends on the index n). Then

$$\Delta\psi_n = Q(\tau_{n+2} - \tau_{n+1}) + \Delta\varphi_n - p\pi \qquad (p = k_n - k_{n+1})$$

and since the period of $\delta(\tau)$ is 2π with respect to τ,

$$\tau_{n+2} - \tau_{n+1} = 2\pi$$

and

$$\Delta\psi_n = 2\pi Q + \Delta\varphi_n - p\pi = 2\pi\left(Q - \frac{p}{2}\right) + \Delta\varphi_n \tag{36}$$

Substituting (35) and (36) in (34) one gets

$$\begin{aligned} \Delta a_n &= -\frac{\epsilon a_n}{2Q}\sin 2\psi_n \\ \Delta\psi_n &= 2\pi\left(Q - \frac{p}{2}\right) - \frac{\epsilon}{2Q}(1 + \cos 2\psi_n) \end{aligned} \tag{37}$$

from which follows

$$\frac{\Delta a_n}{\Delta\psi_n} = \frac{-(\epsilon a_n/2Q)\sin 2\psi_n}{2\pi[Q - (p/2)] - (\epsilon/2Q)(1 + \cos 2\psi_n)} \tag{38}$$

Now, when $Q - (p/2)$ is of order ϵ, one can replace (38), at the approximation of order ϵ, by the differential equation

$$\frac{1}{a_n}\frac{da_n}{d\psi_n} = \frac{-(\epsilon/2Q)\sin 2\psi_n}{2\pi[Q - (p/2)] - (\epsilon/2Q)(1 + \cos 2\psi_n)} \tag{39}$$

which can be immediately integrated to give

$$a_n = \frac{C}{|\, 2\pi[Q - (p/2)] - (\epsilon/2Q)(1 + \cos 2\psi_n)|^{1/2}} \tag{40}$$

C is a constant, independant of index n; i.e., the expression

$$a_n^2\left[2\pi\left(Q - \frac{p}{2}\right) - \frac{\epsilon}{2Q}(1 + \cos 2\psi_n)\right] \tag{41}$$

is an *invariant of the motion*. The value of C is determined by the initial conditions at $\tau = \tau_0$.

2.1.3. Resonances

The function $a_n(\psi_n)$ is unbounded if

$$\left| 2\pi \left(Q - \frac{p}{2}\right) - \frac{\epsilon}{2Q} \right| \leqslant \left| \frac{\epsilon}{2Q} \right|$$

That is, resonances appear for

$$Q = \frac{p}{2} \qquad \text{and} \qquad Q = \frac{p}{2} + \frac{\epsilon}{2\pi Q} \qquad (p, \text{ an integer})$$

$\epsilon/2\pi Q$ is the width of the stopband in the neighborhood of $p/2$. This stopband is narrow, since ϵ is small. On Fig. 6 we have plotted $a_n{}^2$ as a function of ψ_n, both at the resonance $Q = \frac{1}{2}$ and near the resonance $\frac{1}{2}$: $Q \sim \frac{1}{2}$.

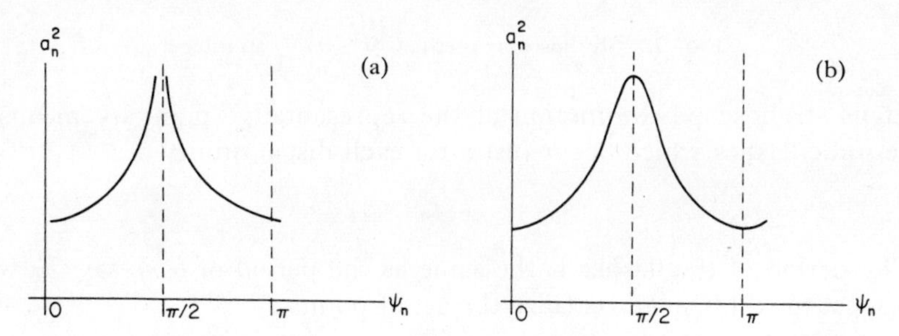

FIG. 6. Orbital stability in a synchrotron: (a) resonance, $Q = \frac{1}{2}$: (b) neighborhood of the resonance, $Q \sim \frac{1}{2}$.

2.1.4. Connection with the Stroboscopic Method†

In vectorial representation the piece of solution $x_n = a_n \cos(Q\tau + \varphi_n)$ is depicted as a circular path, which is described with the angular velocity Q (Fig. 7). The radius of the circle is a_n from $\tau = \tau_n$ to $\tau = \tau_{n+1}$; at $\tau = \tau_{n+1}$ the radius undergoes a sudden jump Δa_n, and the phase angle φ_n undergoes the shift $\Delta\varphi_n$. Between τ_{n+1} and τ_{n+2} another circular path is described, which again undergoes, at $\tau = \tau_{n+2}$, the new change in amplitude Δa_{n+1} and the new shift $\Delta\varphi_{n+1}$, and so on.

Let M_n, M_{n+1}, M_{n+2}, ... be the starting points of these successive circular paths. Obviously it may be convenient to disregard each rotation by arguments similar to those used in the stroboscopic method. That is,

† See [8].

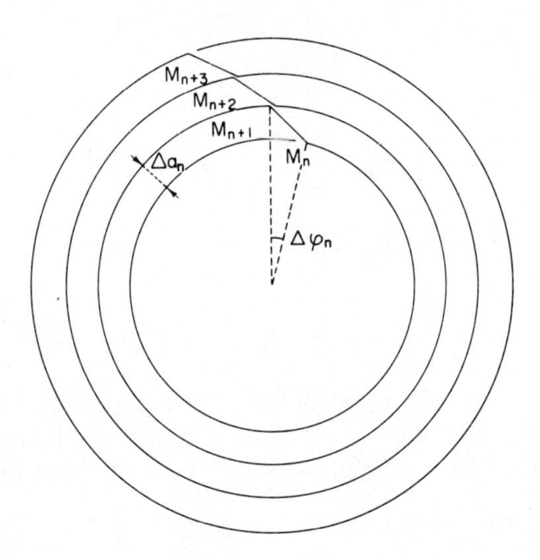

FIG. 7. Stroboscopic method, $Q \sim m$, m an integer.

let us stroboscope the motion of the representative point by means of periodic flashes which occur just after each discontinuity at

$$\tau_0, \tau_1, \ldots, \tau_n, \tau_{n+1}, \ldots$$

The period of the flashes is the same as the period of $\delta(\tau)$, say 2π with respect to τ. Thus we visualize the set of points

$$M_0 \quad M_1 \cdots M_n \quad M_{n+1} \quad M_{n+2} \cdots$$

When the frequency of the circular motion is close to a multiple of the frequency of the flashes, say

$$Q \sim m \qquad (m, \text{ an integer})$$

each point M_i ($i = 0, 1, \ldots, n, n+1, \ldots$) of the set is close to the next one, M_{i+1}. Furthermore, since the small jumps Δa_n and $\Delta \varphi_n$ are slowly varying functions of the index n,

$$M_0 \quad M_1 \cdots M_n \quad M_{n+1} \quad M_{n+2} \cdots$$

can conveniently be approached by a continuous path. This continuous path is the smooth approximation of the stroboscopic locus as defined by Minorsky.

Similarly, when

$$Q \sim \frac{2m+1}{2}$$

the set of points $M_0M_1 \cdots M_nM_{n+1}M_{n+2} \cdots$ is separated into two subsets which are almost symmetric with respect to 0; that is to say, there appear two stroboscopic loci,

$$\begin{matrix} M_0 \quad M_2 \cdots M_n \quad M_{n+2} \cdots \\ M_1 \quad M_3 \cdots M_{n+1} \cdots \end{matrix}$$

each of which can be replaced by a continuous path (Fig. 8).

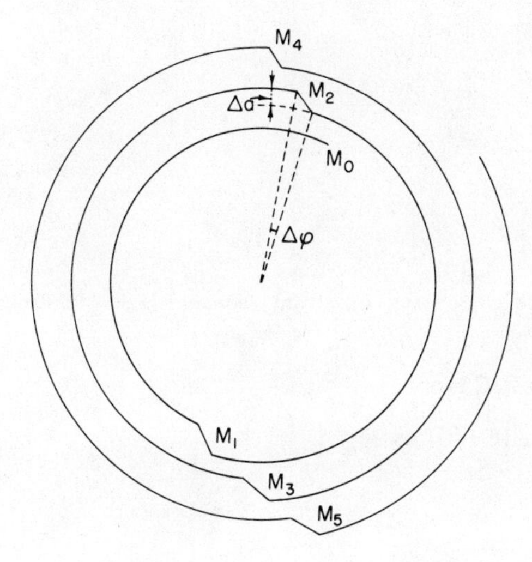

FIG. 8. Stroboscopic method, $Q \sim [(2m + 1)/2]$, m an integer.

The equation of the smoothed stroboscopic locus in polar coordinates ρ_n and Φ_n,

$$\rho_n = a_n^2 \qquad \Phi_n = Q\tau_{n+1} + \varphi_n$$

can be readily deduced from (34). The derivation exhibits many similarities to the one in Section 2.1.2, and one gets

$$\rho_n = \frac{K}{|\,2\pi[Q - (p/2)] - (\epsilon/2Q)(1 + \cos 2\Phi_n)|}$$

where the index n can be dropped. $p = 2m$ or $2m + 1$. When p is even ($p = 2m$), the stroboscopic locus is approached by a continuous path along this curve. When p is odd ($p = 2m + 1$), *each* of the two stroboscopic loci mentioned above is approached by a continuous path along this curve. These stroboscopic loci are shown in Fig. 9a at the resonance $Q = \frac{1}{2}$ and in Fig. 9b near the resonance $\frac{1}{2}$.

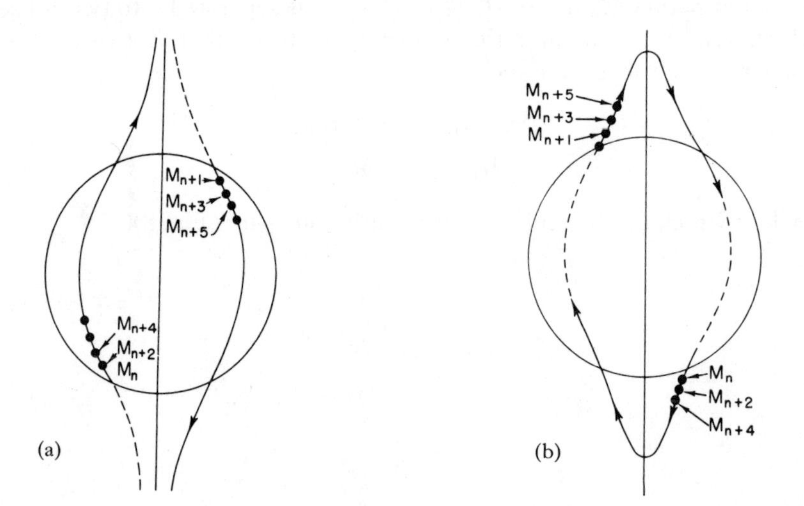

FIG. 9. Stroboscopic locus: (a) at the resonance $Q = \frac{1}{2}$; (b) near the resonance $\frac{1}{2}$.

2.2. Nonlinear Case

Let us rewrite (30) as

$$\frac{d^2x}{d\tau^2} + Q^2x - \beta x^3 = \epsilon\delta(\tau)x^{\nu+1} \tag{42}$$

The method used above can be applied to this nonlinear equation provided a few simplifications are introduced. Only the fundamental terms will generally be taken into consideration, as the higher harmonics raised by nonlinearities are supposed to be negligible. As pointed out earlier, this procedure amounts to assuming that the expected solution $x(\tau)$ is almost sinusoidal, which is ensured by the fact that ϵ and β are small, and it leads to formulating the relation between the angular frequency Ω and the amplitude a of $x(\tau)$. There is no dependency in the case of linear systems.

Since the right side of (42), $\epsilon\delta(\tau)x^{\nu+1}$, is again a periodic impulse function which occurs at $\tau_0, \tau_1, \tau_2, ..., \tau_n, \tau_{n+1}, ...$, the behavior of the particles between these discontinuities is governed by Duffing's equation,

$$\frac{d^2x}{d\tau^2} + Q^2x - \beta x^3 = 0 \tag{43}$$

Say in any open interval $]\tau_n, \tau_{n+1}[$ the solution of (42) is

$$x_n = a_n \cos(\Omega_n\tau + \varphi_n) \tag{44}$$

with

$$\Omega_n = Q - \frac{3\beta}{8Q} a_n^2 \tag{45}$$

This relation between the amplitude and the frequency introduces at the outset a difference with the linear case. We must emphasize this, since the primary aim of this section is to state the most significant features that result from nonlinearities.

Next, since we shall be interested in the small changes Δa_n and $\Delta \varphi_n$ due to the kicks, it seems at first sight that we should also take account of the corresponding changes in Ω_n. However, we have

$$\Delta\Omega_n = -\frac{3\beta}{4Q} a_n \, \Delta a_n \tag{46}$$

Accordingly, when Δa_n is of order $\beta(\Delta a_n \sim \beta)$, $\Delta\Omega_n$ is of order β^2 and can be neglected. Since Δa_n is otherwise of order ϵ, we see that we will be allowed to neglect $\Delta\Omega_n$, provided that

$$\epsilon = \lambda\beta \tag{47}$$

where the only condition imposed on λ is that it be bounded whatever the value of β. We shall make this assumption next.

2.2.1. Introductory Example: Resonance $Q \sim p/3$

To bring out the specific features of the method, when applied to the nonlinear equation (42), let us consider the special case

$$\nu = 1 \qquad \frac{d^2x}{d\tau^2} + Q^2 x - \beta x^3 = \epsilon\delta(\tau)x^2 \tag{48}$$

Here we shall not go into the details of the derivations, as they are similar to the ones we developed in section 2.1. One easily finds

$$\begin{aligned} \Delta a_n &= -\frac{\epsilon}{Q} a_n^2 \sin(\Omega_n\tau_{n+1} + \varphi_n)\cos^2(\Omega_n\tau_{n+1} + \varphi_n) \\ \Delta\varphi_n &= -\frac{\epsilon}{Q} a_n \cos^3(\Omega_n\tau_{n+1} + \varphi_n) \end{aligned} \tag{49}$$

Then

$$\begin{aligned} \Delta a_n &= -\frac{\epsilon a_n^2}{4Q}[\sin 3(\Omega_n\tau_{n+1} + \varphi_n) + \sin(\Omega_n\tau_{n+1} + \varphi_n)] \\ \Delta\varphi_n &= -\frac{\epsilon a_n}{4Q}[\cos 3(\Omega_n\tau_{n+1} + \varphi_n) + 3\cos(\Omega_n\tau_{n+1} + \varphi_n)] \end{aligned} \tag{50}$$

The crux of the discussion relies on the following remarks of Sturrock.

Case 1. When $Q \sim p/3$, p any integer,

$$\Omega_{i+1}\tau_{i+2} + \varphi_{i+1} \sim \Omega_i\tau_{i+1} + \varphi_i + p\,\frac{2\pi}{3} \qquad (i = 0, 1, ..., n, n+1, ...)^\dagger$$

then

$$\sin 3(\Omega_{i+1}\tau_{i+2} + \varphi_{i+1}) \sim \sin 3(\Omega_i\tau_{i+1} + \varphi_i)$$
$$\cos 3(\Omega_{i+1}\tau_{i+2} + \varphi_{i+1}) \sim \cos 3(\Omega_i\tau_{i+1} + \varphi_i)$$

It follows that these terms will pile up, from one kick to the next, and will result in the resonance according to formulas (50).

On the other hand, during the growth of the amplitude due to these terms, $\sin(\Omega_i\tau_{i+1} + \varphi_i)$ and $\cos(\Omega_i\tau_{i+1} + \varphi_i)$ are inefficient, since the small jumps which they produce, sometimes positive and sometimes negative, do not have the same correlation time; i.e., their effect is negligible *in the mean.*

Case 2. When $Q \sim p$, p any integer, the above arguments will hold in favor of the terms $\sin(\Omega_i\tau_{i+1} + \varphi_i)$ and $\cos(\Omega_i\tau_{i+1} + \varphi_i)$, which will produce the resonance, whereas the effects of the terms $\sin 3(\Omega_i\tau_{i+1}+\varphi_i)$ and $\cos 3(\Omega_i\tau_{i+1} + \varphi_i)$ will be negligible *in the mean.*

Accordingly, in Case 1, which we shall discuss as an example, we shall keep only the terms whose frequency is $3\Omega_i$. In Case 2 only the terms whose frequency is Ω_i will be considered.

When $Q \sim p/3$, (50) is rewritten

$$\begin{aligned} \Delta a_n &= -\frac{\epsilon}{4Q}\, a_n{}^2 \sin 3(\Omega_n\tau_{n+1} + \varphi_n) \\ \Delta\varphi_n &= -\frac{\epsilon}{4Q}\, a_n \cos 3(\Omega_n\tau_{n+1} + \varphi_n) \end{aligned} \qquad (51)$$

Then putting $\psi_n = \Omega_n\tau_{n+1} + \varphi_n + k_n(2\pi/3)$, $\rho_n = a_n{}^2$, and taking account of (45), one gets

$$\begin{aligned} \Delta\rho_n &= -\frac{\epsilon}{2Q}\,\rho_n^{3/2} \sin 3\psi_n \\ \Delta\psi_n &= 2\pi\left(Q - \frac{p}{3}\right) - \frac{3\pi}{4Q}\,\beta\rho_n - \frac{\epsilon}{4Q}\,\rho_n^{1/2}\cos 3\psi_n \end{aligned} \qquad (52)$$

From (52) we deduce

$$2\pi\left(Q - \frac{p}{3}\right)\Delta\rho_n - \frac{3\pi}{4Q}\,\beta\rho_n\,\Delta\rho_n - \frac{\epsilon}{4Q}\,\rho_n^{1/2}\cos 3\psi_n\,\Delta\rho_n + \frac{\epsilon}{2Q}\,\rho_n^{3/2}\sin 3\psi_n\,\Delta\psi_n = 0$$

† Since $\Omega_{i+1}(\tau_{i+2} - \tau_{i+1}) \sim 2\pi Q$, $\varphi_{i+1} - \varphi_i \sim \epsilon$.

say

$$\Delta\left[2\pi\left(Q-\frac{p}{3}\right)\rho_n-\frac{3\pi}{8Q}\beta\rho_n{}^2-\frac{\epsilon}{6Q}\rho_n^{3/2}\cos 3\psi_n\right]=0 \tag{53}$$

Finally, we again find an invariant of motion:

$$2\pi\left(Q-\frac{p}{3}\right)\rho_n-\frac{3\pi}{8Q}\beta\rho_n{}^2-\frac{\epsilon}{6Q}\rho_n^{3/2}\cos 3\psi_n=C \tag{54}$$

It leads to a representation of the stroboscopic locus. However, ρ_n is not explicitly obtained as a function of ψ_n.

Let us rewrite (54) in the form

$$F(K,\rho_n)\triangleq\frac{K}{\rho_n^{3/2}}+\frac{Q-(p/3)}{\rho_n^{1/2}}-\frac{3\beta}{16Q}\rho_n^{1/2}=\frac{\epsilon}{6\cdot 2\pi Q}\cos 3\psi_n$$

(K is another constant), whose left side we shall plot as a function of $(\rho_n)^{1/2}$ (for different K). We get the diagrams of Fig. 10 for $Q<p/3$,

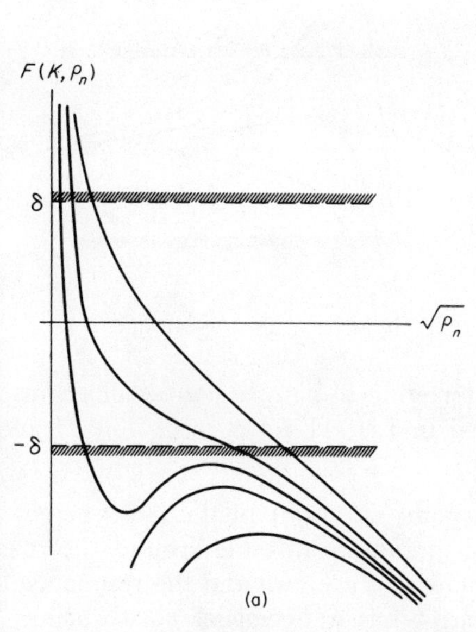

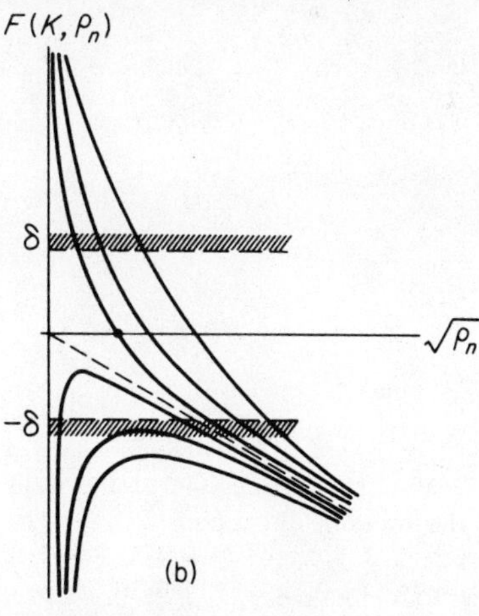

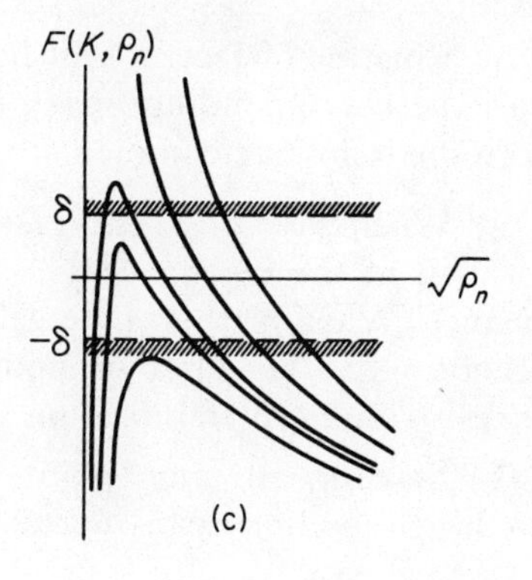

FIG. 10. Discussion of the stability of the motion, illustration 1: (a) $Q<p/3$, $\nu=1$, $\beta\neq 0$; (b) $Q=p/3$, $\nu=1$, $\beta\neq 0$; (c) $Q>p/3$, $\nu=1$, $\beta\neq 0$.

$Q = p/3$, and $Q > p/3$, respectively. Then, since the right side lies in the range $+\delta$, $-\delta$, $\delta = \epsilon/(12\pi Q)$, we can determine the maximum and the minimum value of ρ_n associated with each point of the diagram.

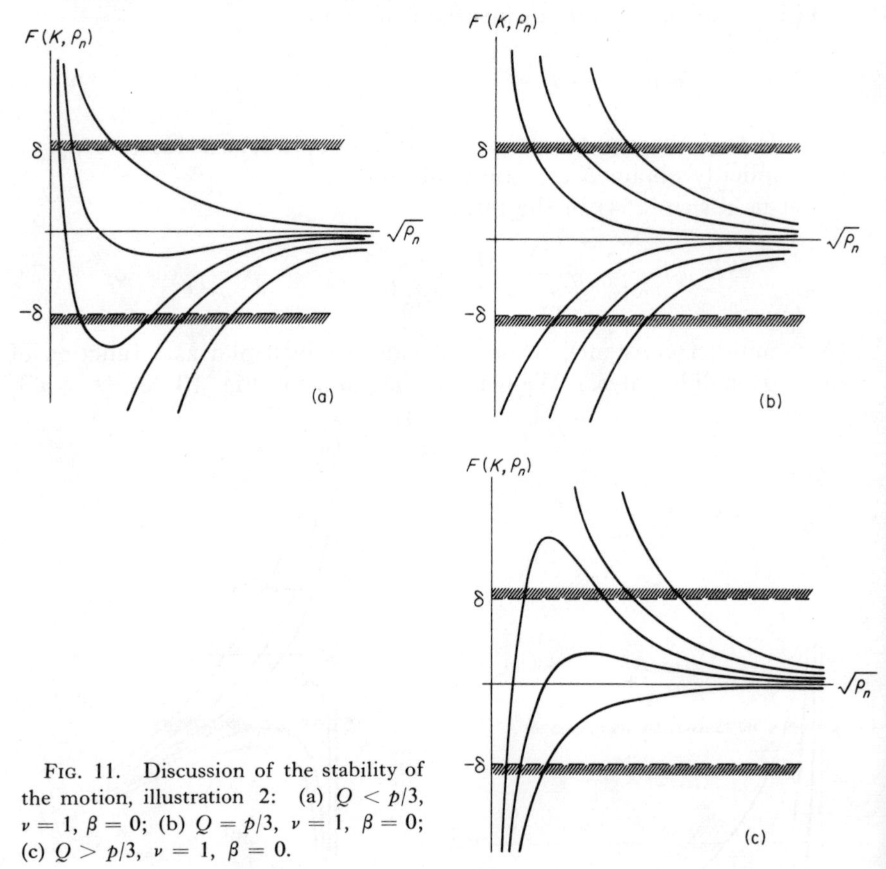

FIG. 11. Discussion of the stability of the motion, illustration 2: (a) $Q < p/3$, $\nu = 1$, $\beta = 0$; (b) $Q = p/3$, $\nu = 1$, $\beta = 0$; (c) $Q > p/3$, $\nu = 1$, $\beta = 0$.

As a matter of fact, it will be interesting to compare these diagrams with the corresponding ones plotted in Fig. 11 for $\beta = 0$. They lead us to the following comments:

(a) When $\beta = 0$, ρ_n increases beyond any limit in the close neighborhood of the resonance $p/3$. The motion is unstable around the resonance. When $\beta \neq 0$, ρ_n remains bounded even around the resonance. Accordingly, this kind of nonlinearity has a beneficial consequence, since it results in stabilization of the motion.

(b) Starting at any point of the diagram where $\sin 3\psi_n \neq 0$, $\cos 3\psi_n \neq \pm 1$, ρ_n can increase or decrease according to the sign of

$\sin 3\psi_n$. For example, we can consider starting points along the axis corresponding to $\cos 3\psi_n = 0$. Then if we are chiefly interested in the growth of the amplitude (whose consequence could be unlucky), we see that, except in the case $Q = p/3$, there exists a critical starting value of ρ_n, say ρ_0, such that: for $\rho_n < \rho_0$, the increase in amplitude is very small; for $\rho_n > \rho_0$, the increase may be rather substantial.

2.2.2. Influence of the Parity of ν

In the above example, ν is odd. Let us consider another one with ν even: for instance, $\nu = 2$; then (49) is replaced by

$$\begin{aligned} \Delta a_n &= -\frac{\epsilon}{Q} a_n{}^3 \sin(\Omega_n \tau_{n+1} + \varphi_n) \cos^3(\Omega_n \tau_{n+1} + \varphi_n) \\ \Delta \varphi_n &= -\frac{\epsilon}{Q} a_n{}^2 \cos^4(\Omega_n \tau_{n+1} + \varphi_n) \end{aligned} \tag{55}$$

from which follows

$$\begin{aligned} \Delta a_n &= -\frac{\epsilon a_n{}^3}{8Q} [\sin 4(\Omega_n \tau_{n+1} + \varphi_n) + 2 \sin 2(\Omega_n \tau_{n+1} + \varphi_n)] \\ \Delta \varphi_n &= -\frac{\epsilon a_n{}^2}{8Q} [\cos 4(\Omega_n \tau_{n+1} + \varphi_n) + 4 \cos 2(\Omega_n \tau_{n+1} + \varphi_n) + 3] \end{aligned} \tag{56}$$

An important difference between (56) and (50) lies in the fact that, in (56), the last term in the expression of $\Delta\varphi_n$ is a constant. Indeed such a constant appears every time ν is even. Its influence on the final result of the calculus can be easily understood if one returns to the arguments which served as the starting point of the above derivations.

Whatever the order of the resonance considered, the constant terms which thus appear in each small jump $\Delta\varphi_n$ will pile up and give a substantial increase of the variable φ_n. For instance, if we study the resonance $Q \sim p/4$, we need keep not only the terms $\sin 4(\Omega_n \tau_{n+1} + \varphi_n)$ and $\cos 4(\Omega_n \tau_{n+1} + \varphi_n)$, but also the constant term, which leads to the equations

$$\begin{aligned} \Delta a_n &= -\frac{\epsilon a_n{}^3}{8Q} \sin 4(\Omega_n \tau_{n+1} + \varphi_n) \\ \Delta \varphi_n &= -\frac{\epsilon a_n{}^2}{8Q} [3 + \cos 4(\Omega_n \tau_{n+1} + \varphi_n)] \end{aligned} \tag{57}$$

Then putting $\psi_n = \Omega_n \tau_{n+1} + \varphi_n + k_n(2\pi/4)$, $\rho_n = a_n^2$, we get

$$\begin{aligned} \Delta \rho_n &= -\frac{\epsilon}{4Q} \rho_n^2 \sin 4\psi_n \\ \Delta \psi_n &= 2\pi \left(Q - \frac{p}{4}\right) - \frac{3\pi}{4Q} \beta \rho_n - \frac{3\epsilon}{8Q} \rho_n - \frac{\epsilon}{8Q} \rho_n \cos 4\psi_n \end{aligned} \tag{58}$$

Finally, from (58) we deduce

$$2\pi \left(Q - \frac{p}{4}\right) \Delta \rho_n - \frac{3\pi}{4Q} \beta \rho_n \Delta \rho_n - \frac{3\epsilon}{8Q} \rho_n \Delta \rho_n - \frac{\epsilon}{8Q} \rho_n \cos 4\psi_n \Delta \rho_n + \frac{\epsilon}{4Q} \rho_n^2 \sin 4\psi_n \Delta \psi_n = 0$$

say

$$\Delta \left[2\pi \left(Q - \frac{p}{4}\right) \rho_n - \frac{3\pi}{8Q} \beta \rho_n^2 - \frac{3\epsilon}{16Q} \rho_n^2 - \frac{\epsilon}{16Q} \rho_n^2 \cos 4\psi_n\right] = 0$$

Accordingly, the invariant of the motion is

$$2\pi \left(Q - \frac{p}{4}\right) \rho_n - \frac{3\pi}{8Q} \beta \rho_n^2 - \frac{3\epsilon}{16Q} \rho_n^2 - \frac{\epsilon}{16Q} \rho_n^2 \cos 4\psi_n = C \tag{59}$$

Not only is the expression of the invariant modified, but also the conclusions that can be drawn from the discussion. Indeed, if we rewrite (59)

$$F(K, \rho_n) \triangleq \frac{K}{\rho_n^2} + \frac{Q - (p/4)}{\rho} - \frac{3\beta}{16Q} = \frac{1}{16} \frac{\epsilon}{2\pi Q} (3 + \cos 4\psi_n)$$

we see that the stability of motion is greatly influenced by the constant which appears in the right side, since the right side will now vary between

$$\delta_1 = \frac{1}{8} \frac{\epsilon}{2\pi Q} \quad \text{and} \quad \delta_2 = \frac{1}{4} \frac{\epsilon}{2\pi Q}$$

The position of this band with respect to the family of curves which represent $F(K, \rho_n)$ is shown in Fig. 12 for the cases $Q < p/4$, $Q = p/4$, $Q > p/4$, respectively.

Obviously the conclusions depend on the sign of ϵ. For $\epsilon > 0$, the variations of ρ_n are always very small, say the motion is stable, even at the resonance $Q = p/4$. It is also stable when $\beta = 0$. For $\epsilon < 0$, the motion can be unstable if

$$-\frac{1}{4} \frac{\epsilon}{2\pi Q} < \frac{3\beta}{8Q} < -\frac{1}{2} \frac{\epsilon}{2\pi Q}$$

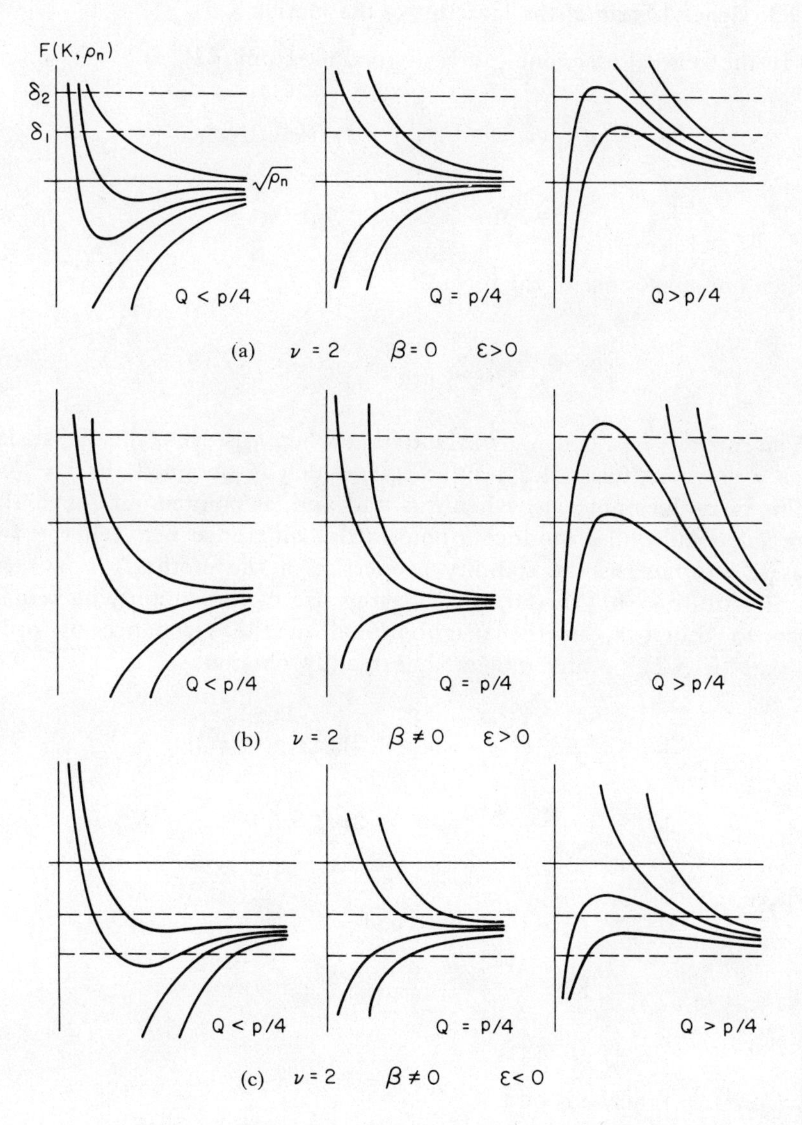

FIG. 12. Stability of motion: (a) $\nu = 2$, $\beta = 0$, $\epsilon > 0$; (b) $\nu = 2$, $\beta \neq 0$, $\epsilon > 0$; (c) $\nu = 2$, $\beta \neq 0$, $\epsilon < 0$.

However, even if this condition is fulfilled, it can be stable for sufficiently small starting amplitudes, provided that $Q \neq p/4$. We see that when $\beta = 0$, the motion is stable.

2.2.3. General Form of the Invariant of the Motion

In the general case, one can easily deduce from (42),

$$\begin{aligned} \Delta a_n &= -\frac{\epsilon}{Q} a_n^{\nu+1} \sin(\Omega_n \tau_{n+1} + \varphi_n) \cos^{\nu+1}(\Omega_n \tau_{n+1} + \varphi_n) \\ \Delta \varphi_n &= -\frac{\epsilon}{Q} a_n^{\nu} \cos^{\nu+2}(\Omega_n \tau_{n+1} + \varphi_n) \end{aligned} \tag{60}$$

Then one makes use of the formula

$$\cos^k \Phi = \frac{1}{2^k} \sum_{j=0}^{k} \frac{k!}{j!(k-j)!} \cos(k-2j)\Phi \tag{61}$$

(k an integer; $j = 0, 1, ..., k$). Note that when k is even, this expansion has a constant term $(1/2^k)\, k!/[(k/2)!]^2$, which is obtained for $j = k/2$. This term does not exist when k is odd and, as pointed out above, this special detail will introduce a noteworthy difference between the two cases, as concerns the stability properties of the motion.

Substituting (61) in (60), and making use of the simplifying remark due to Sturrock, in the neighborhood of the resonance of order $Q = p/(\nu + 2)$,† p any integer, one readily obtains

$$\begin{aligned} \Delta a_n &= -\frac{\epsilon}{2^{\nu+1}} \frac{a_n^{\nu+1}}{Q} \sin[(\nu+2)(\Omega_n \tau_{n+1} + \varphi_n)] \\ \Delta \varphi_n &= -\frac{\epsilon}{2^{\nu+1}} \frac{a_n^{\nu}}{Q} \{\cos[(\nu+2)(\Omega_n \tau_{n+1} + \varphi_n)] + K_\nu\} \end{aligned}$$

where

$$K_\nu = 0 \qquad \text{if } \nu \text{ is odd}$$

$$K_\nu = \frac{1}{2} \frac{(\nu+2)!}{\{[(\nu+2)/2]!\}^2} \qquad \text{if } \nu \text{ is even}$$

Now:

Case 1. When ν is odd:

$$\Delta \rho_n = -\frac{\epsilon}{2^\nu Q} \rho_n^{(\nu+2)/2} \sin(\nu+2)\psi_n$$

$$\Delta \psi_n = 2\pi \left(Q - \frac{p}{\nu+2}\right) - \frac{3\pi}{4Q} \beta \rho_n - \frac{\epsilon}{2^{\nu+1} Q} \rho_n^{\nu/2} \cos(\nu+2)\psi_n$$

† Resonances of different orders can be studied in the same way.

with

$$\rho_n = a_n^2 \qquad \psi_n = \Omega_n \tau_{n+1} + \varphi_n + k_n \frac{2\pi}{\nu + 2} \qquad (k_n \text{ an integer})$$

from which follows

$$\Delta \left[2\pi \left(Q - \frac{p}{\nu + 2} \right) \rho_n - \frac{3\pi}{8Q} \beta \rho_n^2 - \frac{\epsilon}{(\nu + 2) 2^\nu Q} \rho_n^{(\nu+2)/2} \cos(\nu + 2) \psi_n \right] = 0$$

(p an integer). Accordingly, in this case, the invariant of the motion is

$$2\pi \left(Q - \frac{p}{\nu + 2} \right) \rho_n - \frac{3\pi}{8Q} \beta \rho_n^2 - \frac{\epsilon}{(\nu + 2) 2^\nu Q} \rho_n^{(\nu+2)/2} \cos(\nu + 2) \psi_n = C \tag{62}$$

Case 2. When ν is even:

$$\Delta \rho_n = - \frac{\epsilon}{2^\nu Q} \rho_n^{(\nu+2)/2} \sin(\nu + 2) \psi_n$$

$$\Delta \psi_n = 2\pi \left(Q - \frac{p}{\nu + 2} \right) - \frac{3\pi}{4Q} \beta \rho_n - K_\nu \frac{\epsilon}{2^{\nu+1} Q} \rho_n^{\nu/2} - \frac{\epsilon}{2^{\nu+1} Q} \rho_n^{\nu/2} \cos(\nu + 2) \psi_n$$

according to which the invariant of the motion is

$$2\pi \left(Q - \frac{p}{\nu + 2} \right) \rho_n - \frac{3\pi}{8Q} \beta \rho_n^2 - K_\nu \frac{\epsilon}{(\nu + 2) 2^\nu Q} \rho_n^{(\nu+2)/2} - \frac{\epsilon}{(\nu + 2) 2^\nu Q} \rho_n^{(\nu+2)/2} \cos(\nu + 2) \psi_n = C \tag{63}$$

Once the invariant of the motion is embodied, the problem is solved. It is easy enough once you understand how to tackle it, but the "embodying" may appear to be a trick. However, there is no trick involved and, we shall explain why it is possible to obtain an invariant of the motion in this example, on the basis of the hamiltonian representation.

3. HAMILTONIAN REPRESENTATION

3.1. Resonance of a Ballistic Galvanometer

Let us return for a short while to the elementary theory of the ballistic galvanometer, which is the starting point of the above procedure, specifically to the equation

$$\frac{d^2 x}{d\tau^2} + \omega_0^2 x = q \omega_0^2 \delta(\tau) \tag{64}$$

where $\delta(\tau)$ is an impulse which occurs at time τ_i, with $\int_{\Delta\tau} \delta(\tau)\, d\tau = 1$ ($\Delta\tau$, small neighborhood of τ_i), q a small constant, and ω_0 the characteristic angular frequency of the system. The damping of the galvanometer is disregarded.

Let

$$x = a \cos(\omega_0 \tau + \varphi)$$

be the solution of (64) before the kick, and

$$x = (a + \Delta_{\tau_i} a) \cos(\omega_0 \tau + \varphi + \Delta_{\tau_i} \varphi)$$

the solution after the kick. As pointed out earlier, we get $\Delta_{\tau_i} a$ and $\Delta_{\tau_i} \varphi$ by writing

$$\begin{aligned} &\Delta_{\tau_i} x = 0 \\ &\Delta_{\tau_i} p = q\omega_0^2 \qquad p = \frac{dx}{d\tau} \end{aligned} \tag{65}$$

(Δ_{τ_i} means a sudden change at $\tau = \tau_i$). It turns out that (65) can be put in the hamiltonian form

$$\Delta_{\tau_i} x = \frac{\partial H}{\partial p} \qquad \Delta_{\tau_i} p = -\frac{\partial H}{\partial x} \tag{66}$$

with

$$H = -q\omega_0^2 x \tag{67}$$

Now let us use variables ρ and Φ instead of x and p, with

$$x = \sqrt{\rho} \cos \Phi \qquad p = -\omega_0 \sqrt{\rho} \sin \Phi$$

say $\rho = a^2$, $\Phi = \omega_0 \tau + \varphi$. We have

$$\begin{aligned} \frac{\partial x}{\partial \rho} &= \frac{1}{2\sqrt{\rho}} \cos \Phi & \frac{\partial x}{\partial \Phi} &= -\sqrt{\rho} \sin \Phi \\ \frac{\partial p}{\partial \rho} &= -\frac{\omega_0}{2\sqrt{\rho}} \sin \Phi & \frac{\partial p}{\partial \Phi} &= -\omega_0 \sqrt{\rho} \cos \Phi \end{aligned}$$

Hence the Poisson bracket $[x, p]_{\rho,\Phi}$ is

$$P \triangleq [x, p]_{\rho,\Phi} = \frac{\partial x}{\partial \Phi}\frac{\partial p}{\partial \rho} - \frac{\partial x}{\partial \rho}\frac{\partial p}{\partial \Phi} = \frac{\omega_0}{2}$$

It is independent of ρ and Φ. Accordingly, $\Delta_{\tau_i} \rho$ and $\Delta_{\tau_i} \Phi$ can also be obtained from hamiltonian

$$H_0 = \frac{H[x(\rho, \Phi), p(\rho, \Phi)]}{P}$$

This amounts to saying that the transformation from x, p to $P^{1/2}\rho$, $P^{1/2}\Phi$ is a *canonical* one:

$$H = -q\omega_0^2\rho^{1/2}\cos\Phi \qquad H_0 = -2q\omega_0\rho^{1/2}\cos\Phi$$

$$\Delta_{\tau_i}(P^{1/2}\Phi) = \frac{\partial H}{P^{1/2}\partial\rho} \qquad \Delta_{\tau_i}\Phi = \frac{\partial H_0}{\partial\rho} \tag{68}$$

$$\Delta_{\tau_i}(P^{1/2}\rho) = -\frac{\partial H}{P^{1/2}\partial\Phi} \qquad \Delta_{\tau_i}\rho = -\frac{\partial H_0}{\partial\Phi}$$

When a periodic sequence of pulses is applied to the system the benefit of the stroboscopic method is obvious, since it leads to replacing small changes which suddenly occur *at each kick* by continuous changes which are integrated *from one kick to the next kick*. Then (68) can be replaced by a differential equation involving the new stroboscopic variables. For instance, when one considers a periodic sequence of pulses whose frequency is nearly the same as the characteristic frequency of the oscillator, between two successive kicks:

(a) ρ remains constant.

(b) The change in Φ is about 2π.

Accordingly, if we replace Φ by the stroboscopic variable

$$\psi = \Phi - 2\pi$$

we get, *from one kick to the next one*, the change

$$\Delta\psi = 2\pi(\omega_0 - 1) + \Delta_{\tau_i}\Phi$$

and equations (68) are replaced by

$$H_0 = -2q\omega_0\rho^{1/2}\cos\psi$$

$$\Delta\psi = 2\pi(\omega_0 - 1) + \frac{\partial H_0}{\partial\rho} \tag{69}$$

$$\Delta\rho = -\frac{\partial H_0}{\partial\psi}$$

Finally, we get

$$\Delta[2\pi(\omega_0 - 1)\rho + H_0] = 0$$

$$2\pi(\omega_0 - 1)\rho + H_0 = C$$

which is identical to the expression (62) when $\nu = -1$, $\beta = 0$, $Q = \omega_0$, $\epsilon = q\omega_0^2$, and $p = 1$.

This example is a very simple one; however it casts some light upon the close mechanism which raises the invariant of the motion, and the arguments easily fit more intricate situations.

3.2. Hamiltonian Theory of Nonlinear Oscillations

You will note a connection between the above method and the following one, which was developed by Moser [14], by Hagedorn, Hine, and Schoch [11], and by Symon.

Let us keep τ as the azimuthal variable, and consider the equation

$$\frac{d^2x}{d\tau^2} + n(\tau)x = -\frac{\partial V}{\partial x} \tag{70}$$

where $n(\tau)$ is a periodic function whose period is T with respect to τ; next we shall assume that $n(\tau)$ has a constant term Q^2. $V(x)$ is a potential function which describes the distortions of the magnetic field in the close neighborhood of the reference orbit. Moreover $V(x)$ is assumed to depend on τ: $V(x) \triangleq V(x, \tau)$.

Equation (70) can be rewritten in the hamiltonian form by putting

$$H(x, p) = \frac{p^2 + nx^2}{2} + V(x) \qquad p = \frac{dx}{d\tau}, n = n(\tau) \tag{71}$$

Then

$$\frac{dx}{d\tau} = \frac{\partial H}{\partial p} \qquad \frac{dp}{d\tau} = -\frac{\partial H}{\partial x} \tag{72}$$

Next it will be convenient to decompose H into

$$H = H^{(1)} + H^{(2)}$$

with

$$H^{(1)} = \frac{p^2 + nx^2}{2} \qquad H^{(2)} = V(x) = \sum_{k=2}^{\infty} v_k(\tau)x^k$$

$v_k(\tau)$ are coefficients which do not depend on x. $H^{(1)}$ describes the unperturbed linear system, and $H^{(2)}$ represents the perturbational term.

As a matter of fact, the solution for the unperturbed linear system, given by Floquet's theory, assumes the form

$$x = w(\tau)\rho^{1/2} \cos(Q\tau + \varphi) \tag{73}$$

where $w(\tau)$ is a periodic function with respect to τ, whose period is the same as the period of $n(\tau)$, namely T, and ρ and φ are integration constants.

Then we shall take account of the perturbations by letting ρ and φ vary with τ, and we shall use ρ and φ as the new variables inplace of x and p. It may be readily verified that

$$[x, p]_{\rho,\varphi} = 1$$

provided that (73) is properly normalized. Accordingly, the transformation is a canonical one, from which we deduce

$$\begin{aligned} \frac{d\varphi}{d\tau} &= \frac{\partial H^{(2)}[x(\rho, \varphi, \tau), p(\rho, \varphi, \tau), \tau]}{\partial \rho} = \frac{\partial H^{(2)}(\rho, \varphi, \tau)}{\partial \rho} \\ \frac{d\rho}{d\tau} &= -\frac{\partial H^{(2)}[x(\rho, \varphi, \tau), p(\rho, \varphi, \tau), \tau]}{\partial \varphi} = -\frac{\partial H^{(2)}(\rho, \varphi, \tau)}{\partial \varphi} \end{aligned} \tag{74}$$

Next we have to perform the transformation of the variables and write $H^{(2)}(\rho, \varphi, \tau)$ in explicit form. We have

$$H^{(2)} = \sum_{k=2}^{\infty} v_k(\tau) x^k = \sum_{k=2}^{\infty} v_k w^k \rho^{k/2} \cos^k(Q\tau + \varphi)$$

Then using (61),

$$H^{(2)} = \sum_{k=2}^{\infty} \sum_{l+m=k} v_k(\tau) \rho^{k/2} \frac{1}{2^k} \frac{k!}{l!\, m!} w^k(\tau) \cos[(l-m)(Q\tau + \varphi)]$$

Finally, by introducing the Fourier expansion of the function

$$\frac{1}{2^k} \frac{k!}{l!\, m!} v_k(\tau) w^k(\tau)$$

say

$$\frac{1}{2^k} \frac{k!}{l!\, m!} v_k(\tau) w^k(\tau) = \sum_{p=0}^{+\infty} 2 v_{lmp} \cos(p\tau + \varphi_p)$$

one gets

$$\begin{aligned} H^{(2)} = \sum_{k} \sum_{l+m=k} \sum_{p} v_{lmp} \{ &\cos[[(\quad - m)Q + p]\tau + (l-m)\varphi + \varphi_p] \\ &+ \cos[[(l-m)Q - p]\tau + (l-m)\varphi - \varphi_p]\} \rho^{k/2} \end{aligned} \tag{75}$$

In the close neighborhood of resonances of different orders, ρ and φ are slowly varying functions of τ, so that the small perturbations can pile up during a few revolutions, and the amplitude undergoes a substantial increase. We shall again consider such cases, which will lead us to pick out only the slowly varying terms in (75), neglecting the others.

Obviously we again encounter arguments similar to those of Sturrock (Section 2.2.1).

Let us put $l - m = s$, say

$$l = \frac{k+s}{2} \qquad m = \frac{k-s}{2} \qquad (\text{since } l + m = k)$$

and rewrite (75)

$$H^{(2)} = \sum_{k=2}^{\infty} \sum_{s=0}^{k} \sum_{p=0}^{+\infty} 2v_{[(k+s)/2][(k-s)/2]p}\rho^{k/2} \cos[(sQ - p)\tau + s\varphi - \varphi_p] \tag{76}$$

Then in the close neighborhood of

$$sQ - p = 0 \qquad (s, p \text{ given positive integers})$$

we shall keep, in (76):

(a) The constant term corresponding to $s = 0$, $p = 0$.

(b) The term whose angular frequency is $sQ - p$.

Accordingly, $H^{(2)}$ will be replaced by the low-frequency hamiltonian

$$\bar{H}^{(2)} = P_0(\rho) + 2P_1(\rho) \cos[(sQ - p)\tau + s\varphi - \varphi_p]$$

with

$$P_0(\rho) = \sum_k 2v_{(k/2)(k/2)0}\rho^{k/2} \cos \varphi_0 = v^{(1)}\rho + v^{(2)}\rho^{(2)} + \cdots$$

$$P_1(\rho) = \sum_k v_{[(k+s)/2][(k-s)/2]p}\rho^{k/2} = v_{s0p}\rho^{s/2} + \cdots$$

$[v^{(i)} = 2v_{ii0} \cos \varphi_0]$. Then equations (74) are approximated by

$$\begin{aligned} \frac{d\varphi}{d\tau} &= \frac{\partial \bar{H}^{(2)}}{\partial \rho} = \frac{dP_0}{d\rho} + 2\frac{dP_1}{d\rho} \cos[(sQ - p)\tau + s\varphi - \varphi_p] \\ \frac{d\rho}{d\tau} &= -\frac{\partial \bar{H}^{(2)}}{\partial \varphi} = 2P_1 s \sin[(sQ - p)\tau + s\varphi - \varphi_p] \end{aligned} \tag{77}$$

or putting $\psi = [Q - (p/s)]\tau + \varphi - (\varphi_p/s)$, by

$$\begin{aligned} \frac{d\psi}{d\tau} &= \left(Q - \frac{p}{s}\right) + \frac{dP_0}{d\rho} + 2\frac{dP_1}{d\rho} \cos s\psi \\ \frac{d\rho}{d\tau} &= 2P_1 s \sin s\psi \end{aligned} \tag{78}$$

From (78) we deduce an invariant of the motion,

$$\left(Q - \frac{p}{s}\right) \rho + \bar{H}^{(2)}(\rho, \psi) = \left(Q - \frac{p}{s}\right) \rho + P_0 + 2P_1 \cos s\psi = C$$

This conclusion generalizes the ones of earlier sections.

For instance, in the case of resonance of order $p/3$ ($s = 3$), which we considered in Section 2.2.1 from another viewpoint, one finds

$$\left(Q + v^{(1)} - \frac{p}{3}\right) \rho + v^{(2)}\rho^2 + 2v_{30p} \cos 3\psi = C$$

with

$$\psi = \left(Q - \frac{p}{3}\right) \tau + \varphi - \frac{\varphi_p}{3}$$

In practical applications one is interested in the ratio ρ_1/ρ_0, where $(\rho_1)^{1/2}$ is the greatest possible amplitude for a given initial amplitude $(\rho_0)^{1/2}$. This ratio has been plotted by Hagedorn as a function of

$$\xi = \frac{Q + v^{(1)} - (p/3)}{2v_{30p}\rho_0^{1/2}}$$

(the distance in frequency from resonance) with

$$K = \frac{v^{(2)}\rho_0^{1/2}}{2v_{30p}}$$

as a parameter. This is shown in Fig. 13. Hagedorn reaches the conclusion that a value of $K > 10$ is necessary to keep the ratio ρ_1/ρ_0 below 2.

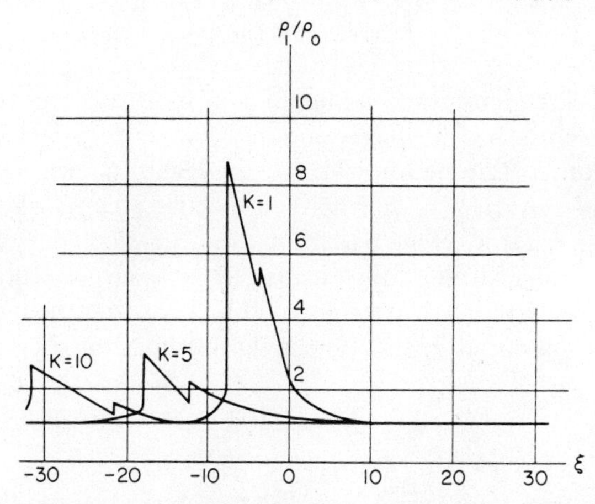

FIG. 13. Ratio ρ_1/ρ_0 vs. ξ. $(\rho_1)^{1/2}$ is the greatest possible amplitude for a given initial amplitude $(\rho_0)^{1/2}$, ξ is the distance in frequency from resonance. The Q value is in the neighborhood of a third-order subresonance.

3.3. Mechanical Analogue for Studying Betatron Oscillations

A mechanical analogue for studying betatron oscillations, in connection with the above theory, say more precisely with (70), has been constructed and described by Barbier [9]. The model is shown in Fig. 14.

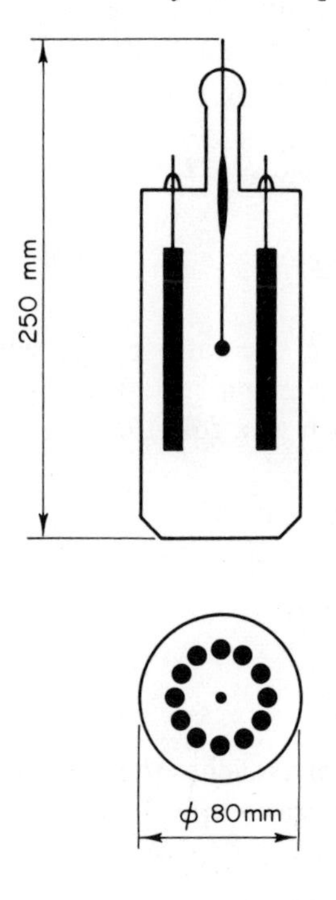

FIG. 14. Mechanical analog for studying the betatron oscillations.

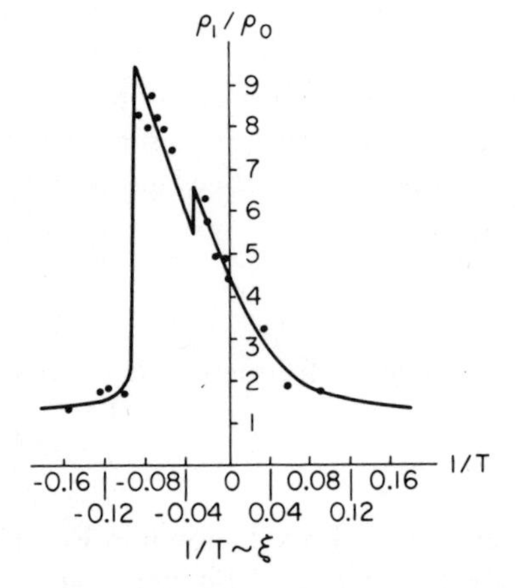

FIG. 15. Third-order subresonance line with constant cubic nonlinearity.

A linear elastic pendulum made of a thin thread of fused quartz, one of whose ends has been given the form of a little ball, oscillates in a vacuum tube. The pendulum is coated with a thin layer of metal, so that the ball can carry a well-defined electric charge when connected to a voltage generator. The ball moves in an electric field produced by a set of cylindrical electrodes arranged on a circle around the pendulum. The electrodes are parallel to the thread of the pendulum.

The electric potential in the close neighborhood of the ball is

$$V(r, \varphi) = \sum_k V_k(t) \left(\frac{r}{R}\right)^k \cos k\varphi$$

where r and φ are cylindrical coordinates of the center of the ball (z nearly constant); R is the distance between the axis of each electrode (the same for all the electrodes) and the supporting line of the thread, and $V_k(t)$ are the coefficients, which are functions of time t only. Accordingly, the forces acting on the ball are

$$F_r = -q\frac{\partial V}{\partial r} = -q\sum_k \frac{kV_k(t)}{R}\left(\frac{r}{R}\right)^{k-1} \cos k\varphi$$

$$F_\varphi = -\frac{q}{r}\frac{\partial V}{\partial \varphi} = q\sum_k \frac{kV_k(t)}{R}\left(\frac{r}{R}\right)^{k-1} \sin k\varphi$$

(q is the charge of the ball).

When a time-varying multipole field, whose frequency is nearly 3, 4, or 5 times the characteristic frequency of the pendulum, is applied to the tube, resonances of order $\frac{1}{3}$, $\frac{1}{4}$, and $\frac{1}{5}$ can be conveniently observed.

The flexibility of this mechanical analogue has enabled many interesting corroborations of the theory. For instance, the above results of Hagedorn (see Fig. 13) have been experimentally confirmed with good accuracy, as can be seen in the curve of Fig. 15, which was plotted by Barbier for the case of third-order subresonance.

4. THE SMOOTH APPROXIMATION

4.1. Smooth Solution to One-Dimensional Motion†

The aim of the smooth approximation is to replace equations with variable (periodic) coefficients by approximate equations with constant coefficients. Indeed, it can be used only if some conditions, which we shall next specify, are fulfilled.

We shall first explain this procedure starting with the equation (see also [6])

$$\frac{d^2x}{ds^2} = (nx + \tfrac{1}{3}ex^3)g(s) \tag{79}$$

which describes the radial motion of the particles along the equilibrium orbit in an alternating-gradient synchrotron with cubic forces. x denotes the radial deviation of a particle, s its curvilinear coordinate along the orbit, $g(s)$ a square wave of unit amplitude and period S, and e is a constant.

First of all, let us disregard the harmonics of $g(s)$ and replace this square wave by its fundamental term, whose angular frequency is

† See [4, 5].

$\omega_0 = 2\pi/S$. That is the first step in smoothing. Equation (79) is rewritten

$$\frac{d^2x}{ds^2} = (nx + \tfrac{1}{3}ex^3)\cos\omega_0 s \tag{80}$$

Then assume that the conditions are such that the solution $x(s)$ can be separated into two terms:

$$x(s) = \xi_1(s) + \xi_2(s)$$

where $\xi_1(s)$ is a slowly varying function of s, and $\xi_2(s) = a\cos\omega_0 s$.

Substituting in (80) and again neglecting the harmonics with frequencies $2\omega_0$ and $4\omega_0$, we get

$$\frac{d^2\xi_1}{ds^2} + \frac{d^2\xi_2}{ds^2} = (n\xi_1 + \tfrac{1}{3}e\xi_1^3 + \tfrac{3}{4}e\xi_1 a^2)\cos\omega_0 s + \left(\frac{na}{2} + \frac{e\xi_1^2 a}{2} + \frac{ea^3}{8}\right) \tag{81}$$

Finally, by identifying separately the low-frequency terms and the terms whose frequency is close to ω_0, we obtain two equations; when a is small,

$$\begin{aligned}\frac{d^2\xi_1}{ds^2} &= \frac{na}{2} + \frac{e\xi_1^2 a}{2}\\ \frac{d^2\xi_2}{ds^2} &= -a\omega_0^2\cos\omega_0 s = (n\xi_1 + \tfrac{1}{3}e\xi_1^3)\cos\omega_0 s\end{aligned} \tag{82}$$

a is determined by the second equation:

$$a(s) = -\frac{n}{\omega_0^2}\xi_1 - \frac{1}{3}\frac{e}{\omega_0^2}\xi_1^3$$

and substituting this expression into the first one we get

$$\frac{d^2\xi_1}{ds^2} = -\frac{n^2}{2\omega_0^2}\left(\xi_1 + \frac{4e}{3n}\xi_1^3 + \frac{e^2}{3n^2}\xi_1^5\right) \tag{83}$$

Equation (83) is the smooth approximation of (79). Its linear approximation gives the well-known value of the angular frequency of betatron oscillations, say

$$\omega_1 = \frac{n}{\omega_0(2)^{1/2}} = \frac{nS}{\pi(8)^{1/2}} \tag{84}$$

Obviously the smooth approximation is valid provided that

$$\omega_1 \ll \omega_0 \qquad \text{say} \quad \omega_0^2 \gg \frac{n}{2^{1/2}}$$

When the nonlinear terms are taken into account, a better approximation is obtained by means of the principle of harmonic balance, which leads to the relation between the amplitude A of $\xi_1(s)$ and the angular frequency of betatron oscillations:

$$\omega = \omega_1 \left(1 + \frac{e}{n} A^2 + \frac{5}{24} \frac{e^2}{n^2} A^4\right)^{1/2} \tag{85}$$

This formula has the correct qualitative behavior.

Other investigations can be carried out starting with the potential energy for motion. From (83) one deduces

$$V(\xi_1) = \frac{n^2}{2\omega_0^2} \left(\frac{1}{2} \xi_1^2 + \frac{1}{3} \frac{e}{n} \xi_1^4 + \frac{1}{18} \frac{e^2}{n^2} \xi_1^6\right)$$

The potential energy and a family of isoenergy closed trajectories in the phase plane are shown for $e > 0$ in Fig. 16 and for $e < 0$ in fig. 17.

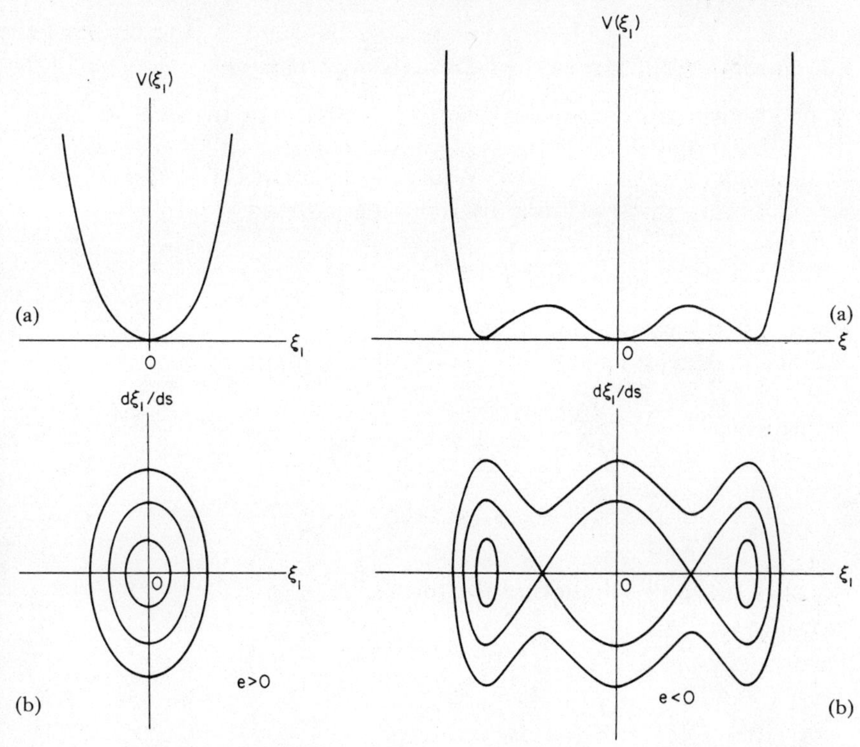

FIG. 16. Smooth approximation to one-dimensional motion: (a) potential energy and (b) isoenergy trajectories in the phase plane, $e > 0$.

FIG. 17. Smooth approximation to one-dimensional motion: (a) potential energy and (b) isoenergy trajectories in the phase plane, $e < 0$.

Formula (85) applies to periodic motions in the central eye in Figs. 16 and 17. Other approximate solutions for motion around the left and right eyes in Fig. 17 can be derived by substituting in (83) the expression

$$\xi_1 = B + A \cos \omega s$$

Then by a straightforward identification one gets

$$B^2 = 2 + \frac{5}{2}\frac{e}{n} A^2 + \left(1 + 4\frac{e}{n} A^2 + \frac{35}{8}\frac{e^2}{n^2} A^4\right)^{1/2}$$

$$\omega = \omega_1 \left(1 + \frac{e}{n} A^2 + \frac{5}{24}\frac{e^2}{n^2} A^4 - 4B^2 + \frac{5}{3} B^4 - \frac{5}{2}\frac{e}{n} A^2 B^2\right)^{1/2}$$

These formulas can be easily discussed.

4.2. Smooth Solution to Two-Dimensional Motion

The above procedure can be easily extended to the case of multi-dimensional motions governed by a set of coupled Hill equations.

As a simple example, let us return to equation (1) of this chapter; we shall begin the smoothing by replacing $n(\theta)$ and $\alpha(\theta)$ by

$$n(\theta) = n \cos \omega_0 \theta \qquad \alpha(\theta) = \bar{\alpha}$$

where $\bar{\alpha}$ is the mean value of $\alpha(\theta)$.

Since in general $| n(\theta) | \gg 1$, (1) will be rewritten

$$\frac{d^2 x_1}{d\theta^2} = n x_1 \cos \omega_0 \theta + \frac{\bar{\alpha}}{2} (x_1{}^2 - x_2{}^2) \tag{86}$$

$$\frac{d^2 x_2}{d\theta^2} = -n x_2 \cos \omega_0 \theta - \bar{\alpha} x_1 x_2 \tag{87}$$

We shall try to fit solutions of the form

$$x_1 = \xi_1(\theta) + \xi_2(\theta) \tag{88}$$

$$x_2 = \eta_1(\theta) + \eta_2(\theta) \tag{89}$$

with

$$\xi_2(\theta) = a \cos \omega_0 \theta \qquad \eta_2(\theta) = b \cos \omega_0 \theta$$

where $\xi_1(\theta)$ and $\eta_1(\theta)$ are slowly varying functions of θ, and the same for $a = a(\theta)$ and $b = b(\theta)$.

By substituting (88) and (89) in (86), we get, when neglecting harmonics of order 2,

$$\frac{d^2\xi_1}{d\theta^2} + \frac{d^2\xi_2}{d\theta^2} = [n\xi_1 + \bar{\alpha}(a\xi_1 - b\eta_1)] \cos \omega_0\theta + \frac{na}{2} + \frac{\bar{\alpha}}{2}({\xi_1}^2 - {\eta_1}^2) + \frac{\bar{\alpha}}{4}(a^2 - b^2) \tag{90}$$

from which, by identifying separately the low-frequency terms and the terms whose frequencies lie in the neighborhood of ω_0, we obtain

$$\frac{d^2\xi_1}{d\theta^2} = \frac{na}{2} + \frac{\bar{\alpha}}{2}({\xi_1}^2 - {\eta_1}^2) + \frac{\bar{\alpha}}{4}(a^2 - b^2)$$

$$\frac{d^2\xi_2}{d\theta^2} = -a{\omega_0}^2 \cos \omega_0\theta = [n\xi_1 + \bar{\alpha}(a\xi_1 - b\eta_1)] \cos \omega_0\theta$$

Then, when assuming that a, b, and $\bar{\alpha}$ are small, say when neglecting a^2, b^2, $\bar{\alpha}a$, $\bar{\alpha}b$, we get

$$a = -\frac{n}{{\omega_0}^2}\xi_1 \tag{91}$$

$$\frac{d^2\xi_1}{d\theta^2} = \frac{na}{2} + \frac{\bar{\alpha}}{2}({\xi_1}^2 - {\eta_1}^2) \tag{92}$$

By similar arguments we get from (87),

$$b = \frac{n}{{\omega_0}^2}\eta_1 \tag{93}$$

$$\frac{d^2\eta_1}{d\theta^2} = -\frac{nb}{2} - \bar{\alpha}\xi_1\eta_1 \tag{94}$$

Finally, by substituting (91) and (93) in (92) and (94), we obtain two coupled equations with constant coefficients with respect to ξ_1 and η_1

$$\begin{aligned} \frac{d^2\xi_1}{d\theta^2} &= -\frac{n^2}{2{\omega_0}^2}\xi_1 + \frac{\bar{\alpha}}{2}({\xi_1}^2 - {\eta_1}^2) \\ \frac{d^2\eta_1}{d\theta^2} &= -\frac{n^2}{2{\omega_0}^2}\eta_1 - \bar{\alpha}\xi_1\eta_1 \end{aligned} \tag{95}$$

This system is the smooth approximation of (1). The procedure is valid provided that $n/\omega_0(2)^{1/2} \ll \omega_0$. These equations have the same form as the ones we considered in Chapter IV, Eq. (7), with $\xi_1 = \bar{x}_1$, $\eta_1 = \bar{x}_2$.

REFERENCES

1. G. Floquet, Sur les équations différentielles linéaires. *Ann. l'École Normale Supérieure* **2 12**, 47–88 (1883).
2. N. W. McLachlan, "Theory and Application of Mathieu Functions." Oxford Univ. Press, New York, 1947.
3. L. Brillouin, A Practical Method for Solving Hill's Equation. *Quart. Appl. Math.* **6**, 167–178 (1948).
4. T. Sigurgeirsson, "Focusing in a Synchrotron with Periodic Field, Perturbation Treatment." CERN/T/TS-3, May 1953.
5. A. Blaquière, Sur les orbites dans le cosmotron à focalisation forte à l'approximation non linéaire. *Compt. Rend.* **239**, 1285–1287 (1954).
6. K. R. Symon, "Smooth Solution to One-dimensional AG Orbits with Cubic Forces." KRS (MURA)-2, July 1954.
7. A. Schoch, "Orbit Stability in a Synchrotron with Non-Linear Restoring Forces." CERN-PS/A. Sch. 2, May 1955.
8. A. Blaquière, Equation de Hill nonlinéaire et méthode stroboscopique de N. Minorsky. *Comp. Rend.* **243**, 1711–1714 (1956).
9. M. Barbier, A Mechanical Analogue for the Study of Betatron Oscillations.†
10. E. D. Courant, Non Linearities in the AG Synchrotron.†
11. R. Hagedorn, M. G. N. Hine, A. Schoch, Non-linear Orbit Problems in Synchrotrons.†
12. A. A. Kolomenski, On the Non-Linear Theory of Betatron Oscillations.†
13. L. J. Laslett and K. R. Symon, Particle Orbits in Fixed field AG Accelerators.†
14. J. Moser, The Resonance Lines for the Synchrotron.†
15. N. N. Bogoliubov and Y. A. Mitropolsky, "Asymptotic Methods in the Theory of Nonlinear Oscillations." Fizmatgiz, Moscow, 1958. (English transl., Gordon and Breach, New York, 1961.)
16. W. J. Cunningham, "Introduction to Nonlinear Analysis." McGraw-Hill, New York, 1958.
17. L. Cesari, Asymptotic Behaviour and Stability Problems in Ordinary Differential Equations. Springer, Berlin, 1959.
18. J. K. Hale, On Differential Equations Containing a Small Parameter. Dept. Math., Univ. California, *Tech. Rept.* 6, Sept. 1960.
19. K. G. Valeev, On Hill's Method in the Theory of Linear Differential Equations with Periodic Coefficients. *J. Appl. Math. Mech.* [Translation of *Prikl. Mat. i Mekhan.* **24**, 1493–1505 (1960).]
20. K. G. Valeev, On Hill's Method in the Theory of Linear Differential Equations with Periodic Coefficients. Determination of the Characteristic Exponents. *J. Appl. Math. Mech.* [Translation of *Prikl. Mat. i Mekhan.* **25**, 460–466 (1961).]
21. C. S. Hsu, On a Restricted Class of Coupled Hill's Equations and Some Applications. *J. Appl. Mech.* **84**, 551–556 (1961).
22. R. A. Struble, The Geometry of the Orbits of Artificial Satellites. *Arch. Rational Mech. Anal.* **7**, 87–104 (1961).
23. R. A. Struble and J. E. Fletcher, General Perturbational Solution of the Harmonically Forced Van der Pol Equation. *J. Math. Phys.* **2**, 880–891 (1961).
24. R. A. Struble and S. M. Yionoulis, General Perturbational Solution of the Harmonicaly Forced Duffing Equation. *Arch. Rational Mech. Anal.*

† *CERN Symp. High Energy Accelerators Pion Phys.*, **1956.**

25. R. A. Struble and J. E. Fletcher, General Perturbational Solution of the Mathieu Equation. *J. Soc. Ind. Appl. Math.* **10** (June 1962).
26. R. A. Struble, "Nonlinear Differential Equations." McGraw-Hill, New York, 1961.
27. C. S. Hsu, On the Parametric Excitation of a Dynamic System Having Multiple Degrees of Freedom. *J. Appl. Mech.* paper **63**, APM-14 (Oct. 11, 1962).
28. C. S. Hsu, On the Stability of Periodic Solutions of Nonlinear Dynamical Systems under Forcing. *Colloq. Intern. Centre Natl. Res. Sci.* (*Marseille*), Sept. 7–12, 1964.

CHAPTER VIII

System Response to Random Inputs

Most physical systems are subjected to random inputs (or noise), which in many cases are sufficiently small to be disregarded in a first analysis of their dynamic behavior. Earlier we discussed this viewpoint, which led us to consider only deterministic laws of motion. However, when the requirements become more severe, owing, for example, to an improvement in the technique of measurement, random fluctuations of the motion need to be taken into account.

These fluctuations can be approached from several different ways:

Approach 1. One can consider the noise as a forcing function which operates upon a deterministic system. (By deterministic system we mean a system governed by deterministic laws when not perturbed by noise.)

Approach 2. One can go deeper into the actual mechanism of the motion, and generally resolve the over-all behavior into a large number of elementary random processes. Then the laws which govern the variables are analyzed from a macroscopic viewpoint based upon the definition of mean values.

Such investigations play an important role, by stating precisely the limit prescribed to the accuracy of the measurements.

Approach 1 is much simplified by the use of Campbell's theorem, which is valid only in the case of linear laws of motion; however, it can be extended to nonlinear systems in the near neighborhood of a steady state if the fluctuations are small enough to permit local linearization. Approach 2 will rely upon the Fokker-Planck-Kolmogorov equation [1, 3–5].

1. CAMPBELL'S THEOREM†

1.1. The Effect of Standard Random Impulses upon a Linear System

Let $\xi(t)$ be the response of a linear system due to an impulse at time $t = 0$, and assume that N impulses, each of which is similar to the latter one, are acting upon the system at random times t_k: $t_0 \leqslant t_k \leqslant t_1$, t_0 and t_1 being given. Moreover, assume that there is no correlation between the random times t_k (Fig. 1). The total effect at time t: $t_0 \leqslant t \leqslant t_1$, is

$$x(t) = \sum_{t_k} \xi(t - t_k)$$

We assume also that N is a random number whose ensemble average is $\langle N \rangle$.

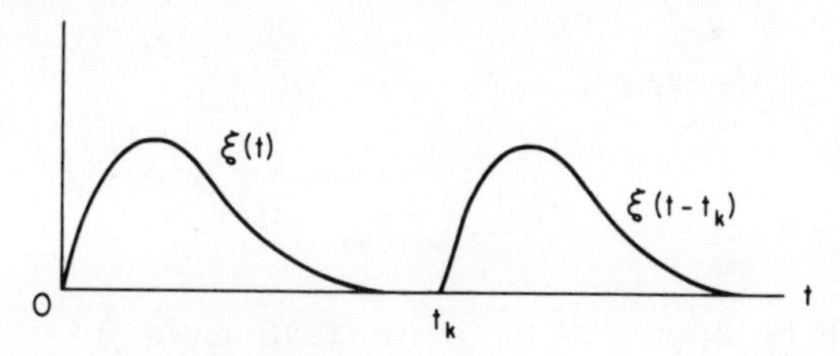

FIG. 1. Response of a linear system due to an impulse, at time $t = 0$ and at time $t = t_k$.

Now let us divide the interval $[t_0, t_1]$ into arbitrarily small subintervals Δt_i at times t_i ($i = 1, 2, ..., m$): $t_0 < t_i < t_1$. The average number of impulses in an interval Δt_i is

$$np(t_i)\,\Delta t_i + o(\Delta t_i) = \frac{\langle N \rangle}{t_1 - t_0}\,p(t_i)\,\Delta t_i + o(\Delta t_i)$$

where the probability $p(t)$ is a given function of t which we shall assume to be continuous on $[t_0, t_1]$. These impulses result in the average response

$$np(t_i)\xi(t - t_i)\,\Delta t_i + o'(t, \Delta t_i)$$

whose magnitude at time t_1 is

$$\eta_i(t_1) \triangleq np(t_i)\xi(t_1 - t_i)\Delta t_i + o'(t_1, \Delta t_i)$$

† See [2, 9].

where $o'(t, \Delta t_i)/\Delta t_i \to 0$ uniformly as $\Delta t_i \to 0$, $\forall\, t : t_0 \leqslant t \leqslant t_1$, from which follows

$$\lim_{\Delta t_i \to 0} [\eta_i(t_1)/\Delta t_i] = np(t_i)\xi(t_1 - t_i)$$

Then by integrating on the interval $[t_0, t_1]$ we get

$$\langle x(t_1)\rangle = n \int_{t_0}^{t_1} p(t_i)\xi(t_1 - t_i)\, dt_i$$

$$= n \int_0^{t_1 - t_0} p(t_1 - t)\xi(t)\, dt$$

where $\langle x(t_1)\rangle$ is the ensemble average value of the output at time t_1 due to the N random impulses, and n the average number of impulses per second.

When $p(t) \equiv 1$, we get

$$\langle x(t_1)\rangle = n \int_0^{t_1 - t_0} \xi(t)\, dt \tag{1}$$

Now we will be interested in the fluctuations $x(t_1) - \langle x(t_1)\rangle$.

Let ν_i be the actual number of impulses in an interval Δt_i whose average value is $\langle \nu_i\rangle = np(t_i)\, \Delta t_i + o(\Delta t_i)$. The corresponding fluctuation of the response, at time t_1, is

$$(\nu_i - \langle \nu_i\rangle)\xi(t_1 - t_i) + o''(\Delta t_i)$$

where $o(\Delta t_i)/\Delta t_i \to 0$, $o''(\Delta t_i)/\Delta t_i \to 0$ uniformly as $\Delta t_i \to 0$.

Now if we assume that ν_i has normal distribution, which is ensured by the central limit theorem when $\langle N\rangle$ is sufficiently large and when the effect of each impulse is sufficiently small, we have

$$\langle(\nu_i - \langle \nu_i\rangle)^2\rangle = \langle \nu_i\rangle$$

Hence the mean square of the fluctuation of the response at time t_1 due to ν_i random impulses in Δt_i is

$$\langle \nu_i\rangle \xi^2(t_1 - t_i) + o'''(\Delta t_i) = np(t_i)\xi^2(t_1 - t_i)\, \Delta t_i + o''''(\Delta t_i)$$

where $o'''(\Delta t_i)/\Delta t_i \to 0$, $o''''(\Delta t_i)/\Delta t_i \to 0$ uniformly as $\Delta t_i \to 0$.

Since the mean squares of the fluctuations relative to the different subintervals are additive, by letting the subintervals tend to zero and integrating on the interval $[t_0, t_1]$ we get

$$\langle [x(t_1) - \langle x(t_1)\rangle]^2\rangle = n \int_{t_0}^{t_1} p(t_i)\xi^2(t_1 - t_i)\, dt_i$$

$$= n \int_0^{t_1 - t_0} p(t_1 - t)\xi^2(t)\, dt$$

e.g., when $p(t) \equiv 1$,

$$\langle [x(t_1) - \langle x(t_1)\rangle]^2\rangle = n \int_0^{t_1 - t_0} \xi^2(t)\, dt \tag{2}$$

Now consider the case where impulses are acting upon the linear system from $-\infty$ to $+\infty$, and let t' be any time, $-\infty < t' < +\infty$. Since impulses which occur at times $t_k > t'$ make no contribution to the value of the output at time t', we get $\langle x(t')\rangle$ and $\langle [x(t') - \langle x(t')\rangle]^2\rangle$ by putting $t_1 = t'$ and letting $t_0 \to -\infty$ in (1) and (2). Then we get Campbell's theorem:

THEOREM. When identical impulses are acting upon a linear system at uncorrelated random times t_k: $-\infty < t_k < +\infty$, with the probability density $p(t_k) \equiv 1$, the average value of the output, and the mean-square value of the corresponding fluctuation about the average, are independent of time and are given by

$$\langle x\rangle = n \int_0^\infty \xi(t)\, dt \tag{3}$$

$$\langle (x - \langle x\rangle)^2\rangle = n \int_0^\infty \xi^2(t)\, dt \tag{4}$$

provided that the integrals on the right sides are defined, and provided that the conditions of applicability of the central limit theorem are fulfilled. [n is the average number of impulses per second and $\xi(t)$ is the response of the system due to an impulse at time $t = 0$.]

In the following, in cases when these integrals are not defined, we shall use formulas (1) and (2) instead of (3) and (4). For instance, these integrals are not defined in the random-walk problem, where $\xi(t)$ is a step function $qY(t)$ (q any scalar, $Y(t)$ the unit step function). Then if the random walk begins at time $t_0 = 0$ and lasts until $t_1 = \tau$,

$$\langle x(\tau)\rangle = nq \int_0^\tau Y(t)\, dt = nq\tau$$

$$\langle [x(\tau) - \langle x(\tau)\rangle]^2\rangle = nq^2 \int_0^\tau Y^2(t)\, dt = nq^2\tau$$

1.2. The Effect of Miscellaneous Random Impulses on a Linear System

In many applications, the random impulses which are acting upon the linear system are not all identical. Next let us assume that the impulses are of different kinds, and let n_1, n_2, ..., n_α be, respectively, the average number per second of impulses of the type

$$\delta_1(t) \qquad \delta_2(t) \cdots \delta_\alpha(t)$$

with

$$\int_{\Delta t} \delta_i(t)\, dt = q_i \qquad (i = 1, 2, ..., \alpha)$$

and

$$n = n_1 + n_2 + \cdots + n_\alpha$$

Assume that the impulses are acting upon the system at uncorrelated random times t_k: $-\infty < t_k < +\infty$, with the probability density $p(t_k) \equiv 1$. Moreover, let us denote by $\xi_0(t)$ the response of the linear system due to the normalized impulse function $\delta(t)$: $\int_{\Delta t} \delta(t)\, dt = 1$, at time $t = 0$. Then if both n_1, n_2, ..., n_α are sufficiently large numbers, and if there is no correlation between the impulses,

$$\langle x \rangle = \sum_{i=1}^{\alpha} \langle x_i \rangle = \sum_{i=1}^{\alpha} n_i q_i \int_0^\infty \xi_0(t)\, dt$$

$$\langle (x - \langle x \rangle)^2 \rangle = \sum_{i=1}^{\alpha} \langle (x_i - \langle x_i \rangle)^2 \rangle$$

$$= \sum_{i=1}^{\alpha} n_i q_i^2 \int_0^\infty \xi_0^2(t)\, dt$$

Now if we define

$$q_{(1)} = \frac{\sum_{i=1}^{\alpha} n_i q_i}{n}$$

and $q_{(2)}$ such that

$$q_{(2)}^2 = \frac{\sum_{i=1}^{\alpha} n_i q_i^2}{n}$$

we get

$$\langle x \rangle = nq_{(1)} \int_0^\infty \xi_0(t)\, dt \tag{5}$$

$$\langle (x - \langle x \rangle)^2 \rangle = nq_{(2)}^2 \int_0^\infty \xi_0^{\,2}(t)\, dt \tag{6}$$

As a matter of fact, it may be that $q_{(1)} = 0$, whereas $q_{(2)} \neq 0$ always; then $\langle x \rangle = 0$.

2. FOKKER-PLANCK-KOLMOGOROV METHOD

The Fokker-Planck-Kolmogorov method is very convenient for studying the behavior of a physical system subjected to random fluctuation.

First of all, we need to consider the variables which will determine the "state" of the system at each given time, the dynamic equations which govern the motion of these variables, and the salient feature of the random terms.

(a) *State variables.* In general we shall designate the state variables by $x_1, x_2, ..., x_n$, and the values of these variables at time t by $x_1(t)$, $x_2(t)$, ..., $x_n(t)$. The physical system will be represented, at any time, by a point in the n-dimensional phase space thus defined, say in R^n $(x_1, x_2, ..., x_n)$. For instance, when the motion of the system is governed by an nth-order differential equation with respect to the variable $x(t)$, it is convenient, as pointed out earlier, to define R^n by putting

$$x_1(t) = x(t) \qquad x_2(t) = \dot{x}(t) \cdots x_n(t) = x^{(n-1)}(t)$$

When it proves convenient we shall assume that the phase space is a plane (x_1, x_2).

(b) *Equations of motion and random terms.* We shall assume that the equations which govern the motion of the representative point in R^n are

$$\dot{x}_i = f_i(x_1, ..., x_n) + F_i(t) \qquad (i = 1, ..., n) \tag{7}$$

where $f_i(x_1, ..., x_n)$ are continuously differentiable functions with respect to the state variables, and $F_i(t)$ are random functions with respect to t. In many practical cases, f_i $(i = 1, ..., n)$ are linear functions,

$$f_i(x_1, ..., x_n) = \sum_j a_{ij} x_j \qquad (i, j = 1, ..., n)$$

where the a_{ij} are constant coefficients.

When the system under consideration is not subjected to random inputs ($F_i(t) \equiv 0$), its representative point follows a continuous deterministic path in R^n. Now we shall assume that the random inputs are always sufficiently small, so that the perturbed motion can be determined by superposing random fluctuations of first-order smallness to a continuous mean trajectory.

2.1. The Fokker-Planck-Kolmogorov Equation

First we shall derive the Fokker-Planck-Kolmogorov equation, following a method developed by Chandrasekhar [8].

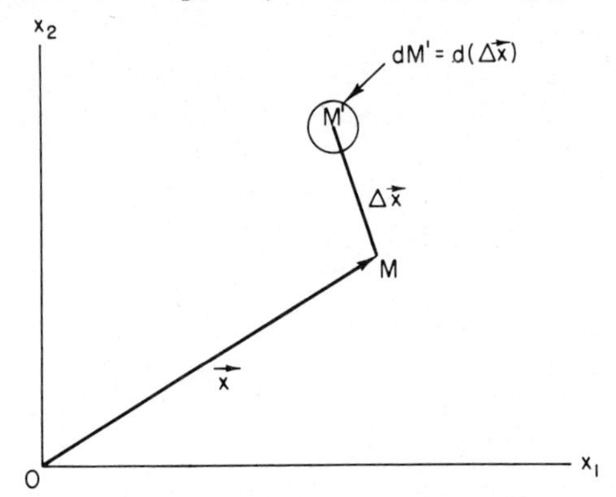

FIG. 2. Calculation of probability density, illustration 1.

Let $p(\mathbf{x}, t)$ be the probability density associated with each position of the representative point in R^n and let $\psi(\mathbf{x}, \Delta\mathbf{x})$ be the probability density associated with the jump $\mathbf{x} \to \mathbf{x} + \Delta\mathbf{x}$ during a given interval Δt (Fig. 2). We shall first compute the probability of finding the re-

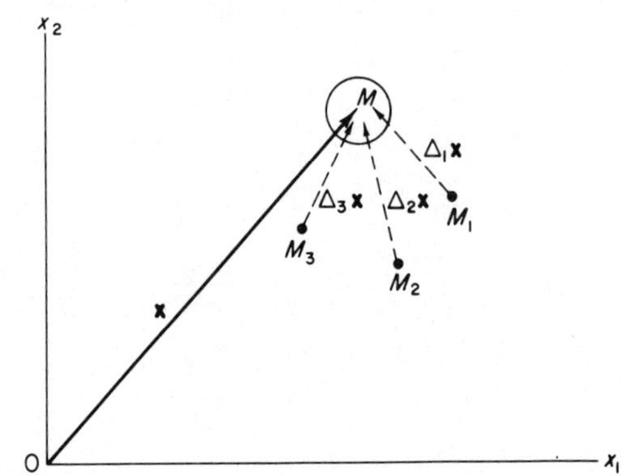

FIG. 3. Calculation of probability density, illustration 2.

presentative point in a small neighborhood $d(\Delta\mathbf{x})$ of $M(\mathbf{x})$, at time $t + \Delta t$; for example, we shall compute

$$p(\mathbf{x}, t + \Delta t)$$

This can be done by specifying the different events which lead to this result, a few of which are listed below (and see Fig. 3):

(a) The representative point is located at $M_1(\mathbf{x} - \Delta_1\mathbf{x})$ at time t; then it executes the jump $\mathbf{M_1M} = \Delta_1\mathbf{x}$, at first-order approximation with respect to $d(\Delta_1\mathbf{x})$, during the interval Δt. The probability density associated with this event is

$$p(\mathbf{x} - \Delta_1\mathbf{x}, t)\psi(\mathbf{x} - \Delta_1\mathbf{x}, \Delta_1\mathbf{x})\, d(\Delta_1\mathbf{x})$$

(b) The representative point is located at $M_2(\mathbf{x} - \Delta_2\mathbf{x})$ at time t; then it executes the jump $\mathbf{M_2M} = \Delta_2\mathbf{x}$, at first-order approximation with respect to $d(\Delta_2\mathbf{x})$, during the interval Δt. The probability density associated with this event is

$$p(\mathbf{x} - \Delta_2\mathbf{x}, t)\psi(\mathbf{x} - \Delta_2\mathbf{x}, \Delta_2\mathbf{x})\, d(\Delta_2\mathbf{x})$$

(c) The representative point is located at $M_3(\mathbf{x} - \Delta_3\mathbf{x})$, etc.

The probability density $p(\mathbf{x}, t + \Delta t)$ will be obtained by summing up the probability densities associated with all possible events (a), (b), (c), ..., from which

$$p(\mathbf{x}, t + \Delta t) = \int_{\Delta\mathbf{x}} p(\mathbf{x} - \Delta\mathbf{x}, t)\psi(\mathbf{x} - \Delta\mathbf{x}, \Delta\mathbf{x})\, d(\Delta\mathbf{x}) \tag{8}$$

Note that this method is valid only if the random process is a *Markoff process*: if the evolution of the system, after any time t, does not depend on its past history.

Now let us assume that Δt and $\Delta\mathbf{x}$ are of first-order smallness and use Taylor's expansion for the probability factors that occur in (8):

$$p(\mathbf{x}, t + \Delta t) = p(\mathbf{x}, t) + \Delta t \frac{\partial p}{\partial t} + o(\Delta t)$$

$$p(\mathbf{x} - \Delta\mathbf{x}, t) = p(\mathbf{x}, t) - \sum_i \Delta x_i \frac{\partial p}{\partial x_i} + \tfrac{1}{2} \sum_i (\Delta x_i)^2 \frac{\partial^2 p}{\partial x_i^2} + \sum_{i<j} \Delta x_i \Delta x_j \frac{\partial^2 p}{\partial x_i \partial x_j} + o'(\Delta x_i\, \Delta x_j)$$

$$\psi(\mathbf{x} - \Delta\mathbf{x}, \Delta\mathbf{x}) = \psi(\mathbf{x}, \Delta\mathbf{x}) - \sum_i \Delta x_i \frac{\partial \psi}{\partial x_i} + \tfrac{1}{2} \sum_i (\Delta x_i)^2 \frac{\partial^2 \psi}{\partial x_i^2} + \sum_{i<j} \Delta x_i\, \Delta x_j \frac{\partial^2 \psi}{\partial x_i \partial x_j} + o''(\Delta x_i\, \Delta x_j)$$

Substituting in (7) we get,

$$p(\mathbf{x}, t) + \frac{\partial p}{\partial t} \Delta t + o(\Delta t)$$

$$= \int_{\Delta \mathbf{x}} \Big[p(\mathbf{x}, t) - \sum_i \frac{\partial p}{\partial x_i} \Delta x_i + \tfrac{1}{2} \sum_i \frac{\partial^2 p}{\partial x_i^2} (\Delta x_i)^2 + \sum_{i<j} \frac{\partial^2 p}{\partial x_i \partial x_j} \Delta x_i \, \Delta x_j + o'(\Delta x_i \, \Delta x_j) \Big]$$

$$\times \Big[\psi(\mathbf{x}, \Delta \mathbf{x}) - \sum_i \frac{\partial \psi}{\partial x_i} \Delta x_i + \tfrac{1}{2} \sum_i \frac{\partial^2 \psi}{\partial x_i^2} (\Delta x_i)^2 + \sum_{i<j} \frac{\partial^2 \psi}{\partial x_i \partial x_j} \Delta x_i \, \Delta x_j + o''(\Delta x_i \, \Delta x_j) \Big] \, d(\Delta \mathbf{x})$$

When calculating the products on the right side, we get such expressions as

$$\langle \Delta x_i \rangle = \int_{\Delta \mathbf{x}} \Delta x_i \psi(\mathbf{x}, \Delta \mathbf{x}) \, d(\Delta \mathbf{x})$$

$$\langle (\Delta x_i)^2 \rangle = \int_{\Delta \mathbf{x}} (\Delta x_i)^2 \psi(\mathbf{x}, \Delta \mathbf{x}) \, d(\Delta \mathbf{x})$$

$$\langle \Delta x_i \, \Delta x_j \rangle = \int_{\Delta \mathbf{x}} \Delta x_i \, \Delta x_j \psi(\mathbf{x}, \Delta \mathbf{x}) \, d(\Delta \mathbf{x})$$

These expressions are the ensemble averages of Δx_i, $(\Delta x_i)^2$, $\Delta x_i \Delta x_j$. Moreover, we have

$$\int_{\Delta \mathbf{x}} \psi(\mathbf{x}, \Delta \mathbf{x}) \, d(\Delta \mathbf{x}) = 1$$

Finally, when all the computations have been made, the Fokker-Planck-Kolmogorov equation is obtained:

$$\frac{\partial p}{\partial t} = \lim_{\Delta t \to 0} \Big\{ - \sum_i \frac{\partial}{\partial x_i} \Big(p \frac{\langle \Delta x_i \rangle}{\Delta t} \Big) + \tfrac{1}{2} \sum_i \frac{\partial^2}{\partial x_i^2} \Big[p \frac{\langle (\Delta x_i)^2 \rangle}{\Delta t} \Big] + \sum_{i<j} \frac{\partial^2}{\partial x_i \, \partial x_j} \Big(p \frac{\langle \Delta x_i \, \Delta x_j \rangle}{\Delta t} \Big) + \frac{o'''(\langle \Delta x_i \, \Delta x_j \rangle)}{\Delta t} \Big\} \tag{9}$$

Next we shall assume that

$$\lim_{\Delta t \to 0} \frac{o'''(\langle \Delta x_i \, \Delta x_j \rangle)}{\Delta t} = 0$$

We shall now use this equation to solve practical problems.

2.2. Explicit Form for the Equation

Let us assume that the physical system under consideration is governed by differential equations (7). By integrating (7) on the time interval Δt, we get

$$\Delta x_i = f_i(x_1, ..., x_n)\,\Delta t + \int_t^{t+\Delta t} F_i(t)\,dt + o(\Delta t) \tag{10}$$

Then

$$\langle \Delta x_i \rangle = f_i(x_1, ..., x_n)\,\Delta t + \left\langle \int_t^{t+\Delta t} F_i(t)\,dt \right\rangle + o(\Delta t)$$

$$\langle (\Delta x_i)^2 \rangle = \left\langle \left[\int_t^{t+\Delta t} F_i(t)\,dt \right]^2 \right\rangle + o'(\Delta t)$$

$$\langle \Delta x_i\,\Delta x_j \rangle = \left\langle \int_t^{t+\Delta t} F_i(t)\,dt \int_t^{t+\Delta t} F_j(t)\,dt \right\rangle + o''(\Delta t)$$

$$(i = 1, ..., n)$$

where

$$\frac{o(\Delta t)}{\Delta t} \to 0 \qquad \frac{o'(\Delta t)}{\Delta t} \to 0 \qquad \frac{o''(\Delta t)}{\Delta t} \to 0 \qquad \text{uniformly, as} \quad \Delta t \to 0$$

Next we shall assume that the following limits exist:

$$\lambda_i = \lim_{\Delta t \to 0} \frac{1}{\Delta t} \left\langle \int_t^{t+\Delta t} F_i(t)\,dt \right\rangle \qquad S_{ii} = \lim_{\Delta t \to 0} \frac{1}{\Delta t} \left\langle \left\{ \int_t^{t+\Delta t} F_i(t)\,dt \right\}^2 \right\rangle$$

$$S_{ij} = \lim_{\Delta t \to 0} \frac{1}{\Delta t} \left\langle \int_t^{t+\Delta t} F_i(t)\,dt \int_t^{t+\Delta t} F_j(t)\,dt \right\rangle \tag{11}$$

When $F_i(t) \triangleq F_i(t, x_1, ..., x_n)$ $(i = 1, ..., n)$ depends not only on t but also on the state variables, and when these functions are nonstationary, the coefficients λ_i, S_{ii}, and S_{ij} are functions of t and of the state vector $\mathbf{x}(x_1, ..., x_n)$:

$$\lambda_i \triangleq \lambda_i(t, \mathbf{x}) \qquad S_{ii} \triangleq S_{ii}(t, \mathbf{x}) \qquad S_{ij} \triangleq S_{ij}(t, \mathbf{x})$$

Finally, substituting in (9), we obtain

$$\frac{\partial p}{\partial t} = -\sum_i \frac{\partial}{\partial x_i} p(f_i + \lambda_i) + \tfrac{1}{2} \sum_i \frac{\partial^2}{\partial x_i^2} pS_{ii} + \sum_{i<j} \frac{\partial^2}{\partial x_i\,\partial x_j} pS_{ij} \tag{12}$$

or, when λ_i, S_{ii}, and S_{ij} do not depend on $\mathbf{x}$

$$\frac{\partial p}{\partial t} + \sum_i \frac{\partial}{\partial x_i} pf_i = -\sum_i \lambda_i \frac{\partial p}{\partial x_i} + \tfrac{1}{2} \sum_i S_{ii} \frac{\partial^2 p}{\partial x_i^2} + \sum_{i<j} S_{ij} \frac{\partial^2 p}{\partial x_i\,\partial x_j} \tag{13}$$

The second form is very convenient in practical applications; we shall refer to it in the next chapter in relation to Berstein's method; it was also used by Crandall [19] in his study of oscillators perturbed by a random force.

2.3. Other Derivation: Physical Meaning of the Equation

Once the physical system has been represented by a point $M(x_1, ..., x_n)$ in phase space R^n, the analysis of its evolution when random forces are acting on it is reduced to the study of a kind of random walk—the brownian motion of point M in R^n.

Then if we consider *at the same time* many copies of *the same* physical system, with different values of the state variables, say

$$
\begin{array}{llll}
M^{(1)}: & x_1^{(1)}(t) \quad x_2^{(1)}(t) \cdots x_n^{(1)}(t) & & \text{for copy 1} \\
M^{(2)}: & x_1^{(2)}(t) \quad x_2^{(2)}(t) \cdots x_n^{(2)}(t) & & \text{for copy 2} \\
 & \vdots & & \\
M^{(k)}: & x_1^{(k)}(t) \quad x_2^{(k)}(t) \quad x_n^{(k)}(t) & & \text{for copy } k \\
 & \vdots & &
\end{array}
$$

we see that the actual problem can be reduced to another one, which deals with the diffusion of a gas in R^n. Indeed, from this viewpoint, $M^{(i)}$ ($i = 1, 2, ..., k, ...$) can be considered a "pseudo particle", and to each position of $M^{(i)}$ in R^n is associated a probability density $p(\mathbf{x}, t)$ which depends on position $\mathbf{x}$ and time t.

Now assume that, at the initial time $t = 0$, all the copies of the physical system are *in the same state*, say

$$M^{(1)}(0) = M^{(2)}(0) = \cdots = M^{(k)}(0) = \cdots = M_0$$

and, starting from this position, let the different systems follow their own history in R^n.

Owing to the incertainty introduced by the random forces $F_1(t), ..., F_n(t)$, the points will gradually be separated in R^n; that is to say, they will generate a *cloud*, which will diffuse in R^n and stretch out in the course of time.

Then $p(\mathbf{x}, t)$ may be thought of either as the local density of this cloud at point $\mathbf{x}$ and time t, or as the density probability defined above. These two concepts will only differ by a constant factor which will have to be normalized if we choose the second meaning. Here we shall prefer the first meaning, and will disregard the constant factor, which does not play a significant role in the theory in any case.

Let $\mathscr{I}(\mathbf{x}, t)$ be the local stream at point $\mathbf{x}$ at time t, in the moving cloud, and use the well-known formula of fluid mechanics,

$$\frac{\partial p}{\partial t} = -\operatorname{div} \mathscr{I} \tag{14}$$

Furthermore, let us separate $\mathscr{I}$ into two terms (Fig. 4),

$$\mathscr{I} = \mathbf{I} + \mathbf{i} \tag{15}$$

where $\mathbf{I}$ is the *global* intensity of the stream and $\mathbf{i}$ is a small *diffusion current* at the point under consideration.

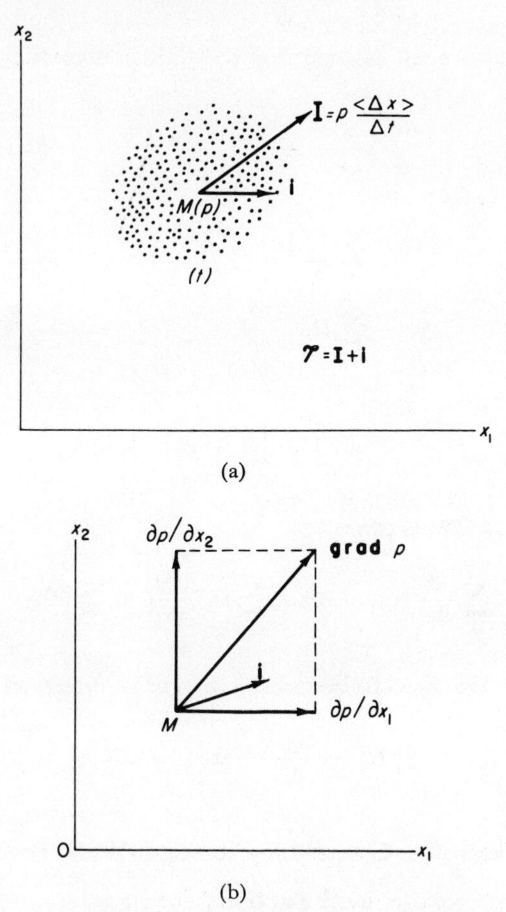

FIG. 4. (a) Local stream, global intensity, and diffusion current in a cloud; (b) vector relation between the diffusion current and the gradient of the density.

According to the usual definition of **I** in fluid mechanics, the components of **I** along the axes of R^n are

$$p \lim_{\Delta t \to 0} \frac{\langle \Delta x_i \rangle}{\Delta t} = p[f_i(\mathbf{x}, t) + \lambda_i] \qquad (i = 1, ..., n) \tag{16}$$

where $\langle \Delta x_i \rangle / \Delta t$ is the mean velocity of the stream at point $\mathbf{x}$, (t) (λ_i has been defined above).

On the other hand, as concerns the diffusion current **i**, we shall make use of the relation

$$\mathbf{i} = -D \,\mathbf{grad}\, p \tag{17}$$

which is the so-called *Fick's law*.

Moreover, we shall assume that the diffusion coefficient D is given in matrix form:

$$D = (D_{ij}) \tag{18}$$

From (17) and (18) we get

$$\begin{aligned} \operatorname{div} \mathbf{i} &= -\sum_{i=1}^{n} \sum_{j=1}^{n} D_{ij} \frac{\partial^2 p}{\partial x_i \, \partial x_j} \\ &= -\sum_i D_{ii} \frac{\partial^2 p}{\partial x_i^2} - \sum_{i<j} (D_{ij} + D_{ji}) \frac{\partial^2 p}{\partial x_i \, \partial x_j} \end{aligned} \tag{19}$$

and from (16),

$$\operatorname{div} \mathbf{I} = \sum_i \frac{\partial}{\partial x_i} p(f_i + \lambda_i)$$

Then formula (14) is rewritten

$$\frac{\partial p}{\partial t} = -\sum_i \frac{\partial}{\partial x_i} p(f_i + \lambda_i) + \sum_i D_{ii} \frac{\partial^2 p}{\partial x_i^2} + \sum_{i<j} (D_{ij} + D_{ji}) \frac{\partial^2 p}{\partial x_i \, \partial x_j} \tag{20}$$

We identify the Fokker-Planck-Kolmogorov equation (12) by putting

$$D_{ii} = \frac{S_{ii}}{2} \qquad D_{ij} + D_{ji} = S_{ij} \tag{21}$$

2.4. First Example: Elementary Random-Walk Problem

Now let us consider more precisely the equation

$$\dot{x} = F(t) \tag{22}$$

This is the law that governs a simple integrator driven by the random force $F(t)$. A mechanical model governed by (22) is shown in Fig. 5. Its only component is a dashpot, one of whose end points is fixed, the

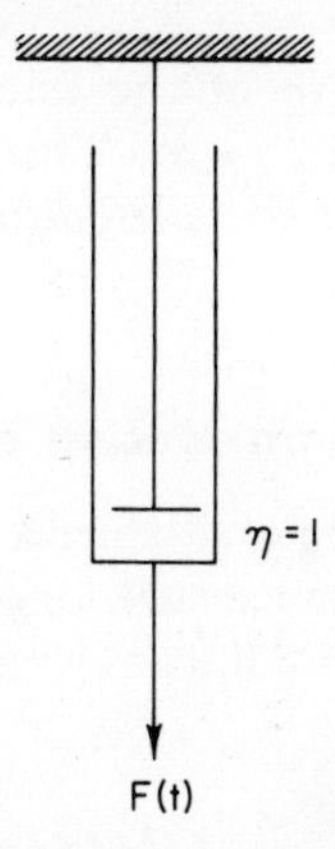

FIG. 5. Simple integrator driven by the random force $F(t)$.

other being moved by the random force $F(t)$. x denotes the translation of the moving end point, and $\eta = 1$ is the viscosity coefficient of the fluid in the dashpot. Obviously there exists no steady motion, since the stretching of the dashpot can increase indefinitely. Note that more realistic models should generally introduce a boundary on x.

Then equation (13) is rewritten

$$\frac{\partial p}{\partial t} = \frac{S}{2} \frac{\partial^2 p}{\partial x^2} \qquad S_{11} = S \tag{23}$$

This equation is well known in many fields concerned with diffusion processes. For instance, it appears in the study of nuclear reactors, for it governs the evolution of a neutron gas which slows down in a moderator; in this field (23) is the so-called equation of the "age," due to Fermi. The "age" is the variable τ which is introduced in (23) by putting

$$\tau = \frac{S}{2} t$$

Then (23) is rewritten

$$\frac{\partial p}{\partial \tau} = \frac{\partial^2 p}{\partial x^2}$$

Its solution is

$$p(x, \tau) = \frac{C}{(4\pi\tau)^{1/2}} \exp\left[-\frac{(x - x_0)^2}{4\tau}\right] = \frac{C}{(2\pi St)^{1/2}} \exp\left[-\frac{(x - x_0)^2}{2St}\right] \tag{24}$$

where x_0 is the value of x at time $t = 0$, and C is a constant of integration.

That is a gaussian law, from which we can easily deduce the mean square of the fluctuation $\langle (x - x_0)^2 \rangle$ at any time t:

$$\langle (x - x_0)^2 \rangle = St \tag{25}$$

We see that the mean square of the fluctuation increases with time beyond any boundary, when $t \to \infty$. This establishes the fact that no steady motion exists. $S/2$ is the so-called *diffusion coefficient* in the classical theory of diffusion processes.

2.5. Second Example: Random Motion of a Nonlinear Oscillator

Let us consider the following oscillator: A unit mass ($m = 1$), mounted on a horizontal string $x'x$, can execute longitudinal vibrations; it is connected to a fixed point by a nonlinear spring (Fig. 6).

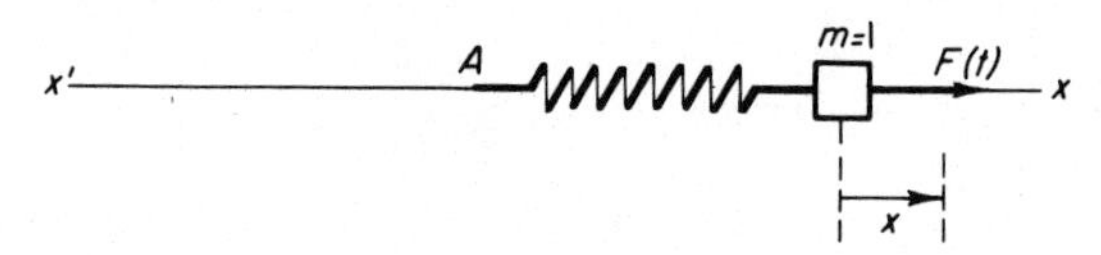

FIG. 6. Oscillation with mass connected to a fixed point.

The force due to the spring is $f(x)$, where x is the translation of the mass with respect to its equilibrium point; the damping force is $-\alpha\dot{x}$ ($\alpha > 0$); furthermore, the random force $F(t)$ is acting on m along $x'x$. We shall assume that $F(t)$ is a stationary random function and that $\langle F(t) \rangle = 0$.

The equation of motion is

$$\ddot{x} + \alpha\dot{x} - f(x) = F(t) \tag{26}$$

For example, by putting $x_1 = x$, $x_2 = \dot{x}$,

$$\dot{x}_1 = x_2 \qquad \dot{x}_2 = f(x_1) - \alpha x_2 + F(t) \tag{27}$$

In this example we have

$$f_1(x_1, x_2) \equiv x_2 \qquad f_2(x_1, x_2) \equiv f(x_1) - \alpha x_2$$
$$F_1(t) \equiv 0 \qquad F_2(t) \equiv F(t)$$

Accordingly, (13) is rewritten

$$\frac{\partial p}{\partial t} + x_2 \frac{\partial p}{\partial x_1} + f(x_1) \frac{\partial p}{\partial x_2} = \alpha p + \alpha x_2 \frac{\partial p}{\partial x_2} + \frac{S}{2} \frac{\partial^2 p}{\partial x_2^{\,2}} \tag{28}$$

with

$$\lambda_1 = \lambda_2 = 0 \qquad S_{11} = 0 \qquad S = S_{22} = \lim_{\Delta t \to 0} \frac{1}{\Delta t} \left\langle \left[\int_t^{t+\Delta t} F(t)\, dt \right]^2 \right\rangle$$

Next we shall consider only stationary motion, which is governed by

$$\frac{\partial p}{\partial t} = 0$$

$$x_2 \frac{\partial p}{\partial x_1} + f(x_1) \frac{\partial p}{\partial x_2} = \alpha p + \alpha x_2 \frac{\partial p}{\partial x_2} + \frac{S}{2} \frac{\partial^2 p}{\partial x_2{}^2} \tag{29}$$

2.6. Maxwell-Boltzmann Solution of the Equation of Motion

The oscillator described above is a nonconservative system:

(a) The friction (represented by the damping term) dissipates energy.

(b) Because of the random force, the oscillator can be considered to exchange energy with an outside source.

When these two phenomena are small enough to be disregarded, the system can be considered a conservative one, the equation of which is

$$\ddot{x} - f(x) = 0 \tag{30}$$

Equation (30) is a first approximation of (26).

It turns out that the Maxwell-Boltzmann distribution has two advantages. It describes correctly the behavior of a set of systems, or a set of points, when there is no energy exchanged between them, say when each of them is conservative. This is a well-known property in the kinetic theory of perfect gases. It also describes correctly the behavior of a set of systems, or a set of particles, when there is energy exchanged between them, but in this case it is valid only if the set has reached the steady state, i.e., when the systems (or the particles) have reached "thermal equilibrium."

We shall verify these properties using (29). First of all, let us assume that $\alpha = 0$, $S = 0$; then (29) becomes

$$x_2 \frac{\partial p}{\partial x_1} = -f(x_1) \frac{\partial p}{\partial x_2} \tag{31}$$

We have only to verify that

$$p(x_1, x_2) = A(x_1) \exp\left(-\frac{x_2{}^2}{\theta}\right) \tag{32}$$

is a solution of (31). We have

$$\frac{\partial p}{\partial x_1} = \frac{dA}{dx_1} \exp\left(-\frac{x_2^2}{\theta}\right) \qquad \frac{\partial p}{\partial x_2} = -\frac{2A}{\theta} x_2 \exp\left(-\frac{x_2^2}{\theta}\right)$$

Then, by substituting in (31), we get

$$x_2 \frac{dA}{dx_1} = f(x_1) \frac{2A}{\theta} x_2$$

say

$$A(x_1) = K \exp\left[\frac{2}{\theta} \int_0^{x_1} f(x)\, dx\right]$$

where K is a constant of integration. Finally, we find that

$$p(x_1, x_2) = K \exp\left[-\frac{x_2^2}{\theta} + \frac{2}{\theta} \int_0^{x_1} f(x)\, dx\right] \tag{33}$$

Until now, θ has been an arbitrary constant. The meaning is as follows: When there is no energy exchanged between the particles of a gas, "thermal equilibrium" cannot be defined; more precisely, the temperature of the gas is undetermined. In the second case, however, it will be possible to compute the "temperature" of the gas in the steady state.

When $\alpha \neq 0$, $S \neq 0$, we get by substituting (33) in (29),

$$\frac{\partial p}{\partial x_2} = -\frac{2x_2}{\theta} p$$

$$\frac{\partial^2 p}{\partial x_2^2} = -\frac{2}{\theta} p + \frac{4x_2^2}{\theta^2} p$$

$$\alpha p + \alpha x_2 \frac{\partial p}{\partial x_2} + \frac{S}{2} \frac{\partial^2 p}{\partial x_2^2} = \alpha p - \frac{2\alpha x_2^2}{\theta} p - \frac{S}{\theta} p + \frac{2S x_2^2}{\theta^2} p = 0 \tag{34}$$

since $x_2(\partial p/\partial x_1) - f(x_1)(\partial p/\partial x_2) = 0$ is identically verified. And from (34) we get

$$\theta = \frac{S}{\alpha} \tag{35}$$

Finally,

$$p(x_1, x_2) = K \exp\left\{-\frac{\alpha}{S}\left[x_2^2 - 2\int_0^{x_1} f(x)\, dx\right]\right\} \tag{36}$$

2.7. Nyquist's Formula for the Thermal Noise of Radioelectric Oscillators

In the example above we have considered a mechanical oscillator, but obviously the theory can be applied to radioelectric oscillators. For instance, let us consider (Fig. 7) the usual resonant circuit, excited

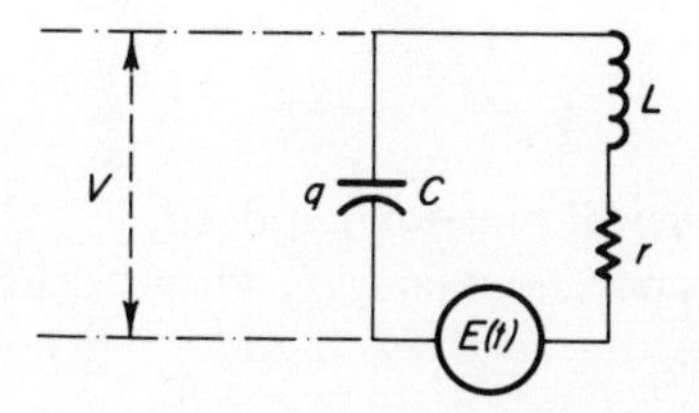

FIG. 7. Resonant circuit with thermal noise source.

by a random forcing function, whose equation is

$$LC\ddot{V} + rC\dot{V} + V = E(t) \tag{37}$$

where r is the series resistance of the circuit, L the inductance of the coil, C the capacitance, V the voltage between the end points of the capacitance, and $E(t)$ a random emf which we shall assume to be generated by the thermal noise of the resistance r.

Let us introduce the reduced variable

$$x = LCV$$

Equation (37) becomes

$$\ddot{x} + \frac{r}{L}\dot{x} + \frac{1}{LC}x = E(t) \tag{38}$$

By comparing (38) with the equation of the above example, we see that

$$\alpha = \frac{r}{L} \qquad f(x) = -\frac{x}{LC} \qquad \int_0^{x_1} f(x)\,dx = -\frac{x^2}{2LC}$$

$$p(x_1, x_2) = K \exp\left[-\frac{1}{\theta}\left(x_2^2 + \frac{x_1^2}{LC}\right)\right] \tag{39}$$

Now let us return to the earlier variable V. We have

$$x_1 = LCV = Lq \qquad x_2 = LC\dot{V} = Li$$

where $q = CV$ is the electric charge of the capacitance and $i = dq/dt$ is the intensity through the circuit. Then (36) is rewritten

$$p(q, i) = K \exp\left[-\frac{2L}{\theta}\left(\frac{Li^2}{2} + \frac{q^2}{2C}\right)\right] \tag{40}$$

In this formula we have brought forth the expression

$$\frac{1}{2}Li^2 + \frac{1}{2}\frac{q^2}{C}$$

which has a simply physical meaning. Indeed, $q^2/2C$ is the energy stored in the capacitance; it is potential energy, whereas the term $\frac{1}{2}Li^2$ is similar to kinetic energy—it is due to the self-inductance of the coil.

Let us put

$$W = \frac{1}{2}Li^2 + \frac{1}{2}\frac{q^2}{C}$$

and rewrite (40) in the more usual form

$$p(q, i) = K \exp\left(-\frac{W}{kT}\right) \tag{41}$$

which is the Maxwell-Boltzmann formula, when the physical quantities involved are: T, the absolute temperature of the circuit, and k, the Boltzmann constant: 1.37×10^{-23} watt-sec/deg. This gives the expression of θ in terms of T and k, say

$$\theta = 2LkT \tag{42}$$

from which, by substituting in (35), we get

$$S = \lim_{\Delta t \to 0} \frac{1}{\Delta t}\left\langle\left[\int_t^{t+\Delta t} E(t)\, dt\right]^2\right\rangle = \alpha\theta = 2rkT \tag{43}$$

Finally, if the thermal noise of the resistance is considered to be a sequence of impulses $\delta(t - t_j)$, which occur at random times t_j, with

$$\int_{\Delta t} \delta(t)\, dt = q$$

One easily obtains

$$\frac{1}{\Delta t}\left\langle\left[\int_t^{t+\Delta t} E(t)\, dt\right]^2\right\rangle = nq^2$$

(n, average number of impulses per second) and formula (43) is rewritten

$$nq^2 = 2rkT \tag{44}$$

This is the well-known formula for the thermal noise generated by resistance r, when the absolute temperature of this resistance is T.

When the power spectrum of the noise is considered, formula (44) can be easily transformed, and the Nyquist formula can be obtained:

$$\langle \mathscr{E}^2 \rangle = 4kTr\,\Delta\nu \tag{45}$$

where $\mathscr{E}$ is the rms value of the thermal-noise voltage, and $\Delta\nu$ is the bandwidth of the measuring system (in cycles/sec). Equation (45) shows that the thermal noise has a uniform distribution of power throughout the frequency spectrum. $4kTr$ is the power of the noise per unit bandwidth.

3. SOLUTION OF THE FOKKER-PLANCK-KOLMOGOROV EQUATION BASED ON CAMPBELL'S THEOREM†

In some problems, where the system under consideration is a linear one, or a nonlinear one whose law of motion can be approximately linearized in the close neighborhood of a steady state, the solution of the Fokker-Planck-Kolmogorov equation can be obtained by means of Campbell's theorem.

We shall explain this method by considering the simple two-dimensional problem in which a physical system (M) is governed by the differential equations

$$\dot{x}_1 = -\alpha x_1 + F_1(t) \tag{46}$$

$$\alpha, \beta > 0$$

$$\dot{x}_2 = -\beta x_2 + F_2(t) \tag{47}$$

which are assumed to be independent of one another; i.e., $F_1(t)$ and $F_2(t)$ have no intercorrelation function, say $S_{12} = 0$. The method can be easily extended to more complicated situations.

Consider at time $t = 0$ a set of copies of the system (M) *in the same initial state*, which state is represented by the point M_0. At a later time t, because of the random forces $F_1(t)$ and $F_2(t)$, the representative points of the different copies are scattered in the neighborhood of a mean position $\langle M \rangle$ (Fig. 8). In this problem the solution of the corresponding

† See [23].

Fokker-Planck-Kolmogorov equation, $p(M_0, t, M)$, is the density of the cloud at point M, and time t, *given* M_0; indeed, the cloud moves with time t, and gradually spreads throughout the plane.

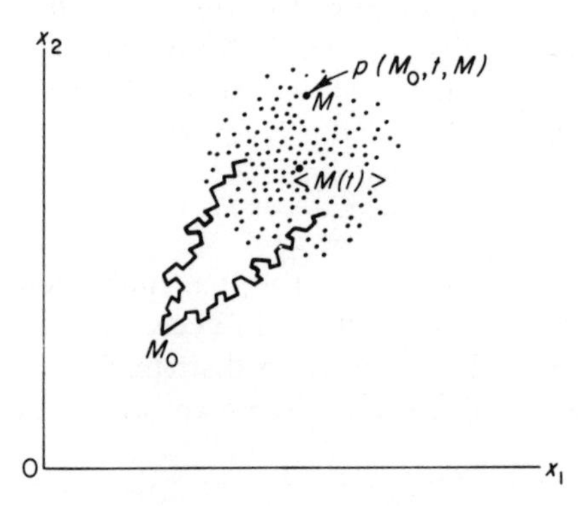

FIG. 8. Diffusion process.

Instead of searching for an explicit form of the function $p(M_0, t, M)$, we shall wish to determine

(a) The center of mass of the cloud, say $\langle M(t)\rangle$.

(b) The first moment of the probability distribution $p(M_0, t, M)$, say the spread of the cloud.

These two quantities can be easily obtained by means of Campbell's theorem.

3.1. Motion of the Center of Mass $\langle M(t)\rangle$

The law of motion of the center of mass $\langle M(t)\rangle$ is determined by formula (1). Indeed, the random forces $F_1(t)$ and $F_2(t)$ can be considered as sequences of impulses which occur at random times t_j. First, we shall make the assumption that there is no correlation between them. We can also assume, without any loss of generality, that these impulses are all identical to one another, respectively, in $F_1(t)$ and $F_2(t)$; i.e., let

$$\int_{\Delta t} \delta_1(t)\, dt = q_1 \qquad \text{be associated with} \quad F_1(t)$$

$$\int_{\Delta t} \delta_2(t)\, dt = q_2 \qquad \text{be associated with} \quad F_2(t)$$

n_1 and n_2 will designate the numbers per second relative to $F_1(t)$ and $F_2(t)$, respectively.

When the random forces are disregarded, the trajectory of the representative point is determined by

$$x_1(t) = x_1{}^0 e^{-\alpha t} \qquad x_2(t) = x_2{}^0 e^{-\beta t} \tag{48}$$

where $x_1{}^0$ and $x_2{}^0$ are the coordinates of the starting point M_0.

When random forces are considered, there is a drift of the representative point with respect to the unperturbed trajectory. We shall compute the drift as follows: Let us consider, for example, the variable $x_1(t)$:

(a) The instantaneous deviation which is produced by each impulse is q_1, and the corresponding transient is

$$q_1 e^{-\alpha t}$$

(b) The average value of the deviation due to all the impulses which occur between $t = 0$ and any time t, given by formula (1), is

$$\langle \Delta x_1(t) \rangle = n_1 q_1 \int_0^t e^{-\alpha t}\, dt = n_1 q_1 \frac{1 - e^{-\alpha t}}{\alpha}$$

Finally, the average value of $x_1(t)$ is

$$\langle x_1(t) \rangle = x_1{}^0 e^{-\alpha t} + n_1 q_1 \frac{1 - e^{-\alpha t}}{\alpha} \tag{49}$$

By similar arguments we get

$$\langle x_2(t) \rangle = x_2{}^0 e^{-\beta t} + n_2 q_2 \frac{1 - e^{-\alpha t}}{\beta} \tag{50}$$

$\langle x_1(t) \rangle$ and $\langle x_2(t) \rangle$ are the coordinates of $\langle M(t) \rangle$.

3.2. Mean Square of the Fluctuations

As a matter of fact, the representative point, which started at M_0 at time $t = 0$, is not, in general, in coincidence with the center of mass $\langle M(t) \rangle$ at time t. The deviation between the actual values $x_1(t)$ and $x_2(t)$, and the average values $\langle x_1(t) \rangle$ and $\langle x_2(t) \rangle$, respectively, can be either positive or negative; say the mean value of the fluctuation is zero.

As for the mean square of the fluctuation, it is given by formula (2), which leads to the following expressions:

$$\langle \Delta_1{}^2 \rangle = n_1 q_1{}^2 \int_0^t e^{-2\alpha t}\, dt = n_1 q_1{}^2 \frac{1 - e^{-2\alpha t}}{2\alpha} \tag{51}$$

$$\langle \Delta_2{}^2 \rangle = n_2 q_2{}^2 \int_0^t e^{-2\beta t}\, dt = n_2 q_2{}^2 \frac{1 - e^{-2\beta t}}{2\beta} \tag{52}$$

with

$$\Delta_1 = x_1(t) - \langle x_1(t) \rangle \qquad \Delta_2 = x_2(t) - \langle x_2(t) \rangle$$

3.3. Solution of the Fokker-Planck-Kolmogorov Equation in the Linear Case

When the number of impulses during the time interval $(0, t)$ is large enough, the distribution of points about $\langle M(t) \rangle$, in the cloud, is given by the *central limit theorem.* This theorem states that the points obey a "normal distribution," whose center is determined by $\langle x_1(t) \rangle$ and $\langle x_2(t) \rangle$. This distribution is

$$p(x_1, x_2, t) = \frac{1}{2\pi(\alpha_{11}\alpha_{22})^{1/2}} \exp\left(-\frac{1}{2}\left\{\frac{[x_1 - \langle x_1(t) \rangle]^2}{\alpha_{11}} + \frac{[x_2 - \langle x_2(t) \rangle]^2}{\alpha_{22}}\right\}\right) \tag{53}$$

with

$$\alpha_{11} = \langle \Delta_1^2 \rangle = \frac{n_1 q_1^2}{2\alpha}(1 - e^{-2\alpha t})$$

$$\alpha_{22} = \langle \Delta_2^2 \rangle = \frac{n_2 q_2^2}{2\beta}(1 - e^{-2\beta t})$$

Equation (53) is the solution of (13).

When $F_1(t)$ and $F_2(t)$ have zero mean value, we have

$$\langle x_1(t) \rangle = x_1^0 e^{-\alpha t} \qquad \langle x_2(t) \rangle = x_2^0 e^{-\beta t}$$

Now in the case where $\beta = 0$, $\alpha \neq 0$, one finds

$$\langle x_2(t) \rangle = x_2^0 \qquad \langle \Delta_2^2(t) \rangle = n_2 q_2^2 t$$

Accordingly, (53) becomes

$$p(x_1, x_2, t) = \frac{1}{2\pi(\alpha_{11}\alpha_{22})^{1/2}} \exp\left\{-\frac{1}{2}\left[\frac{(x_1 - x_1^0 e^{-\alpha t})^2}{\alpha_{11}} + \frac{(x_2 - x_2^0)^2}{\alpha_{22}}\right]\right\} \tag{54}$$

with

$$\alpha_{11} = \frac{n_1 q_1^2}{2\alpha}(1 - e^{-2\alpha t}) \qquad \alpha_{22} = n_2 q_2^2 t$$

$$S_1 = n_1 q_1^2 \qquad S_2 = n_2 q_2^2$$

For instance, this method can be readily applied to the example of section 2.3. One gets

$$p(x, t) = \frac{1}{(2\pi\alpha)^{1/2}} \exp\left[-\frac{(x - x_0)^2}{2\alpha}\right] \qquad \text{with } \alpha = nq^2t,\ x_0 = x(0)$$

which is identical to expression (24), with $\alpha = 2\tau$.

REFERENCES

1. A. Einstein, Zur Theorie der Brownschen Bewegung. *Ann. Phys.* **19**, 371 (1906).
2. A. Campbell, *Proc. Cambridge Phil. Soc.* **15**, 117–136, 310–328 (1909).
3. A. D. Fokker, "Die mittlere Energie rotierender Elektrischer Dipole im Strahlungsfeld." 1914.
4. M. Planck, Über einen Satz der statischen Dynamik und seine Erweiterung in der Quantentheorie. *Sitzber. Preuss. Akad. Wiss., Phys.-Math. Klasse* **324** (1917).
5. A. Kolmogorov, *Math. Ann.* **104** (1931).
6. J. M. Whittaker, *Proc. Cambridge Phil. Soc.* **33**, 451–458 (1937).
7. H. A. Kramers, Brownian Motion in a Field of Force and the Diffusion Model of Chemical Reactions. *Physica* **7**, 284–303 (1940).
8. S. Chandrasekhar, Stochastic Problems in Physics and Astronomy. *Rev. Mod. Phys.* **15** (1943).
9. S. O. Rice, Mathematical Analysis of Random Noise. *Bell System Tech. J.* **23**, 282–332 (1944); **24**, 46–156 (1945).
10. T. C. Wang and G. E. Uhlenbeck, On the Theory of the Brownian Motion. *Rev. Mod. Phys.* **17**, 323–342 (1945).
11. A. Blaquière, Effet du bruit de fond sur la fréquence des autooscillateurs à lampes. Précision ultime des horloges radioélectriques. *Comp. Rend.* **234**, 419–421 (1952); Effet du bruit de fond sur l'amplitude des oscillateurs entretenus. *Comp. Rend.* **234**, 710–712 (1952); Effet du bruit de fond sur la fréquence des autooscillateurs à lampes. Précision ultime des horloges radioélectriques. *Ann. Radioelec.* **8** (1953).
12. R. C. Booton, Jr., Nonlinear Control Systems with Random Inputs. *IRE Trans. Circuit Theory* **1**, 9–18 (1954).
13. P. N. Nikiforuk, Response of a Particular Nonlinear Control System to Random Signals. *Trans. AIEE* **75**, Part 2, Applications and Industry, 419–422 (1956).
14. T. K. Caughey, Response of a Nonlinear String to Random Loading. *J. Appl. Mech.* **26**, 341–344 (1959); Response of Van der Pol's Oscillator to Random Excitation. *J. Appl. Mech.* **26**, 345–348 (1959).
15. S. H. Crandall, Random Vibration. *Appl. Mech. Rev.* **12**, 739–742 (1959).
16. T. K. Caughey, Random Excitation of a Loaded Non-Linear String. *J. Appl. Mech.* **27**, 575–578 (1960); Random Excitation of a System with Bilinear Hysteresis. *J. Appl. Mech.* **27**, 649–652 (1960).
17. R. H. Lyon, Equivalent Linearization of the Hard Spring Oscillator. *J. Acoust. Soc. Am.* **32**, 1161–1162 (1960).
18. J. C. West, "Analytical Techniques for Non-Linear Control Systems." English Universities Press, London, 1960.
19. S. H. Crandall, "Random Vibration of Systems with Nonlinear Restoring Forces." Contract 49(638)-564, Air Force Office of Scientific Research, 1961.

20. A. Blaquière and R. Pachowska, Le Bruit neutronique des réacteurs nucléaires. *Bull. Inform. Sci. Tech.* **52**, June 1961; "Fluctuations d'un système dépendant de plusieurs paramètres aléatoires, application aux réacteurs nucléaires." *Rept. CEA 2115*, 1962.
21. A. Blaquière, Théorie de la réaction de fission en chaine (Chap. 9) in "Bibliothèque des sciences et techniques nucléaires." Presses Universitaires de France, Paris, 1962.
22. J. M. Dolique, Thèses, Collisions dans les plasmas à ions chauds, Paris, 1962.
23. A. Blaquière, Résolution de l'équation de Fokker-Planck au moyen des théorèmes de Campbell. *Compt. Rend.* **256**, 1084–1086 (1963); *Rept. CEA* 2275.
24. A. Blaquière, Fluctuations et largeur de raie d'un auto-oscillateur : équivalence des théories non-linéaires et des méthodes de linéarisation. *Colloq. Intern. Centre Natl. Rech. Sci.* (*Marseille*) Sept. 7–12, 1964.
25. T. K. Caughey, On the Response of a Class of Nonlinear Oscillators to Stochastic Excitation. *Colloq. Intern. Centre Natl. Rech. Sci.* (*Marseille*), Sept. 7–12, 1964.
26. S. H. Crandall, Random Forcing of Nonlinear Systems. *Colloq. Intern. Centre Natl. Rech. Sci.* (*Marseille*), Sept. 7–12, 1964.

CHAPTER IX

Random Fluctuations of Self-Oscillators†

INTRODUCTION

Evaluation of the effects of random noise on the behavior of self-oscillators has recently received a renewed interest with the discovery of atomic clocks and lasers. These devices produce an extremely pure sine wave, which raises the problem of explaining why quartz clocks presently appear to be inferior to quantum devices.

The subject is a broad one, and to simplify it, we shall disregard its practical aspects and try only to weigh the basic physical arguments. We shall narrow the discussion still further by concentrating our attention primarily on "line-width" problems.

First, let us note that experimental data are very scarce in this domain, and up to now result mainly from comparison of quartz clocks with atomic standards (as is done, for example, at the National Bureau of Standards, Boulder, Colorado).

The chief difference between atomic and tube clocks is apparent at first sight, and resides in the different nature of the active element. In quantum oscillators we have a medium with negative susceptibility, which shows a very low intrinsic noise level, even if it is not maintained at a very low temperature, as is ordinarily done in a maser amplifier. In ordinary clocks the active element is a tube, which suffers in principle from shot noise provided by a "hot source," the cathode at some 1000K.

But if we look closer at the tube oscillator, we observe that contrary to the amplifier case, shot noise plays only a negligible role, as long as one deals with thermal high-frequency noise and if one chooses the right structure for the oscillator. Indeed, when the tank circuit is connected to the grid of the tube (Fig. 1), the shot noise is acting through the feedback coupling M, which may be made very small for a high-Q

† The work developed in this chapter was performed in collaboration with Prof. P. Grivet.

circuit. The shot noise then becomes negligible, as can be shown by a simple calculation.

If R_e is the noise-equivalent resistance of the tube, the mean-square value of the noise emf on the grid is

$$\langle E^2 \rangle = 4kTR_e \, \Delta\nu$$

(kT is the Boltzmann factor and $\Delta\nu$ is the elementary frequency bandwidth). On the other hand, near the resonant frequency, the Nyquist noise of the circuit appears on the grid with magnitude E_c,

$$\langle E_c^2 \rangle = 4kTR_p \, \Delta\nu$$

and for a good crystal, $R_p \gg R_e$,

$$R_p = Q^2 r = 10^{14}$$

($Q = 10^6$, $r = 100$, Q is the quality factor, and r is the series resistance).

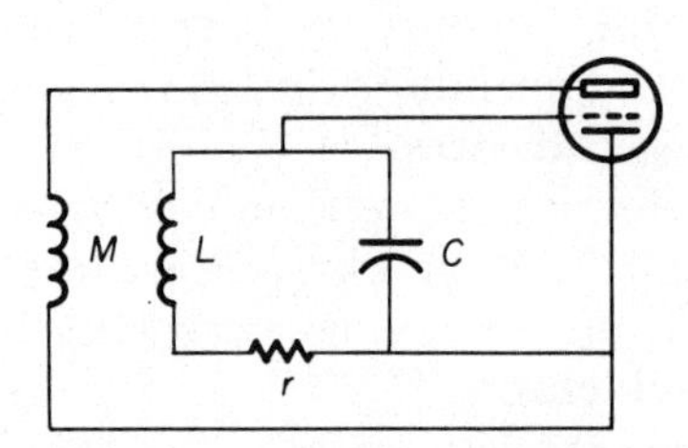

FIG. 1. Vacuum-tube oscillator, tank circuit connected to the grid.

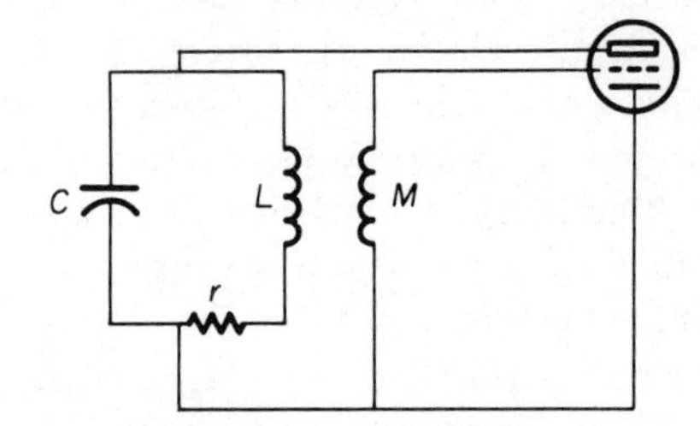

FIG. 2. Vacuum-tube oscillator, tank circuit connected to the plate—a noisy schema for a clock.

Shot noise appears then to be negligible in the useful band of frequencies, which, owing to good average stability of the oscillator, is a very narrow one near the resonance frequency of the tank.

On the other hand, it is not difficult to give an example of a bad scheme. If we connect the tank circuit to the plate (Fig. 2), the coupling of shot noise to the circuit becomes fixed and at a high value. Accordingly, we shall next disregard the shot noise of the tube.

1. BERSTEIN'S METHOD†

1.1. Diagram of the Self-Oscillator

One of the earliest analyses concerning the random fluctuations of self-sustained oscillators is due to Berstein, who used the Fokker-

† See [3].

Planck-Kolmogorov method for describing the phenomenon and for investigating its salient features. The diagram of the oscillator which is considered is the one of a classical vacuum-tube oscillator with the tank circuit connected to the anode, with regenerative coupling by mutual inductance M between the grid circuit and the tank. In the following, owing to the above arguments, we shall assume that the tank circuit is connected to the grid of the tube.

Next we shall refer to the more usual diagram, shown in Fig. 3.

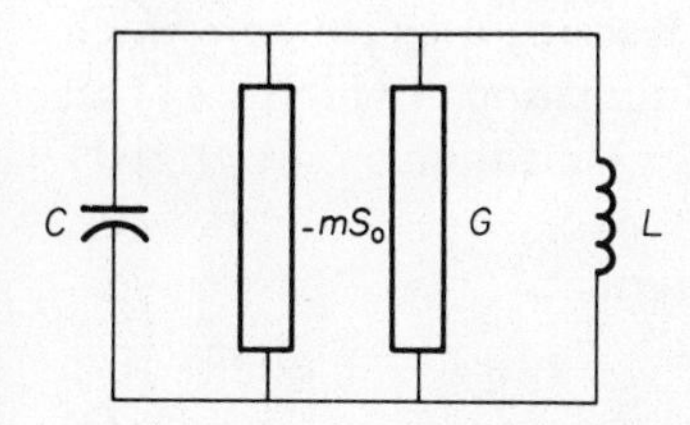

FIG. 3. Simplified diagram of the self-oscillator, including the clock case and using the concept of "negative resistance."

$G = 1/R$ is the parallel conductance of the resonant tank circuit; L and C are the self-inductance and the capacitance of the tank; and $-mS_0$ is the parallel negative conductance of the self-oscillator: m is a coefficient which depends on the actual degree of coupling; for instance, in the model considered by Berstein, m is proportional to the mutual inductance M. S_0 is the slope of the vacuum tube at the operating point; we suppose that the tube has a very high internal resistance. The variable which we consider is U, the oscillating voltage between the terminals of the condensor C.

Next we shall take account of the nonlinearity due to the characteristic curve of the tube, by replacing S_0 by $S_0 + S_2U^2$. This representation was discussed in Chapter 2.

Then it may easily be seen that the equation of motion of the oscillator, when thermal noise is taken into account, is

$$LC\ddot{U} + L(G - mS_0 - 3mS_2U^2)\dot{U} + U = E(t) \tag{1}$$

where $E(t)$ is the random forcing function which represents the thermal noise.

1.2. The Source of Noise

In Fig. 3 we have shown the parallel resistance R of the tank circuit. However, we must keep in mind the fact that the actual source of noise is the series resistance r of the tank. Accordingly, we shall also introduce r, and take account of the well-known relation $R = Q^2r$, $Q = L\omega_0/r$, the quality factor of the resonant tank circuit.

1.3. Equations of Motion in the Phase Plane

Let us introduce the new variables

$$x = U(t)$$

$$y = \frac{1}{\omega_0} \dot{U}(t) \quad \text{with} \quad \omega_0{}^2 = \frac{1}{LC}$$

Hence the state of the system, at any time t, is represented by a point in phase space R^2 (x, y), and this point will describe a trajectory which will be determined by superposing random fluctuations (which we shall now assume to be of first-order smallness) to a continuous mean trajectory, as explained in Chapter 8.

Then (1) is replaced by the differential system

$$\begin{aligned} \dot{x} &= \omega_0 y \\ \dot{y} &= -\omega_0 x - \frac{1}{C}(G - mS_0 - 3mS_2x^2)y + \omega_0 E(t) \end{aligned} \tag{2}$$

Following Berstein, let us put

$$\Gamma(x, y) = -\frac{1}{C}(G - mS_0 - 3mS_2x^2)y \tag{3}$$

$$f_2(t) = \omega_0 E(t) \tag{4}$$

We get

$$\begin{aligned} \dot{x} &= \omega_0 y \\ \dot{y} &= -\omega_0 x + \Gamma(x, y) + f_2(t) \end{aligned} \tag{5}$$

More generally, Berstein analyzes the behavior of oscillators governed by the equations

$$\dot{x} = \omega_0 y + F(x, y) + f_1(t)$$

$$\dot{y} = -\omega_0 x + \Gamma(x, y) + f_2(t)$$

However, in this chapter we shall restrict our study to equations (5), the arguments can be easily extended.

1.4. Differential Equations of the Oscillator in Polar Coordinates

For the sake of simplification, we shall next assume that the amplitude of the oscillator is a slowly varying function of time. This will enable us to use the smooth-approximation principle—to quickly replace varying variables by their time average over an interval of time which does not exceed, in practice, a few cycles of the oscillation.

Then polar coordinates are very convenient. Let us put

$$x = \rho \cos \theta \qquad y = \rho \sin \theta$$

according to which (5) is replaced by

$$\dot{\rho} = \Gamma(\rho \cos \theta, \rho \sin \theta) \sin \theta + f_2(t) \sin \theta$$

$$\dot{\theta} = -\omega_0 + \frac{1}{\rho} \Gamma(\rho \cos \theta, \rho \sin \theta) \cos \theta + \frac{1}{\rho} f_2(t) \cos \theta$$

Finally, taking account of the above remarks, we have after smoothing

$$\begin{aligned} \dot{\rho} &= \Phi(\rho) + f_2(t) \sin \theta \\ \dot{\theta} &= -\omega_0 + \psi(\rho) + \frac{1}{\rho} f_2(t) \cos \theta \end{aligned} \tag{6}$$

with

$$\begin{aligned} \Phi(\rho) &= \frac{1}{2\pi} \int_0^{2\pi} \Gamma(\rho \cos \theta, \rho \sin \theta) \sin \theta \, d\theta \\ \psi(\rho) &= \frac{1}{2\pi\rho} \int_0^{2\pi} \Gamma(\rho \cos \theta, \rho \sin \theta) \cos \theta \, d\theta \end{aligned} \tag{7}$$

1.5. Perturbations in the Neighborhood of Stationary Motion

Let ρ_0 be the stationary amplitude of the oscillator (which we assume to exist and to be stable) when the effect of noise is neglected. Next we shall take account of the noise, the effect of which we shall consider as a small perturbation in the neighborhood of the theoretical stationary oscillation.

As a matter of fact, the stationary oscillation is the generating approximation, and the small perturbation can be obtained by replacing equations (6) by their quasi-linear approximation.

Let us put

$$\rho = \rho_0 + z \qquad (z \ll \rho_0)$$

and replace variable θ by phase angle φ in the left side of (6), with

$$\theta = -\omega t + \varphi$$

We get, at first-order approximation,[†]

$$\dot{z} = \left(\frac{\partial \Phi}{\partial \rho}\right)_{\rho_0} z + f_2(t) \sin \theta$$
$$\dot{\varphi} = \frac{1}{\rho_0} f_2(t) \cos \theta \tag{8}$$

or

$$\dot{z} + sz = F_1(t)$$
$$\dot{\varphi} = F_2(t) \tag{9}$$

with

$$\left(\frac{\partial \Phi}{\partial \rho}\right)_{\rho_0} = -s \tag{10}$$

$$F_1(t) = f_2(t) \sin \theta = \omega_0 E(t) \sin \theta$$
$$F_1(t) = \frac{1}{\rho_0} f_2(t) \cos \theta = \frac{\omega_0 E(t)}{\rho_0} \cos \theta \tag{11}$$

1.6. The Fokker-Planck-Kolmogorov Equation for the Perturbed Self-Oscillator

Now starting with the perturbational equations (9), one can easily derive the Fokker-Planck-Kolmogorov equation, which governs the amplitude and phase fluctuations of the oscillator. This equation, whose general form is that of Eq. (12), Chapter VIII, is, in the present problem,

$$\frac{\partial p}{\partial t} - \frac{\partial}{\partial z}(psz) = \frac{S_{11}}{2} \frac{\partial^2 p}{\partial z^2} + \frac{S_{22}}{2} \frac{\partial^2 p}{\partial \varphi^2} + S_{12} \frac{\partial^2 p}{\partial z \, \partial \varphi} \tag{12}$$

with

$$S_{11} = \lim_{\Delta t \to 0} \frac{1}{\Delta t} \left\langle \left[\int_t^{t+\Delta t} F_1(t)\, dt\right]^2 \right\rangle$$
$$S_{22} = \lim_{\Delta t \to 0} \frac{1}{\Delta t} \left\langle \left[\int_t^{t+\Delta t} F_2(t)\, dt\right]^2 \right\rangle \tag{13}$$
$$S_{12} = \lim_{\Delta t \to 0} \frac{1}{\Delta t} \left\langle \int_t^{t+\Delta t} F_1(t)\, dt \int_t^{t+\Delta t} F_2(t)\, dt \right\rangle$$

[†] The equation relative to the variable φ is obtained by taking account of the relation

$$\omega = \omega_0 - \psi(\rho_0 + z)$$

which gives ω as a function of the deviation z in the neighborhood of the stationary generating approximation.

By substituting expressions (11) for $F_1(t)$ and $F_2(t)$ in these coefficients, one gets

$$S_{11} = \lim_{\Delta t \to 0} \frac{1}{2\,\Delta t} \left\langle \left[\int_t^{t+\Delta t} f_2(t)\, dt \right]^2 \right\rangle$$

$$S_{22} = \lim_{\Delta t \to 0} \frac{1}{2\rho_0^2\,\Delta t} \left\langle \left[\int_t^{t+\Delta t} f_2(t)\, dt \right]^2 \right\rangle \tag{14}$$

$$S_{12} = 0$$

from which, taking account of (4),

$$S_{11} = \lim_{\Delta t \to 0} \frac{\omega_0^2}{2\,\Delta t} \left\langle \left[\int_t^{t+\Delta t} E(t)\, dt \right]^2 \right\rangle$$

$$S_{22} = \lim_{\Delta t \to 0} \frac{\omega_0^2}{2\rho_0^2\,\Delta t} \left\langle \left[\int_t^{t+\Delta t} E(t)\, dt \right]^2 \right\rangle \tag{15}$$

Finally, the random emf $E(t)$, which is caused by the thermal noise, will be specified. Indeed, we have obtained in Chapter VIII, Section 2.6 the physical relation

$$\lim_{\Delta t \to 0} \frac{1}{\Delta t} \left\langle \left[\int_t^{t+\Delta t} E(t)\, dt \right]^2 \right\rangle = 2kTr \tag{16}$$

where r is the resistance which generates the noise, T the absolute temperature of the noise source, and k is Boltzmann's constant. By substituting in (15) we get

$$S_{11} = kTr\omega_0^2 \tag{17}$$

$$S_{22} = \frac{kTr\omega_0^2}{\rho_0^2} \tag{18}$$

1.7. Amplitude and Phase Fluctuations

The solution of (12) has been obtained in Chapter VIII (Section 3.3); it is

$$p(z_0, \varphi_0, t, z, \varphi) = \frac{1}{2\pi(\alpha_{11}\alpha_{22})^{1/2}} \exp\left\{ -\frac{1}{2} \left[\frac{(z - z_0 e^{-st})^2}{\alpha_{11}} + \frac{(\varphi - \varphi_0)^2}{\alpha_{22}} \right] \right\}$$

with

$$\alpha_{11} = \frac{S_{11}}{2S}(1 - e^{-2st}) \qquad \alpha_{22} = S_{22} t$$

z_0 and φ_0 are the initial values of z and φ at time $t = 0$. Accordingly, the mean square of the phase fluctuations is

$$\langle(\varphi - \varphi_0)^2\rangle = S_{22}t = \frac{kTr\omega_0^2}{\rho_0^2}\, t \tag{19}$$

and the mean square of the amplitude fluctuations, at the stationary limit, say when $t \to \infty$, is

$$\langle z^2\rangle = \frac{S_{11}}{2s} = \frac{kTr\omega_0^2}{2s} = \frac{kTLG}{2Cs}\,\omega_0^2 \tag{20}$$

Moreover, s can be computed from (3), (7), and (10); one gets

$$s = -\frac{G - mS_0}{C} \tag{21}$$

1.8. Berstein-Blaquière's Method

The mean squares of the amplitude and phase fluctuations can be more readily obtained using Campbell's theorem. This simplification is made possible because the assumptions which underlie the derivation of Campbell's theorem and of the Fokker-Planck-Kolmogorov equation are the same.

Let us synthesize the stochastic excitation function $E(t)$ by an infinite series of pulses of uniform strength randomly distributed on the t axis with average density n. The individual strengths of the pulses which synthesize $F_1(t)$ and $F_2(t)$ are ϵq_1 and ϵq_2 ($\epsilon = \pm 1$), respectively.

For an individual pulse, the responses relative to z and φ are

$$\delta z(t) = \epsilon q_1 e^{-st} \qquad \delta\varphi(t) = \epsilon q_2$$

At this state we may use Campbell's theorem, which gives the mean values for z and φ, and also the mean-square values of these variables.

First, because ϵ has equal probability of being positive or negative, we get

$$\langle z(t)\rangle = n\langle\epsilon\rangle \int_0^t q_1 e^{-st}\, dt = 0$$

$$\langle\varphi(t)\rangle = n\langle\epsilon\rangle \int_0^t q_2\, dt = 0$$

On the other hand, the mean squares of z and φ are

$$\langle z^2(t)\rangle = n \int_0^t \epsilon^2 q_1^2 e^{-2st}\, dt = \frac{nq_1^2}{2s}(1 - e^{-2st}) \tag{22}$$

$$\langle\varphi^2(t)\rangle = n \int_0^t \epsilon^2 q_2^2\, dt = nq_2^2 t \tag{23}$$

By using the definition of $F_1(t)$ and $F_2(t)$, we get

$$\langle F_1^2(t)\rangle = \frac{\omega_0^2}{2}\langle E^2(t)\rangle$$

$$\langle F_2^2(t)\rangle = \frac{\omega_0^2}{2\rho_0^2}\langle E^2(t)\rangle$$

The result is that q_1 and q_2 are related to q, the strength of the pulses for $E(t)$, by

$$q_1^2 = \frac{\omega_0^2}{2} q^2 \qquad q_2^2 = \frac{\omega_0^2}{2\rho_0^2} q^2$$

q is defined by Nyquist's law,

$$nq^2 = 2kTr$$

One then concludes that

$$\langle z^2\rangle = \frac{kTr}{2s}\omega_0^2 \qquad \langle \varphi^2\rangle = \frac{kTr\omega_0^2}{\rho_0^2} t$$

These are formulas (19) and (20) obtained in Section 1.7.

2. BLAQUIÈRE'S METHOD†

Now we shall summarize the method we developed earlier in our analysis of the fluctuations of radioelectric clocks. For a more complete study of this question, we refer the reader to the references at the end of the chapter.

2.1. [a] and [φ] Components of a Pulse

First let us consider the effect of an individual pulse on the behavior of the oscillator. Let q be the strength of the pulse, and assume that the unperturbed oscillation, say the oscillation before time t_j at which the pulse occurs, is correctly represented by

$$U = a_0 \sin(\omega_0 t + \varphi)$$

We know that at time t_j, $\dot{U}(t)$ exhibits the discontinuity $\Delta\dot{U}$, whereas $U(t)$ is continuous. $\Delta\dot{U}$ is readily deduced from (1); that is, $\Delta\dot{U} = q\omega_0^2$.

† See [6].

It follows that

$$\Delta U = \Delta a \sin(\omega_0 t_j + \varphi) + a_0 \, \Delta\varphi \cos(\omega_0 t_j + \varphi) = 0$$

$$\frac{1}{\omega_0} \Delta\dot{U} = \Delta a \cos(\omega_0 t_j + \varphi) - a_0 \, \Delta\varphi \sin(\omega_0 t_j + \varphi) = q\omega_0$$

Accordingly, the changes in amplitude and phase angle due to the individual pulse are

$$\Delta a = q\omega_0 \cos(\omega_0 t_j + \varphi) \tag{24}$$

$$\Delta\varphi = -\frac{q\omega_0}{a_0} \sin(\omega_0 t_j + \varphi) \tag{25}$$

Now let us remark that if time t_j is such that

$$\omega_0 t_j + \varphi = 2k\pi \qquad (k \text{ integer})$$

then the pulse results in the change in amplitude

$$\Delta a = q\omega_0$$

whereas the phase angle is not modified.

On the other hand, if time t_j is such that

$$\omega_0 t_j + \varphi = \frac{\pi}{2} + 2k\pi$$

then the pulse results in the small phase shift

$$\Delta\varphi = -\frac{q\omega_0}{a_0}$$

and the amplitude is not affected. Pulses of the first kind are called [a] pulses" and pulses of the second kind "[φ] pulses."

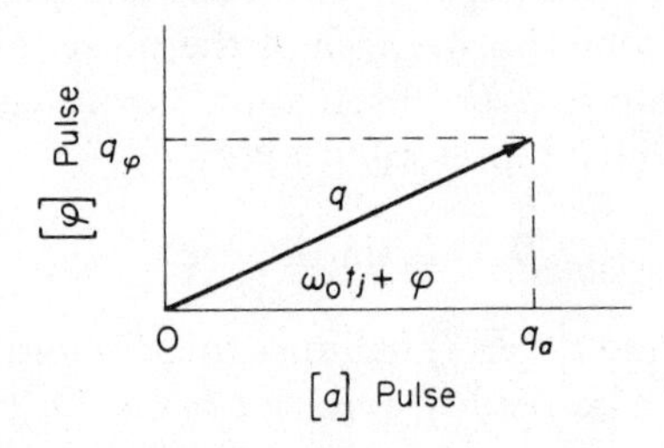

FIG. 4. The use of the [a] and [φ] pulses as vector components.

Now, from (24) and (25) we deduce a useful property: Any impulse can be split into two components, one of which is an [a] pulse and the other one a [φ] pulse. The strengths of these components are, respectively,

$$q_a = q\cos(\omega_0 t_j + \varphi) \qquad q_\varphi = q\sin(\omega_0 t_j + \varphi) \tag{26}$$

These formulas suggest the vectorial representation shown in Fig. 4. Each pulse is portrayed by a vector whose length is q and whose direction is determined with respect to the a axis by the angle $\omega_0 t_j + \varphi$. Its component along the a axis represents the [a] pulse, and its component along the φ axis represents the corresponding [φ] pulse. From (26) we get

$$\langle q_a{}^2\rangle = \langle q_\varphi{}^2\rangle = \frac{q^2}{2}$$

Let n be the average density of the pulses which perturb the oscillator; since each pulse of any random sequence can be decomposed into an [a] pulse and a [φ] pulse, it follows that the average density of pulses of each kind is also n.

Accordingly, we can replace the actual pulses either by [a] and [φ] pulses with the average density n, and with the mean-square strength $q^2/2$, the same for each kind; or by sets of [a] and [φ] pulses with the average density $n/2$, and with the mean-square strength q^2, the same for each kind.

Next we shall approach the problem from this viewpoint; say we consider two sets of pulses: [a] pulses with average density $n/2$, [φ] pulses with average density $n/2$. The pulses of each kind will be standard pulses with strength q.

2.2. Mean-Square Value of Amplitude Fluctuations

Let τ be the effective time constant, governing the experimental law of return to the steady state after a small perturbation. The amplitude response of the oscillator to an individual [a] pulse is

$$\delta a(t) = q\omega_0 e^{-t/\tau}$$

Hence from Campbell's theorem we deduce the mean-square value of the amplitude fluctuations:

$$\langle \Delta a^2\rangle = \frac{n}{2}\int_0^\infty q^2\omega_0{}^2 e^{-2t/\tau}\,dt = \frac{nq^2\omega_0{}^2\tau}{4}$$

By taking account of Nyquist's law,

$$nq^2 = 2kTr$$

we get†

$$\langle \Delta a^2 \rangle = \frac{kTr\omega_0^2 \tau}{2} = \frac{kTr}{2s} \omega_0^2 \qquad \text{with} \quad \tau = \frac{1}{s} \tag{27}$$

2.3 Mean-Square Value of the Phase Fluctuations

Likewise the small phase shift which is inflicted on the oscillation by an individual [φ] pulse is

$$\delta\varphi = -\frac{q\omega_0}{a_0}$$

and from Campbell's theorem we deduce

$$\langle (\Delta\varphi)^2 \rangle = \frac{n}{2} \int_0^t \frac{q^2\omega_0^2}{a_0^2} \, dt = \frac{nq^2\omega_0^2}{2a_0^2} t$$

say†

$$\langle (\Delta\varphi)^2 \rangle = \frac{kTr\omega_0^2}{a_0^2} t \tag{28}$$

We see that the phase fluctuations never tend toward a stationary process.

2.4. Second-Order Approximation

The above results have been obtained in both methods starting with a perturbational equation in which the effects of second-order smallness due to each pulse have been disregarded. A simple example will clearly show that they need be taken into account when one wishes to analyze more accurately some properties of the perturbed oscillator.

Consider, for instance, a linear undamped oscillator, whose unperturbed equation of motion is

$$\ddot{x} + \omega_0^2 x = 0$$

Then let a pulse of strength q act upon the system at any time corresponding to the maximum deflection; consider a [φ] pulse (Fig. 5).

Since the system is linear, we can use the "superposition principle"

† Equations (27) and (28) are identical to (20) and (19), respectively, when we return to the earlier notations $z = \Delta a$, $\rho_0 = a_0$.

and obtain the perturbed motion by superposing on the steady unperturbed oscillation

$$a_0 \sin \omega_0 t$$

the sinusoidal response due to the pulse,

$$q\omega_0 \cos \omega_0 t$$

The phase shift between the latter one and the unperturbed oscillation is $\pi/2$, since the pulse occurs at a maximum deflection point.

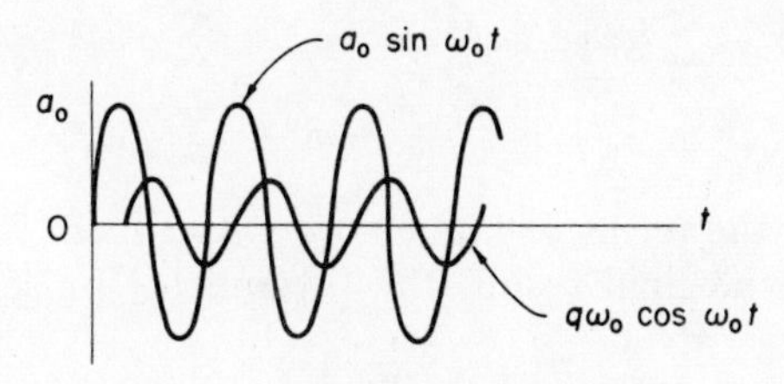

FIG. 5. Effect of a [φ] pulse on the amplitude, application of the "superposition principle."

Assume that $q\omega_0$ is very small with respect to a_0, and let us refer to the vectorial representation shown in Fig. 6. We see that the above pulse results in:

(a) The phase perturbation $\quad \delta\varphi = q\omega_0/a_0$

(b) The amplitude perturbation $\quad \delta a = q^2\omega_0{}^2/2a_0$

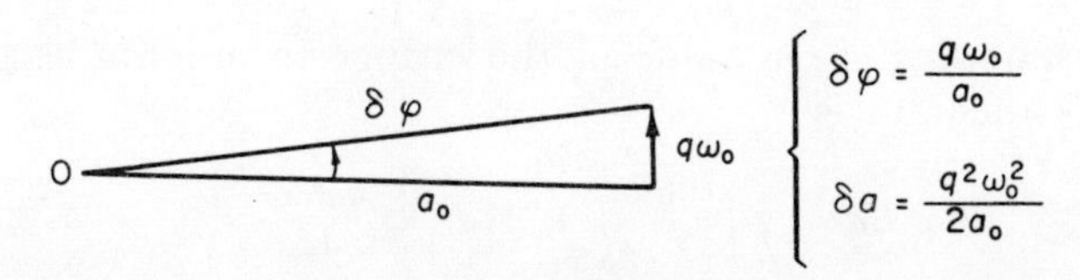

FIG. 6. Calculation of the second-order effect of the [φ] pulse.

Indeed δa is of order q^2, so that it was negligible in our earlier derivations. However, next we shall wish to take account of such second-order terms, since they are all positive and can pile up, thus leading to a nonnegligible effect in the problem we shall now consider.

For instance, if a random sequence of [a] and [φ] pulses disturbs the above oscillator†:

† We again assume that the individual strengths of the pulses are ϵq, $\epsilon = \pm 1$. In this example ϵ is a random parameter.

(a) Each [a] pulse will result in the amplitude perturbation

$$\delta a = \epsilon q \omega_0$$

(b) Each [φ] pulse will result in the amplitude perturbation

$$\delta a = \frac{q^2 \omega_0^2}{2a_0}$$

Then the mean-square value of the amplitude of the oscillation, at any time t, is

$$\left\langle \left[a_0 + \sum (\epsilon q \omega_0) + \sum (q^2 \omega_0^2 / 2a_0)\right]^2 \right\rangle = a_0^2 + \sum q^2 \omega_0^2 + 2a_0 \sum (q^2 \omega_0^2 / 2a_0)$$
$$= a_0^2 + \sum q^2 \omega_0^2 + \sum q^2 \omega_0^2$$

when neglecting the terms of higher-order smallness with respect to q. We see that, in the final result, the [a] and [φ] pulses have the same contribution.

Now let us return to actual oscillators, which have an effective time constant τ, governing the experimental law of return to the steady state after each small perturbation. Then each [φ] pulse will result in the small deviation

$$\delta a = \frac{q^2 \omega_0^2}{2a_0}$$

which will originate the transient

$$\frac{q^2 \omega_0^2}{2a_0} e^{-t/\tau}$$

The corresponding mean value of the change in amplitude, given by Campbell's theorem is

$$\Delta a = \frac{n}{2} \int_0^\infty \frac{q^2 \omega_0^2}{2a_0} e^{-t/\tau} \, dt = \frac{n q^2 \omega_0^2}{4a_0} \tau$$

$n/2$ is the mean density of the [φ] pulses along the t axis, say (with $nq^2 = 2kTr$)

$$\Delta a = \frac{kTr\omega_0^2}{2a_0} \tau \tag{29}$$

Finally, it turns out that the amplitude fluctuations, given by the earlier first-order approximation, occur in the neighborhood of the amplitude (Fig. 7).

$$a_0 + \frac{kTr\omega_0^2}{2a_0} \tau \tag{30}$$

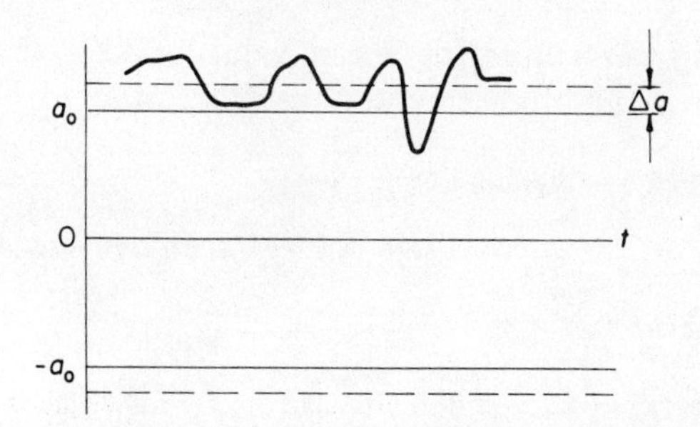

FIG. 7. Second-order effect of the [φ] pulses on the amplitude.

As a matter of fact, in the first-order approximation we have neglected the small change Δa, which more accurately defines the central value of the amplitude probability distribution, but could be disregarded in the first part of the analysis. Furthermore, note that terms of second-order smallness with respect to q do not occur in the study of phase fluctuations.

2.5. The Line Width of Good Oscillators

2.5.1. Autocorrelation Function of the Signal

Let us consider the perturbed oscillation $U(t)$ beginning at time $t = 0$ and for which $U(0) = 0$ (Fig. 8). We shall disregard the amplitude

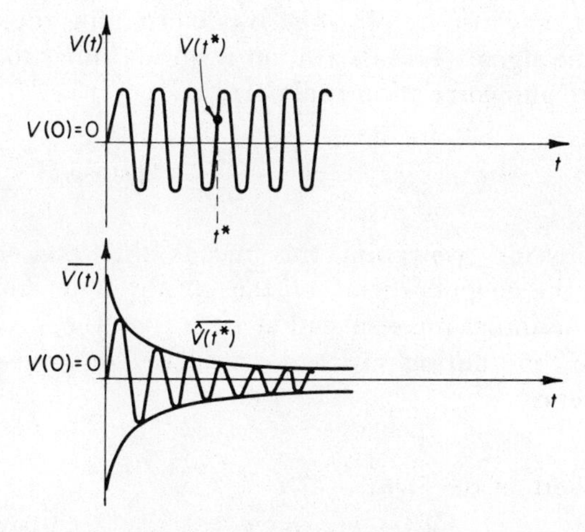

FIG. 8. Evaluation of the mean value of the ordinate at time t^*, allowing for stochastic displacement of the curve parallel to the t axis.

fluctuations and determine the mean value of $U(t)$ at time t^*, taking into account only the phase fluctuations. The value of $U(t)$ at time t^* is

$$U(t^*) = a_0 \sin(\omega_0 t^* + \Delta\varphi)$$
$$= a_0 \cos \Delta\varphi \sin \omega_0 t^* + a_0 \sin \Delta\varphi \cos \omega_0 t^*$$

and its mean value is

$$\langle U(t^*)\rangle = a_0 \langle \cos \Delta\varphi \rangle \sin \omega_0 t^* + a_0 \langle \sin \Delta\varphi \rangle \cos \omega_0 t^*$$

Now it is convenient to write, as a first approximation

$$\langle \cos \Delta\varphi \rangle \simeq \exp[-\tfrac{1}{2}\langle(\Delta\varphi)^2\rangle]$$

whence

$$\langle U(t^*)\rangle \simeq a_0 \exp[-\tfrac{1}{2}\langle \Delta\varphi)^2\rangle] \sin \omega_0 t^*$$

Finally, using (28), we get

$$\langle U(t^*)\rangle = a_0 \exp\left(-\frac{kTr\omega_0^2}{2a_0^2} t^*\right) \sin \omega_0 t^* \tag{31}$$

We see that the perturbed oscillation, when considered in the mean, behaves like a damped oscillation, the time constant of which is

$$t_0^* = \frac{2a_0^2}{kTr\omega_0^2} \tag{32}$$

This result may also be obtained by determining the autocorrelation function of the signal. This derivation is quite similar to the one above. It leads to the autocorrelation function

$$\psi(t^*) = \tfrac{1}{2}a_0^2 \exp\left(-\frac{kTr\omega_0^2}{2a_0^2} t^*\right) \sin \omega_0 t^* \tag{33}$$

From the physical viewpoint this means that the oscillator, when operated in the neighborhood of the steady state, slowly loses any memory of the initial phase it had at time $t = 0$. t_0^* is the correlation time, say the time during which the oscillator undergoes a substantial loss of coherence.

2.5.2. Line Width of the Signal

When the amplitude fluctuations are disregarded, the Fourier spectrum of the signal is readily obtained. We may either take the

Fourier transform of $\psi(t^*)$ or write down directly the power spectrum of the damped oscillation (31), whose well-known expression is proportional to

$$\frac{1}{(\omega^2 - {\omega_0}^2)^2 + D^2\omega^2} \tag{34}$$

where

$$D = \frac{kTr{\omega_0}^2}{{a_0}^2} \tag{35}$$

We find for the total line width $2\delta\omega$ on the ω scale,

$$D = 2\delta\omega = \frac{kTr{\omega_0}^2}{{a_0}^2} \tag{36}$$

If $2\delta\omega_r$ is the total line width of the resonant circuit between the half-power points, the following circuit relations hold:

$$\omega_0 = 2Q\,\delta\omega_r \qquad P = \frac{{a_0}^2}{2Q^2r}$$

Q is the quality $L\omega_0/r$ of the circuit and P is the power in the circuit. Inserting these values, we get

$$\delta\omega = \frac{kT}{P}(\delta\omega_r)$$

or returning to frequencies ν,

$$2\delta\nu = \frac{4\pi kT(\delta\nu_r)^2}{P} \tag{37}$$

2.5.3. The Time Constants of the Oscillator

As we have pointed out in Chapter VI, Section 5.6, the law of variation with time of the amplitude of oscillation, in the neighborhood of the steady state, is characterized by time constants τ_1 and τ_2 (Fig. 9). τ_1 is the local time constant at the amplitude $a_0 + \Delta a$ ($\Delta a > 0$ or $\Delta a < 0$). τ_2 is the time constant which governs the law of return to the steady state after any small perturbation Δa. This constant is identical with parameter τ, which was introduced in the derivations above.

From the results of Chapter 6 it turns out that

$$\frac{\tau_2}{\tau_1} = \frac{\Delta a}{a_0} \tag{38}$$

where Δa is any deviation from the steady state. This relation will be useful in connection with (29).

$$\Delta a = \frac{kTr\omega_0^2}{2a_0} \tau_2 \tag{39}$$

is the mean deviation from the unperturbed amplitude a_0, due to the noise, and we shall wish to compute the corresponding local time constant τ_1; from (38) and (39) we get

$$\tau_1 = \frac{a_0 \tau_2}{\Delta a} = \frac{2a_0^2}{kTr\omega_0^2}$$

This expression is significant, for it is identical with the correlation time (32). It leads to the following conclusions.

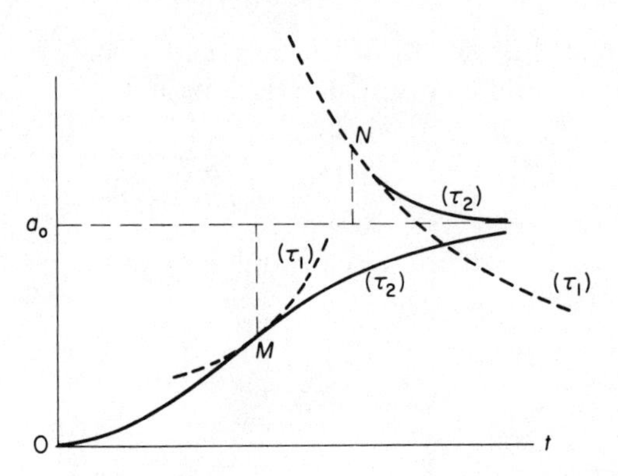

FIG. 9. Distinction between two approximations of the time constant of the oscillator.

2.5.4. Introduction to the Quasi-Linear Method

As a matter of fact, up to now we have met with two kinds of arguments:

(a) In the nonlinear theory, the loss of coherence of the oscillation was measured by the correlation time t_0^*, and the line width was explained by this loss of coherence,

$$2\delta\omega = \frac{2}{t_0^*} \tag{40}$$

(b) *In the mean* the actual oscillator behaves like a *linear oscillator*, which exhibits random fluctuations in the neighborhood of amplitude

$$a_0 + \frac{kTr\omega_0^2}{2a_0} \tau_2 \tag{41}$$

The time constant of this linear oscillator is

$$\tau_1 = t_0{}^* \tag{42}$$

and the line width between the half-power points is

$$2\delta\omega = \frac{2}{\tau_1} \tag{43}$$

i.e., the line width is determined by the lifetime of the transients of the linear oscillator thus introduced.

This second viewpoint will serve as a starting point for the quasi-linear method, which we shall now explain.

3. LERNER'S QUASI-LINEAR METHOD†

The line width of a maser oscillator was determined by Gordon, Zeiger, and Townes [8] by means of a quasi-linear method first introduced by Lerner in the case of classical radioelectric oscillators.

We shall now explain the quasi-linear method from this viewpoint.

(a) The oscillator is replaced by an amplifier with a very narrow bandwidth.

(b) The input signal to the amplifier is the noise.

(c) The output signal is a quasi-sinusoidal wave whose amplitude and phase are random functions of time.

Indeed, this model is not so bad, since it is difficult to perceive any difference between the output of this selective amplifier and the signal which is produced by the perturbed self-sustained oscillator.

(d) Moreover, let us assume that the feedback loop of the amplifier involves a nonlinear component whose gain depends not only on the frequency but also on the amplitude of the signal.

The amplifier is shown in Fig. 10. Assume, for instance, that the amplifier is a very selective resonant circuit, governed by the differential equation

$$LC\ddot{U} + r^*C\dot{U} + U = E(t) \tag{44}$$

where $E(t)$ is the noise and r^* is a very small resistance which is a function of the amplitude of the oscillation. Let us also assume that the noise obeys Nyquist's law

$$\langle E^2(t) \rangle = 4kTr\, d\nu \tag{45}$$

† See [5].

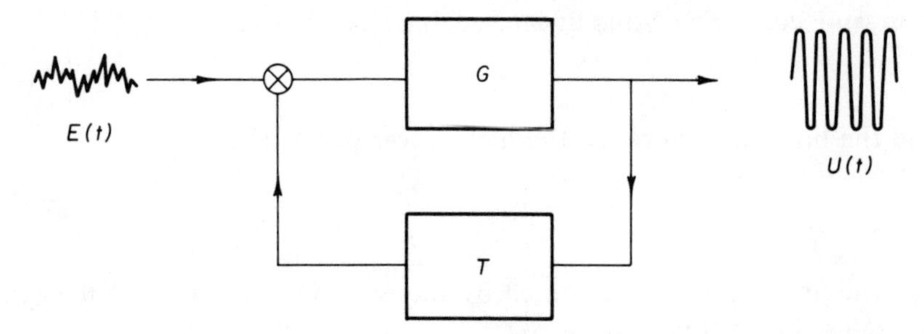

FIG. 10. Selective amplifier model for a noisy oscillator. $E(t)$, input noise; $U(t)$, quasi-sinusoidal output signal.

where r is the resistance which generates noise ($r \neq r^*$). By integrating upon the bandwidth of the circuit, we get

$$\langle U^2 \rangle = \frac{1}{2\pi} \int_0^\infty \frac{4kTr\, d\omega}{(1 - LC\omega^2)^2 + (r^*)^2 C^2 \omega^2} = \frac{kTr}{r^* C} \tag{46}$$

where

$$\langle U^2 \rangle \simeq \frac{a_0^2}{2}$$

On the other hand, the time constant of the circuit, governed by (44), is

$$\tau_1 = \frac{2L}{r^*} \tag{47}$$

where r^* is determined by the value of the amplitude of oscillation during any small interval of time, i.e., by $a \simeq a_0$. The corresponding line width of the signal is

$$2\delta\omega = \frac{2}{\tau_1} = \frac{r^*}{L} \tag{48}$$

Now from (46) we get

$$r^* = \frac{2kTr}{Ca_0^2}$$

and by substituting in (48), the line width of the oscillation is obtained. We find

$$2\delta\omega = \frac{2kTr}{LCa_0^2} = \frac{2kTr\omega_0^2}{a_0^2} \tag{49}$$

Obviously there is a discrepancy between (36) and (49): The line width given by the above quasi-linear method is twice the correct value given by the nonlinear theory. This leads us to analyze the above method more carefully.

As a matter of fact, the mean value of the amplitude of the output oscillation slightly differs from a_0. Lerner ascribes parameter r^* in (44) to this small deviation and puts

$$r^* = K\frac{\Delta a}{a_0} \qquad \Delta a = \langle a \rangle - a_0$$

where $\langle a \rangle$ is the mean value of the actual amplitude and K is a factor of proportionality. By substituting in (46) one gets

$$\frac{\langle a^2 \rangle}{2} \simeq \frac{a_0^2}{2} = \frac{kTra_0}{KC\,\Delta a}$$

say

$$\frac{\Delta a}{a_0} = \frac{2kTr}{KCa_0^2}$$

On the other hand, let us compute K. From (47) we get

$$\tau_1 = \frac{2L}{r^*} = \frac{2L}{K}\frac{a_0}{\Delta a}$$

Then by taking account of (38): $(\tau_2/\tau_1) = (\Delta a/a_0)$, we find

$$K = \frac{2L}{\tau_2}$$

Finally, we obtain the mean value of the deviation Δa:

$$\Delta a = \frac{2kTr}{KCa_0} = \frac{kTr\omega_0^2}{a_0}\tau_2$$

Again we remark that the value of the deviation thus obtained is twice the correct value of (39), and this enables us to explain the earlier discrepancy between the line widths, given by the different procedures.

Indeed as was pointed out in section 2.4, the mean deviation Δa, given by nonlinear theory, is due to $[\varphi]$ pulses only. Moreover, the density of $[\varphi]$ pulses along the t axis is one-half the actual density of the pulses which synthesize the stochastic excitation $E(t)$.

Accordingly, the quasi-linear method will lead to the correct line width provided that only one-half the actual noise is in (46). That is, (45) should be replaced by

$$2kTr\,d\nu$$

This statement can be readily verified.

4. FLICKER NOISE†

4.1. The Flicker Problem

In the previous sections we dealt with high-frequency noise from 10 kHz upward, and we had to consider thermal noise only. But in the low-frequency range, say from 5000 Hz downward, it is necessary to take into account the flicker effect of the tube, whose spectrum is sketched in Fig. 11 for a number of usual values.

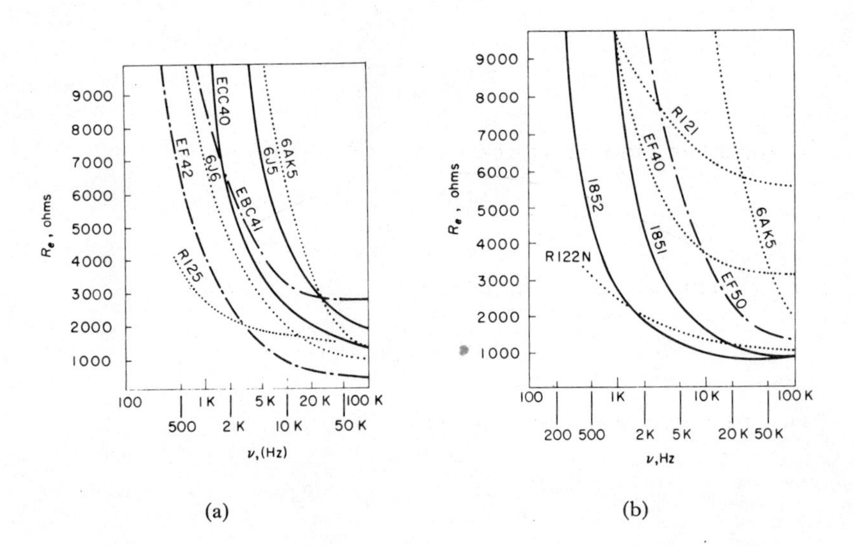

FIG. 11. (a) Flicker noise of triodes; (b) flicker noise of pentodes.

The spectrum possesses a branch which is nearly infinite at low frequencies, so that in the very low range of frequencies, the tube noise is no longer negligible in comparison with the thermal effect, even if we take into account the marked attenuation by the circuit coupling, which we have considered at the begining of this chapter.

On the other hand, it must be realized first that no kind of linear filtering will be able to get rid of the flicker components. The components appear on the first tube plate as mixed with the nominal clock frequency, which acts as a "carrier." These "modulation frequencies" are so near

† See [22, 31].

the carrier that there is no hope of filtering them out by any linear process.

This mechanism is very similar to the standard harmonic generation process, which is known to be dependent on the curvature of the tube characteristic, e.g., on the coefficient S_1 of

$$i_a = S_0 U + S_1 U^2 + S_2 U^3 \tag{50}$$

This may help us understand intuitively why for the theory of the flicker effect we need (50) in full, including the $S_1 U^2$ term.

Of course, we also have to introduce the equivalent noise resistance of the tube R_e (which we could disregard in the earlier sections). Generally speaking, R_e is much larger here, because it represents the flicker noise, but its size can only be properly defined in the frequency domain. This is postponed because we start the theory in the time domain (Fig. 12).

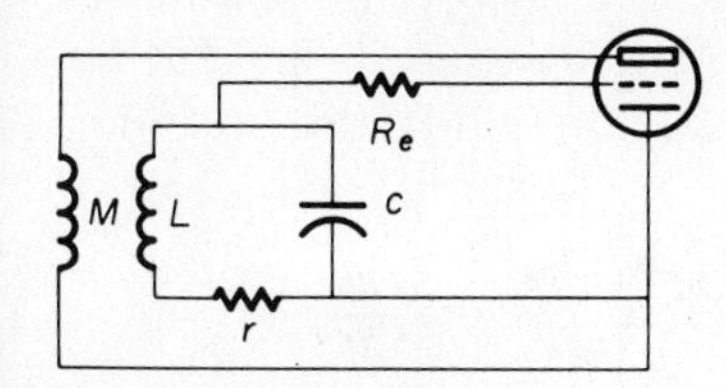

FIG. 12. Description of flicker noise, using the equivalent noise resistance of the tube, R_e.

The calculations will be notably more complicated because:

(a) We have included the $S_1 U^2$ term in (50).

(b) We have a new source of noise R_e, outside the tank circuit.

The coupling between noise source and circuit is more complicated at the outset, because E_e originates in R_e. U, the grid potential of the theory, is no longer a convenient variable, easy to measure. To get significant results we must write the equation in terms of a new physical variable u,

$$u = U - E_e \tag{51}$$

We shall not go into detail here but simply note the important steps of the calculation. The interested reader may refer to Blaquière and Grivet [22] for a detailed treatment of Colpitt's oscillator along this line of reasoning.

Summing up, we first find, by standard circuit analysis, the coupling between noise source and circuit, in terms of the original variable U.

It reads

$$\frac{d^2U}{dt^2} + \omega_0^2(rC - MS_0 - 2MS_1U - 3MS_2U^2)\frac{dU}{dt} + \omega_0^2U = \left(\frac{d^2}{dt^2} + 2C\omega_0^2\frac{d}{dt} + \omega_0^2\right)E_e \quad (52)$$

Going to the physical variable u defined by (51), we get

$$\frac{d^2u}{dt^2} + \omega_0^2(rC - MS_0)\frac{du}{dt} + \omega_0^2u + M\omega_0^2\left[S_0\frac{dE_e}{dt} - S_1\frac{d(u+E_e)^2}{dt} - S_2\frac{d(u+E_e)^3}{dt}\right] = 0 \quad (53)$$

We now proceed to the "smoothing" of this equation. Here we must choose a direct procedure different from the general Berstein-Blaquière theory of Section 1.8. Flicker noise is a slow process compared to thermal noise. We can "smooth" the excitation function itself (e.g., the noise) instead of smoothing only the solution. This is done easily in (53) by neglecting the various time derivatives of E_e, including the first one, dE_e/dt. Equation (53) now reads

$$\frac{d^2u}{dt^2} + (rC - MS_0)\omega_0^2\frac{du}{dt} + \omega_0^2u - M\omega_0^2\left(S_1\frac{du^2}{dt} + S_2\frac{du}{dt}\right) = M\omega_0^2\left(2S_1E_e\frac{du}{dt} + 6S_2E_eu\frac{du}{dt} + 3S_2E_e^2\frac{du}{dt}\right) \quad (54)$$

Finally, we smooth the solution of this equation, using the first-harmonic approximation. Let us set $u = \hat{u}\sin\omega_0t$. We place this expression in (54). Balancing the first harmonic, it leads to

$$(rC - MS_0)\hat{u}\omega_0\cos\omega_0t - \frac{3MS_2}{4}\hat{u}^3\omega_0\cos\omega_0t = 2MS_1E_e\hat{u}\omega_0\cos\omega_0t + 3MS_2E_e^2\hat{u}\omega_0\cos\omega_0t \quad (55)$$

Hence, neglecting E_e^2, we get

$$\hat{u}(t) = 2\left[\frac{rC - MS_0 - 2MS_1E_e(t)}{3MS_2}\right]^{1/2} \quad (56)$$

Looking at this expression, we observe that $\hat{u}$ is a slowly varying function, like $E_e(t)$. By differentiating, we get $\delta\hat{u}$, whose mean square (time average) is

$$\overline{(\delta\hat{u})^2} = \frac{4MS_1^2}{3S_2(rC - MS_0)}\overline{dE_e^2} \quad (57)$$

This formula shows that the flicker effect depends strongly on the curvature of the i_a, U characteristics measured by coefficient S_1. The flicker noise can be suppressed, or at least its influence considerably reduced, by choosing a quiescent point on the characteristic for which the coefficient S_1 vanishes. This remark is important in the case of some oscillators, such as, for example, the Robinson [17] low-noise oscillator, in which this condition is naturally fulfilled. This kind of oscillator has been sudied by Blaquière and Grivet in [22, 31].

4.2. Comparison of the Flicker and Thermal Effect

Starting with (57), we easily obtain the spectrum of the amplitude perturbation due to the flicker effect:

$$P_f(\nu)\, d\nu = \frac{4MS_1^2}{3S_2(rC - MS_0)} 4kTR_e\, d\nu \tag{58}$$

But for the sake of clarity of argument, we find it advisable to introduce two others parameters:

(a) The actual degree of coupling n^* is defined by

$$n^* = \frac{M - M_0}{M_0} \tag{59}$$

M_0 being the threshold value for barely sustained oscillations.

(b) The quality factor for the oscillator Q^* is defined by

$$Q^* = \frac{Q}{n^*} \tag{60}$$

These parameters were first introduced by Buyle-Bodin [16] and Hasenjäger [14]. Equation (58) may be rewritten

$$P_f(\nu)\, d\nu = -\frac{4S_1^2}{3S_2S_0} \frac{1 + n^*}{n^*} R_e(4kT\, d\nu) \tag{61}$$

Equations (34) and (35) describe the spectrum due to the thermal effect and may be written with the new parameters as

$$\begin{aligned} P_t(\nu)\, d\nu &= \frac{r(4kT\, d\nu)}{(\omega/\omega_0)^2 + (1/Q^*)^2} \simeq (Q^*)^2 r(4kT\, d\nu) \\ &= \frac{R_p}{(n^*)^2}(4kT\, d\nu) \end{aligned} \tag{62}$$

For an order-of-magnitude comparison of (61) and (62) we may disregard the common factor $(4kT\,d\nu)/n^*$ and consider only the coefficients

$$\frac{4S_1^2}{3S_2S_0} R_e \qquad \text{for flicker noise}$$

$$\frac{R_p}{n^*} \qquad \text{for thermal noise} \tag{63}$$

where R_p is the shunt resistance of the tank circuit at the clock frequency. This parameter is usually a small number in clocks, as may be seen from its link with the steady voltage amplitude $\hat{u}$ of the tank circuit, representing, for example, the quartz crystal. Indeed (55), when one sets $E_e \equiv 0$, gives $\hat{u}$ as

$$\hat{u}^2 = 4\,\frac{rC - MS_0}{3MS_2} = \frac{4S_0}{3S_2}\,\frac{n^*}{1 + n^*} \tag{64}$$

so that approximately

$$n^* = \frac{3S_2}{4S_0}\,\hat{u}^2 \tag{65}$$

As R_p may lie in the megohm range, (63) shows that the flicker noise is ordinarily negligible, except at very low frequencies, from 0.1 Hz downward. This is true as long as the factor S_1^2/S_2S_0 assumes normal values of the order of 10.

5. ERROR IN FREQUENCY MEASUREMENT USING A FINITE TIME t′

There exist many techniques for determining a frequency. Here we shall not go into the details of the discussion, which may sometimes be rather intricate, since the definition of a frequency is not always an easy matter. As a matter of fact, there are in general many possible definitions and many possible techniques of measurement [9, 32].

We shall pay attention to an important conclusion which may be illustrated by the so-called *zero-crossing method*, namely by the method in which the frequency of an approximately periodic (fluctuating) signal is obtained by counting the number of zero crossings on a given interval of time t'. Again we shall assume that the fluctuations of the signal are due to thermal noise, whose effect has been analyzed in the above sections.

Let $\langle(\delta\varphi)^2\rangle$ and $\langle(\Delta\varphi)^2\rangle$ be, respectively, the mean-square values of the phase fluctuations of the signal per cycle and after an elapsed time t'. Moreover, assume that t' is equal to n cycles of the unperturbed signal, say

$$t' = \frac{n}{\nu_0} \tag{66}$$

where $\nu_0 = \omega_0/2\pi$ is the frequency of the unperturbed signal. Then

$$\langle\Delta\varphi\rangle = 0 \qquad \text{and} \qquad \langle(\Delta\varphi)^2\rangle = n\langle(\delta\varphi)^2\rangle$$

On the other hand, it may easily be seen that the frequency fluctuations, relative to the experimental time t', are linked to the phase fluctuations $\Delta\varphi$ by

$$\frac{\Delta\nu}{\nu} = \frac{\Delta\varphi}{2\pi n} \tag{67}$$

It follows that

$$\langle(\Delta\nu)^2\rangle = \nu_0^2 \frac{\langle(\Delta\varphi)^2\rangle}{(2\pi n)^2} = \nu_0^2 \frac{\langle(\delta\varphi)^2\rangle}{n(2\pi)^2} \tag{68}$$

or, by taking account of (66)

$$\langle(\Delta\nu)^2\rangle = \frac{1}{(2\pi)^2} \frac{\langle(\delta\varphi)^2\rangle}{t'} \nu_0 \tag{69}$$

Finally, we can deduce the mean-square value of the phase fluctuations per cycle from formula (28), by replacing t by $1/\nu_0$. We get

$$\langle(\delta\varphi)^2\rangle = (2\pi)^2 \frac{kTr}{a_0^2} \nu_0 \tag{70}$$

Then (69) reads

$$\left[\frac{\langle(\Delta\nu)^2\rangle}{\nu_0^2}\right]^{1/2} = \left[\frac{kTr}{a_0^2 t'}\right]^{1/2} \tag{71}$$

It is the uncertainty in a frequency measurement of duration t'. This uncertainty decreases when t' increases. Note that $\langle(\Delta\nu)^2\rangle$ is an average value and not a "line width," as the one which is given by the Fourier transformation. These remarks will be complemented by the discussion of section 6.3.

6. APPLICATION TO MASERS†

6.1. Relation with Classical Oscillator Theory

One can see easily that formula (71) is not specific for the zero-crossing method but applies to all techniques in which the frequency measurements use a finite time t'. It can be applied to quartz clocks, as shown by our earlier investigations. It can also be applied to masers, in which field it has raised renewed interest. Here we shall consider the example of masers.

At the outset let us emphasize a few similarities between classical radioelectric oscillators and masers, as shown in Fig. 13. Figure 13a

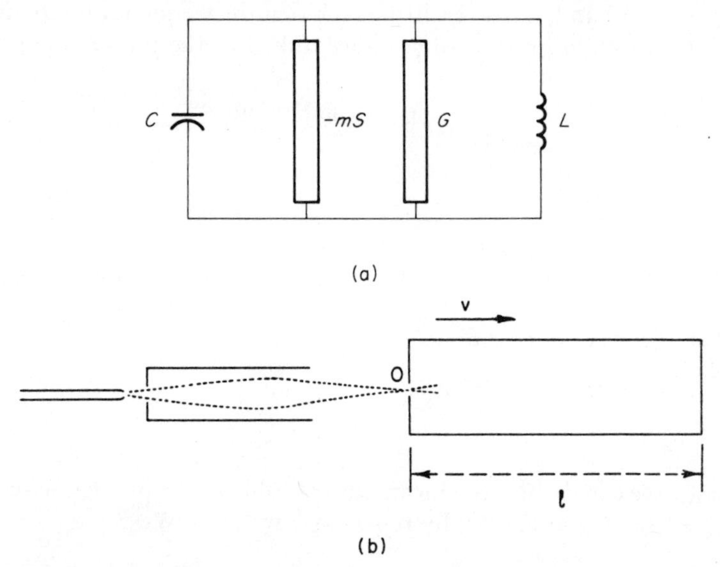

FIG. 13. (a) Classical self-oscillator; (b) simplified schema for a maser. The beam of particles enters the cavity through 0.

shows a classical self-oscillator. We assume that the tank circuit (L, C, G) has a very high quality factor, say its damping-time constant $\tau = 2L/r$ (r is the series resistance of the tank: $r = LG/C$) is very large, so that the frequency of the self-sustained oscillator is very close to the characteristic frequency of the tank.

Figure 13b refers to a maser. The beam of particles enters the cavity through 0. These particles can be thought of as quantum oscillators with a very high quality factor. Their characteristic frequency is

† See [10, 24, 28, 31].

$\nu_0 = \omega_0/2\pi$ and their time constant is τ. They *basically* perform the same role as the tank circuit of Fig. 13a.†

More precisely, the time constant τ of the quantum oscillator is the average interval of time during which a particle undergoes a transition from initial level W_1 to final level W_2.

On the other hand, in a classical oscillator, the effective loss of energy per second is

$$\frac{a_0^2}{2R} = \frac{a_0^2}{2Q_l^2 r}$$

where a_0 is the amplitude of the signal, $R = 1/G$ the parallel resistance of the tank circuit, and Q_l the quality factor of the tank

$$Q_l = \frac{L\omega_0}{r} = \frac{R}{L\omega_0}$$

The corresponding loss in a maser oscillator is the mean power which the beam yields to the cavity. Let us designate this power by P. Then if we identify P with its classical analogue, formula (71) can be rewritten in terms of the new parameters relative to the maser. One readily gets

$$\left[\frac{\langle(\Delta\nu)^2\rangle}{\nu_0^2}\right]^{1/2} = \left(\frac{kTR}{a_0^2 Q_l^2 t'}\right)^{1/2} = \frac{1}{Q_l}\left(\frac{kT}{2Pt'}\right)^{1/2} \tag{72}$$

Moreover, if one defines the quality factor of the quantum resonator by

$$Q_l = \frac{\nu_0}{\Delta\nu_r} \tag{73}$$

where $\Delta\nu_r$ is the resonance line width at half-height (the so-called "optical line width") one gets, by substituting in (72),

$$[\langle(\Delta\nu)^2\rangle]^{1/2} = \left(\frac{kT}{2Pt'}\right)^{1/2} \Delta\nu_r \tag{74}$$

This formula can also be applied to ordinary circuits.

6.2. Saturation Effect of the Quantum Oscillator‡

To compare classical clocks and masers we have described masers in macroscopical terms; we have emphasized the links between the theory

† The corresponding parameters do not have the same order of magnitude in actual devices.

‡ See [10].

of masers and the classical theory of clocks. It is possible to go deeper into this similarity, while keeping in mind the limitations of such arguments, by considering a more realistic model, shown in Fig. 14. Resonant circuit 1 depicts the cavity. Let Q and ω_c be its quality factor and its characteristic frequency, respectively. Resonant circuit 2 depicts the quantum oscillator, whose quality factor and characteristic frequency are Q_l and ω_0.

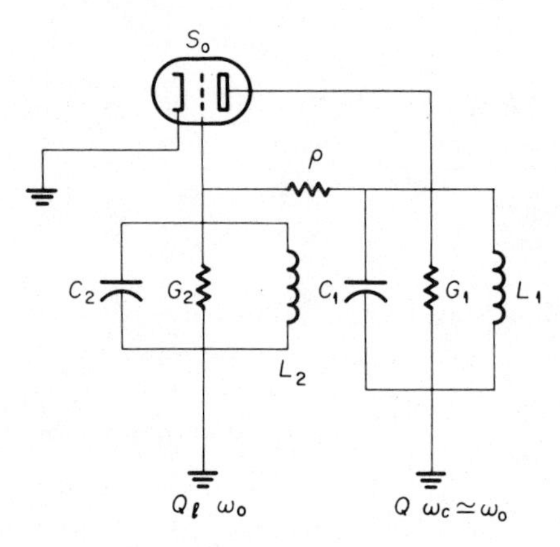

FIG. 14. Equivalent circuit for a maser.

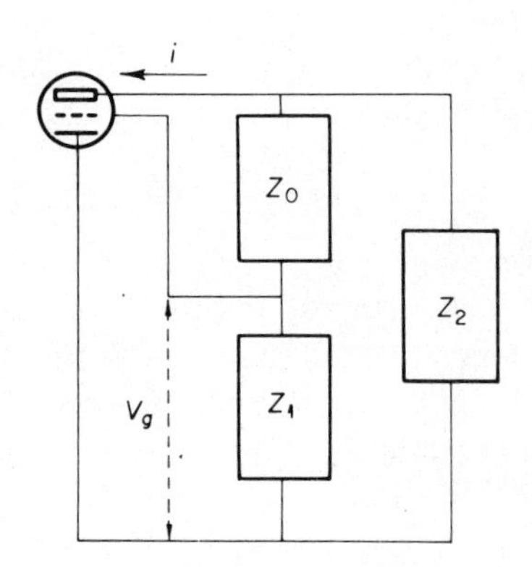

FIG. 15. General scheme for a self-oscillator.

Assume that $Q_l \gg Q$. The coupling between the quantum oscillator and the cavity is represented by resistance ρ. Let us also assume that ω_c is very close to ω_0. However, since in general ω_c cannot be strictly equal to ω_0, the frequency of the self-sustained oscillator is different from ω_c and ω_0, but it is also very close to ω_c and ω_0.

This model can be readily reduced to the general scheme (Fig. 15) which we have considered in Chapter II, section 1.2. Here we will refer to the corresponding equation

$$-Z_1 Z_2 S_0 = Z_0 + Z_1 + Z_2 \tag{75}$$

Impedances Z_0, Z_1, and Z_2 are shown in Fig. 15. S_0 is the slope of the characteristic curve of the tube at the operating point. Furthermore, assume that $Z_0 \gg Z_1, Z_2$, according to which (75) reduces to

$$-Z_1 Z_1 S_0 = Z_0 \tag{76}$$

Next we shall express Z_0, Z_1, and Z_2, in terms of the parameters of the circuits.

Following the notation of Fig. 14 one gets

$$Z_1 = \frac{jL_1\omega}{1 - L_1C_1\omega^2 + j\omega L_1G_1} = \frac{jL_1\omega}{[(\omega_0^2 - \omega^2)/\omega_0^2] + [j(\omega/\omega_0)(1/Q)]}$$

$$Z_2 = \frac{jL_2\omega}{1 - L_2C_2\omega^2 + j\omega L_2G_2} = \frac{jL_2\omega}{[(\omega_0^2 - \omega^2)/\omega_0^2] + [j(\omega/\omega_0)(1/Q_l)]}$$

$$Z_0 = \rho$$

Let us put $G_0 = 1/\rho$ and assume that ρ is a very high resistance. Then

$$Z_1 \simeq \frac{jL_1\omega_c}{2[(\omega_c - \omega)/\omega_c] + [j(1/Q)]}$$

$$Z_2 \simeq \frac{jL_2\omega_0}{2[(\omega_0 - \omega)/\omega_0] + [j(1/Q_l)]}$$

By substituting in (76) we get the following relation:

$$\left(2\frac{\omega_0 - \omega}{\omega_0} + j\frac{1}{Q_l}\right)\left(2\frac{\omega_c - \omega}{\omega_c} + j\frac{1}{Q}\right) = S_0L_1L_2\omega_0\omega_cG_0$$

say, since $(\omega_0 - \omega)/\omega_0$ and $(\omega_c - \omega)/\omega_c$ are very small,

$$\frac{1}{Q} - 2j\frac{\omega_c - \omega}{\omega_c} \simeq -G_0S_0L_1L_2\omega_0\omega_cQ_l\left(1 + 2jQ_l\frac{\omega_0 - \omega}{\omega_0}\right)$$

from which we deduce

$$G_0S_0L_1L_2\omega_0\omega_cQ_lQ = -1 \tag{77}$$

$$\frac{\omega - \omega_c}{\omega_c} = \frac{Q_l}{Q}\frac{\omega_0 - \omega}{\omega_0} \tag{78}$$

As a matter of fact, condition (77) can be fulfilled only if the slope S_0 of the tube, at the operating point, is negative.

Condition (78) is similar to the one obtained in the quantum theory of masers. When the amplitude of the oscillation is small, say when the threshold condition is nearly fulfilled, one gets

$$\frac{\omega - \omega_0}{\omega_0} = 1.07\frac{Q}{Q_l}\frac{\omega_c - \omega}{\omega_c} \tag{79}$$

However, quantum theory is absolutely necessary when one wishes to take account of the saturation effect of the quantum oscillator. Then condition (79) is replaced by

$$\frac{\omega - \omega_0}{\omega_0} = \frac{Q}{Q_l}\frac{1}{0.89\pi}\frac{1 - \cos 2\theta}{1 - \dfrac{\sin 2\theta}{2\theta}}\frac{\omega_c - \omega}{\omega_c} \tag{80}$$

where θ is the ratio of the actual power P to a reference level P_c, given by quantum theory. Formula (80) enables us to borrow a semiclassical treatment by improving the above model.

Assume that the quality factor Q_l of circuit 2 depends on the power of the oscillation, and that it is governed by the law

$$Q_l = Q_{l_0} 0.95\pi \frac{1 - (\sin 2\theta/2\theta)}{1 - \cos 2\theta} \simeq Q_{l_0}\pi \frac{1 - (\sin 2\theta/2\theta)}{1 - \cos 2\theta}$$

Q_{l_0} being the value of Q_l at very low power.

Then the uncertainty in the measurement of the frequency, given by formula (72), is

$$\left[\frac{\langle(\Delta\nu)^2\rangle}{{\nu_0}^2}\right]^{1/2} = \frac{1}{\pi} \frac{1 - \cos 2\theta}{1 - (\sin 2\theta/2\theta)} \frac{1}{Q_{l_0}} \left(\frac{kT}{2Pt'}\right)^{1/2}$$

Moreover, if the cavity of the maser is tuned in such a way that the transition time τ of the quantum oscillator is nearly equal the average transit time of the particles

$$\tau = \frac{l}{v}$$

where l is the length of the cavity in the direction of the beam and v the average velocity of the particles in this direction, one gets

$$\left[\frac{\langle(\Delta\nu)^2\rangle}{{\nu_0}^2}\right]^{1/2} = \frac{1}{\pi} \frac{2v}{l\omega_0} \frac{2\theta \sin^2\theta}{2\theta - \sin 2\theta} \left(\frac{2kT}{Pt'}\right)^{1/2} \tag{81}$$

which was deduced from quantum theory by Shimoda, Wang, and Townes.

6.3. Optical Line Width of the Signal†

6.3.1. Accuracy of Frequency Measurements

Let us return to formula (71) and assume that t' is gradually increased. We see that the uncertainty of the measurement tends to zero when $t' \to \infty$, following a law in $1/\sqrt{t'}$, which is a straightforward consequence of the so-called "law of large numbers," which is well known in the theory of stochastic processes.

It will be interesting to pay attention to the special case in which the experimental time t' is equal to the correlation time t_0^* of the signal. As pointed out previously, t_0^* is the time during which the oscillator

† See [6, 8, 9, 26, 31].

undergoes a substantial loss of coherence (say of phase memory). The value of $t_0{}^*$ has been computed in Section 2.5.1:

$$t_0{}^* = \frac{2a_0{}^2}{kTr\omega_0{}^2} \tag{82}$$

Then one finds

$$\left[\frac{\langle(\Delta\nu)^2\rangle}{\nu_0{}^2}\right]^{1/2} = \left(\frac{kTr}{a_0{}^2 t_0{}^*}\right)^{1/2} = \frac{1}{\sqrt{2}}\,\frac{kTr\omega_0}{a_0{}^2} \tag{83}$$

Now, recalling formula (36),

$$\frac{2\delta\omega}{\omega_0} = \frac{2\delta\nu}{\nu_0} = \frac{kTr\omega_0}{a_0{}^2} \tag{84}$$

which represented the uncertainty in the frequency measurement, using a spectroscopic technique, based on the Fourier transformation of the signal and the measurement of the line width, we see that (83) is nearly equal to the optical line width of the signal. Note that the factor $1/\sqrt{2}$, which introduces a distinction between (83) and (84), is of little consequence, for it is merely due to some arbitrariness in the definition of the correlation time $t_0{}^*$. One can easily find a definition of the correlation time $t_0{}^*$ such that the factor $1/\sqrt{2}$ does not appear in formula (83). As a matter of fact, the definition depends on what is meant by "a substantial loss of coherence."

We arrive at the following conclusions†:

(a) Let $(\Delta\nu/\nu_0)_l = [\langle(\Delta\nu)^2\rangle/\nu_0{}^2]^{1/2}$ be the uncertainty relative to *a technique in which the frequency measurement uses a finite time t'* [formula (71)], and $(\Delta\nu/\nu_0)_s = 2\delta\nu/\nu_0$ be the uncertainty defined by the optical line width of the signal [formula (84)], say the uncertainty relative to *a spectroscopic method* (in which no finite time t' is explicitly specified); then from (71) and (84) we deduce [9]

$$\left(\frac{\Delta\nu}{\nu_0}\right)_l = \frac{1}{(2\pi\nu_0 t')^{1/2}}\left(\frac{\Delta\nu}{\nu_0}\right)_s^{1/2} \tag{85}$$

(b) If $t' \ll t_0{}^*$, where $t_0{}^*$ is the correlation time of the signal, the uncertainties relative to both techniques are absolutely different, even *they do not have the same order of magnitude*:

$$\left(\frac{\Delta\nu}{\nu_0}\right)_s \ll \left(\frac{\Delta\nu}{\nu_0}\right)_l$$

† Here we shall use standard notation to simply comparision of the different formulas.

It is interesting to note that $(\Delta\nu/\nu_0)_s$ can be deduced from a measurement which uses a finite time t', provided that one refers to relation (85). An experimental method for this purpose has been described by Berstein [3, 4].

(c) If $t' \sim t_0{}^*$, both techniques have the same accuracy.

(d) If $t' \gg t_0{}^*$, the technique which uses a finite time t', for instance the zero-crossing technique, is much more accurate than the spectroscopic techniques, since then one takes advantage of the law in $1/\sqrt{t'}$. However, the counterpart of the gain in accuracy is the increase in the time of measurement, which may appear to be a serious disadvantage.†

6.3.2. Optical Line Width of a Maser Oscillator

Formula (84) can be applied to maser oscillators. Let $\Delta\omega_r$ be the *full* resonance line width at half-height, on the ω scale, of the quantum oscillator, i.e.,

$$\frac{1}{Q_l} = \frac{\Delta\omega_r}{\omega_0}$$

Then (84) reads

$$\frac{2\delta\omega}{\omega_0} = \left(\frac{\Delta\omega}{\omega_0}\right)_s = \frac{kTr\omega_0{}^2}{a_0{}^2} = \frac{kTQ_l{}^2r(\Delta\omega_r)^2}{a_0{}^2} \tag{86}$$

say, by taking account of

$$P = \frac{a_0{}^2}{2R} = \frac{a_0{}^2}{2Q_l{}^2r}$$

(P is the power the beam delivers to the cavity).

$$(\Delta\omega)_s = \frac{kT(\Delta\omega_r)^2}{2P} \tag{87}$$

To compare this formula with the one obtained by Gordon, Zeiger, and Townes, let

$$(\Delta\omega)_s = 4\pi(\delta\nu)_s \qquad \Delta\omega_r = 4\pi\delta\nu_r$$

where $(\delta\nu)_s$ and $\delta\nu_r$ are the *half-widths* at half-height of the output signal and of the resonance line, respectively. Then (87) reads (see also [8, 31])

$$2\delta\nu_s = \frac{4\pi kT(\delta\nu_r)^2}{P} \tag{88}$$

† However in most of the practical cases, the technique of measurement is determined by the physical nature of the phenomenon observed.

From the above discussion we conclude that (74) and (88) are not so different as they might appear to be at first sight, only that they do not refer to the same definition of the frequency uncertainty. In (74) one considers a measurement which uses finite time t', whereas (88) is a spectroscopic formula. The formulas are linked together by (85). They become identical when t' is the correlation time of the maser.

REFERENCES

1. P. Langevin, *Compt. Rend.* **146**, 530 (1908).
2. A. Andronov and S. Chaikin, Theory of Non-Linear Oscillations, Moscow, 1937. (English transl. by S. Lefschetz, A. Andronov, and S. Chaikin, Princeton Univ. Press, Princeton, N.J., 1949.)
3. I. L. Berstein, On Fluctuations in the Neighborhood of Periodic Motion of an Auto-Oscillating System (in English). *C. R. Acad. Sci. USSR, Mat.-Fiz.* **20** (1938); Fluctuations of Auto-Oscillatory Systems. *J. Tech. Phys.* (*USSR*) **11** (1941); On the Fluctuations of Tube Generators. *Dokl. Akad. Nauk USSR*, **68**(1949); Amplitude and Phase Fluctuations of Tube Generators. *Izv. Acad. Nauk USSR Ser. Fiz.* **14** (1950).
4. G. S. Gorelik, On the Technical and Natural Line-Width of a Tube Generator. *J. Exptl. Theoret. Phys., Akad. Nauk USSR* **20** (1950); Nonlinear Oscillations, Interference and Fluctuations. *Izv. Akad. Nauk SSSR, Ser. Fiz.* **14** (1950).
5. R. M. Lerner, The Effect of Noise on the Frequency Stability of a Linear Oscillator *Proc. Natl. Electron. Conf.* **7**, 275–280 (1951).
6. A. Blaquière, Effet du bruit de fond sur la fréquence des autooscillateurs à lampes. Précision ultime des horloges radioélectriques. *Compt. Rend.* **234**, 419–421 (1952); Effet du bruit de fond sur l'amplitude des oscillateurs entretenus. *Compt. Rend.* **234**, 710–712 (1952); Effet du bruit de fond sur la fréquence des autooscillateurs à lampes. Précision ultime des horloges radioélectriques. *Ann. Radioelec.* **8**, 36–81 (1953); Spectre de Puissance d'un oscillateur non linéaire perturbé par le bruit. *Ann. Radioelec.* **8**, 153–179 (1953); Influence du bruit de la lampe dans les oscillateurs. *Compt. Rend.* **237**, 1316–1318 (1953).
7. N. G. Basov and A. M. Prokhorov, The Theory of a Molecular Oscillator and a Molecular Power Amplifier. *Discussions Faraday Soc.* **19**, 96–99 (1955).
8. J. P. Gordon, H. J. Zeiger, and C. H. Townes, The Maser, a New Type of Microwave Amplifier, Frequency Standard and Spectrometer. *Phys. Rev.* **99**, 1264–1274 (1955).
9. A. Blaquière, Limite imposée par le bruit de fond à la précision des horloges radioélectriques. *Ann. Franc. Chronométrie*, 2nd ser., pp. 15–43 (1956).
10. K. Shimoda, T. C., Wang, and C. H. Townes, Further Aspects of the Theory of the Maser. *Phys, Rev.* **102**, 1308–1321 (1956).
11. R. V. Pound, Spontaneous Emission and the Noise Figure of Maser Amplifier. *Ann. Phys.* **1**, 24–32 (1957).
12. K. Shimoda, H. Takahasi, and C. H. Townes, Fluctuations in Amplification of Quanta with Application to Maser Amplifiers. *J. Phys. Soc. Japan* **12**, 686–700 (1957).
13. A. Blaquière and P. Grivet, "Le Bruit de fond." Masson, Paris, 1958.
14. H. J. Hasenjäger, Contribution à l'étude de l'oscillateur HF en vue de son application en spectroscopie hertzienne. Thèse, Faculté des Sciences de Grenoble, France, 1958.

15. S. Kurochkin, On the Theory of the Spin Oscillator. *Radiotekhn. i Elektron.* **3**, 198–201 (1958).
16. M. Buyle-Bodin, Sensibilité et fidélité des oscillateurs autodynes en spectroscopie hertzienne. *J. Phys. Radium* **20**, 159 (1959).
17. F. N. H. Robinson, Nuclear Resonance Absorption Circuit. *J. Sci. Instr.* **36**, 481–487 (1959).
18. A. Blaquière, Mécanique non linéaire, les oscillateurs à régimes quasi-sinusoidaux. "Mémorial des sciences mathématiques," fasc. 141. Gauther-Villars, Paris, 1960.
19. M. J. E. Golay, Monochromaticity and Noise in a Regenerative Electrical Oscillator. *Proc. IRE* (1960).
20. P. Grivet, Mesure des champs magnétiques faibles du type champ terrestre. *Bull. Ampère* **9**, 567–620 (1960).
21. E. L. Tolnas, Measurement of NH_3 Maser-Oscillator Frequency and Noise. *Bull. Am. Phys. Soc.* **5**, 342–343 (1960).
22. A. Blaquière and P. Grivet, L'effet non linéaire du bruit blanc et du bruit de scintillation dans les spectromètres à résonance nucléaire, du type oscillateur marginal. *Arch. Sci. Geneva, Spec. Issue* (1961).
23. G. Bonnet, Validité de la représentation par modèle électrique des effets de RMN. *J. Phys. Radium* **22**, 204–214 (1961); Propriétés statistiques du bruit de fond en RMN. *Bull. Ampère* **10**, 297–304 (1961).
24. A. Blaquière and P. Grivet, Spectre d'un oscillateur maser, relation avec la théorie des autooscillateurs non linéaires classiques. *Compt. Rend.* **255**, 2929–2931 (1962).
25. A. Blaquière and R. Pachowska, Fluctuations d'un système dépendant de plusieurs paramètres aléatoires, application aux réacteurs nucléaires. *Rept. CEA 2115*, 1962.
26. A. Blaquière, Théorie de la réaction de fission en chaine. "Bibliothèque des sciences et techniques nucléaires." Presses Universitaires de France, Paris, 1962; Largeur de raie d'un oscillateur laser considéré comme le siège d'une réaction en chaine. *Compt. Rend.* **255**, 3141–3143 (1962).
27. G. Bonnet, Possibilités nouvelles des magnétomètres à protons. *Ann. Geophys.* **18**, 62–91, 150–178 (1962).
28. D. Kleppner, H. M. Goldenberg, and N. F. Ramsey, Theory of the Hydrogen Maser. *Phys. Rev.* **126**, 603–615 (1962).
29. D. K. C. Macdonald, "Noise and Fluctuations," pp. 37 and 107. Wiley, New York, 1962.
30. C. H. Townes, "Masers," pp. 39–67. Academic Press, New York, 1962.
31. A. Blaquière and P. Grivet, "Masers and Classical Oscillators Symposium on Optical Masers." Brooklyn, New York, 1963; Nonlinear Effects of Noise in Electronic Clocks. *Proc. IEEE* **51** (1963).
32. A. Blaquière, G. Bonnet, and P. Grivet, Maser and Magnetometers and Spin-Oscillators. *Third Quantum Electronics Symp., Paris,* 1963.
33. A. Javan, F requency Stability of He-Ne Lasers. *Third Quantum Electronics Symp., Paris,* 1963.
34. A. Blaquière, Fluctuations et largeur de raie d'un autooscillateur: Equivalence des théories non-linéaires et des méthodes de linéarisation. *Colloq. Intern. Centre Natl. Rech. Sci. (Marseille)*, Sept. 7–12, 1964.
35. T. K. Caughey, On the Response of a Class of Nonlinear Oscillators to Stochastic Excitation. *Colloq. Intern. Centre Natl. Rech. Sci. (Marseille)*, Sept. 7–12, 1964.
36. S. H. Crandall, Random Forcing of Nonlinear Systems. *Colloq. Intern. Centre Natl. Rech. Sci. (Marseille)*, Sept. 7–12, 1964.

APPENDIX

Sinusoidal Modes of Electromagnetic Resonators

F. Bertein

Faculty of Sciences Orsay, Institute of Electronics
Orsay (S. O.), France

1. EQUATION FOR LINEAR OSCILLATIONS

We shall briefly examine the nonlinear problem which occurs in the analysis of free oscillations of an electromagnetic cavity filled with a dielectric material.

Let us suppose that the cavity, K, is empty and the losses are disregarded; the number of "modes," say, the number of oscillatory sinusoidal states, is infinite. In this case K exhibits similarities with a set of oscillatory circuits composed of inductances and capacitances. Next we shall see that the dielectric material introduces a coupling between the modes and also produces a nonlinear effect. As a consequence of both properties, a number of phenomena can occur, particularly the well-known synchronization effect.

First, we shall briefly derive the equations of oscillation. We shall use mks units and complex notations. Spatial coordinates will be denoted by x.

Let us suppose that the dielectric is isotropic, but heterogeneous, and that it is described by susceptibility $\chi(j\omega, x)$. That is, in the case of sinusoidal oscillations, the polarization, $\mathbf{M}(x, t)$, at every point, is linked to the field, $\mathbf{E}(x, t)$, by the relation

$$\mathbf{M}(x, t) = \chi(j\omega, x)\mathbf{E}(x, t) \tag{1}$$

Accordingly, in the first part of this account, the susceptibility will be assumed to be independent of the field amplitude. On the other hand, we shall make use of the restrictive assumption: $|\chi| \ll 1$.

Let us introduce the so-called *modes*, vector functions $\mathbf{E}_i(x)$, which are the spatial parts of sinusoidal electric fields $\mathbf{E}_i(x)\exp(j\omega_i t)$ in the empty cavity, with losses disregarded. Moreover, we shall assume that these functions have been normalized, and orthogonalized, in the volume K:

$$\int \mathbf{E}_i{}^*\mathbf{E}_k\,d\tau = \delta_{ik} \tag{2}$$

According to the theory of resonant cavities, fields $\mathbf{E}$ and $\mathbf{M}$ can be expanded as follows:

$$\begin{aligned} \mathbf{E}(x, t) &= \sum q_i \mathbf{E}_i \\ \mathbf{M}(x, t) &= \sum m_i \mathbf{E}_i \end{aligned} \tag{3}$$

q_i and m_i are the "coordinates" of $\mathbf{E}$ and $\mathbf{M}$, respectively. They are functions of time t. Accordingly, (1) is rewritten

$$\sum_i m_i \mathbf{E}_i = \chi \sum_i q_i \mathbf{E}_i$$

Let us multiply both sides by $\mathbf{E}_k{}^*$ and integrate over the cavity. By taking account of (2), one gets a set of equations in which the m_i are linear functions of the q_i. In the following it will be convenient to write these equations in the form

$$-\omega^2 m_i = \sum_k C_{ik} q_k \tag{4}$$

$$C_{ik} = -\omega^2 \int \mathbf{E}_i{}^*\mathbf{E}_k \chi(j\omega, x)\,d\tau \tag{5}$$

Now we shall apply Maxwell's equations to the electromagnetic field. By using expansions (3) for $\mathbf{E}$ and $\mathbf{M}$, these equations reduce to the subsequent equalities between the coordinates, following a classical derivation in the theory of cavities. Dissipation in the walls is taken into account by additional terms in the usual way:

$$\ddot{q}_i + \frac{\omega_i}{Q_i}\dot{q}_i + \omega_i{}^2 q_i + \ddot{m}_i = 0$$

In the case of sinusoidal oscillations,

$$\left(-\omega^2 + j\frac{\omega_i \omega}{Q_i} + \omega_i{}^2\right) q_i - \omega^2 m_i = 0 \tag{6}$$

Finally, by taking account of relation (4), we get

$$\left(-\omega^2 + j\frac{\omega_i \omega}{Q_i} + \omega_i^2\right) q_i + \sum_k C_{ik} q_k = 0 \qquad C_{ik}: \quad (5)$$

These are algebraic linear equations, which are similar to the ones of coupled circuits, with ω_i, characteristic frequencies, and C_{ik}, coupling coefficients.

Let us rewrite the above equations in the form

$$(\omega_i^2 - \omega^2) q_i + \sum_k C_{ik} q_k = 0 \tag{7}$$

by incorporating coefficients C_{ii} with $j(\omega_i \omega / Q_i)$. Let us recall that linearity is due to the fact that we have assumed the susceptibility to be independent of the applied field; this assumption is valid when the field is sufficiently small.

Then the well-known property of coupled circuits holds—one can observe oscillations such that field expansion (3) actually reduces to a single mode, for instance, to $\mathbf{E}_i(x)(q_k = 0 \text{ if } K \neq i)$. As a matter of fact, this situation occurs when resonance frequency ω_i is sufficiently far from the other frequencies.

2. NONLINEAR OSCILLATIONS: SINGLE MODE

Now let us consider the case where polarization **M** is linked to field **E** by a susceptibility which depends on the amplitude of this latter one. This situation will lead us to nonlinear equations.

In agreement with the usual experimental results, we shall assume that the relation between the polarization and the field can be written [instead of (1)]

$$\mathbf{M}(x, t) = [\chi(j\omega, x) + \xi(j\omega, x)\hat{E}^2(x)]\mathbf{E}(x, t) \tag{8}$$

where $\hat{E}(x)$ is the maximum value of the field at point x. Then the susceptibility is

$$\chi(j\omega, x) + \xi(i\omega, x)\hat{E}^2(x)$$

where $\chi(j\omega, x)$ is the linear approximation, when the field is sufficiently weak.

Relations (4) and (5), between the coordinates of **E** and **M**, must be replaced by the following ones:

$$-\omega^2 m_i = \sum \tilde{C}_{ik} q_k \tag{9}$$

$$\tilde{C}_{ik} = -\omega^2 \int \mathbf{E}_i{}^* \mathbf{E}_k (\chi + \xi \hat{E}^2)\, d\tau = C_{ik} - \omega^2 \int \xi \mathbf{E}_i{}^* \mathbf{E}_k \hat{E}^2\, d\tau \tag{10}$$

Peak values, $K_i = \hat{q}_i$, of the coordinates of $\hat{E}$, and volume integrals will appear in the last term. Volume integrals will be denoted by

$$c_{ik}^{ln} = -\omega^2 \int \xi(j\omega, x) \mathbf{E}_i{}^* \mathbf{E}_k \mathbf{E}_l{}^* \mathbf{E}_n\, d\tau \tag{11}$$

To obtain the equations of the sinusoidal oscillations one must substitute expressions (10) into (6), the latter ones being deduced from Maxwell's equations; coefficients $\tilde{C}_{ii}$ and C_{ii} will be incorporated with $j(\omega_i \omega / Q_i)$.

First of all, let us consider an oscillation which is reduced to a single mode $\mathbf{E}_i(x)$, i.e., the usual motion of a simple oscillator. As pointed out earlier, this situation occurs only when frequency ω_i is sufficiently far from the other frequencies.

In this case the equation of motion is

$$(\omega_i{}^2 - \omega^2 + \tilde{C}_{ii}) q_i = 0$$

say,

$$(\omega_i{}^2 - \omega^2 + C_{ii} + c_i K_i{}^2) q_i = 0 \tag{12}$$

$$c_i = c_{ii}^{ii} = -\omega^2 \int \xi |\, E_i \,|^4\, d\tau \tag{13}$$

The equation has two possible solutions:

1. $q_i = 0$, no oscillation; this solution is not admissible, since it represents an unstable state which cannot persist because of the random noise.

2. Stable oscillation, determined by

$$\omega_i{}^2 - \omega^2 + C_{ii} + c_i K_i{}^2 = 0$$

which is equivalent to two equations involving real quantities, the frequency $\omega_0 i$ and the amplitude $K_0 i$ of the steady oscillation. In other words,

$$\omega_i{}^2 - \omega_{0i}^2 + C_{ii} + c_i K_{0i}^2 = 0 \tag{14}$$

3. SYNCHRONIZATION OF TWO MODES, SPATIALLY SEPARATED, IN THE NONLINEAR REGION

Now let us consider an oscillation involving two modes:

$$\mathbf{E} = q_1\mathbf{E}_1 + q_2\mathbf{E}^2$$

which requires that frequencies ω_1 and ω_2 be sufficiently separate from the other ones.

We shall investigate the condition for a sinusoidal oscillation of frequency ω to exist, or, in other words, the condition for the *synchronization of two modes*. Indeed this problem is similar to the one concerning coupled circuits [1].

The equations which determine an oscillation of frequency ω are

$$\begin{aligned} (\omega_1{}^2 - \omega^2 + \tilde{C}_{11})q_1 + \tilde{C}_{12}q_2 = 0 \\ \tilde{C}_{21}q_1 + (\omega_2{}^2 - \omega^2 + \tilde{C}_{22})q_2 = 0 \end{aligned} \tag{15}$$

For the sake of simplification we shall assume that, at every point of the spatial nonlinear region ($\xi \neq 0$), one of the two modes has an amplitude whose magnitude is very large with respect to the amplitude of the other one.

One can see easily that the $\tilde{C}_{ik}$ are practically reduced to the following:

$$\begin{aligned} \tilde{C}_{11} = C_{11} + c_1K_1{}^2 \qquad \tilde{C}_{22} = C_{22} + c_2K_2{}^2 \\ \tilde{C}_{12} = C_{12} \qquad \tilde{C}_{21} = C_{21} \end{aligned} \tag{16}$$

Furthermore, let us make use of amplitude K_{0i} and frequency ω_{0i}, determined by (14). These quantities characterize the oscillations of each mode, when these modes are assumed to be separate.

Equations (14) for the synchronized motion are rewritten

$$\begin{aligned} [\omega_{01}^2 - \omega^2 + c_1(K_1{}^2 - K_{01}^2)]q_1 + C_{12}q_2 = 0 \\ C_{21}q_1 + [\omega_{02}^2 - \omega^2 + c_2(K_2{}^2 - K_{02}^2)]q_2 = 0 \end{aligned} \tag{17}$$

They are equivalent to four equations involving real quantities—frequency ω, amplitudes K_1 and K_2, phase shift $\varphi(q_1 = K_1e^{j\omega t}$, $q_2 = K_2e^{j(\omega t - \varphi)})$. Solving these equations, one will obtain the quantities which characterize the motion.

Next we shall put

$$\begin{aligned} &\Delta\omega = \omega_{02} - \omega_{01} \\ &\delta_1 = \omega - \omega_{01} \qquad \delta_2 = \omega - \omega_{02} \text{ with } \omega_{02} > \omega_{01} \end{aligned}$$

In practical cases, frequencies ω_i, ω_{0i}, and ω are very close to one another. Accordingly, one can replace $\omega_{0i}^2 - \omega^2$ by $-2\omega_{0i}\delta_i$ and ω by ω_1 or ω_2 in expressions (13) for c_i.

Then the resolution of (17) exhibits different cases, depending on the values of the parameters. Here we shall restrict the discussion to the simplest case, the one in which

$$\begin{aligned} C_{12} &= C_{21} = j\Gamma \qquad && \Gamma \text{ real} \\ c_1 &= c_2 = j\gamma && \gamma \text{ real} > 0 \\ K_{01} &= K_{02} = K_0 \end{aligned} \tag{18}$$

Such conditions are met in some physical cases of considerable importance; they are obtained, for instance, in the case of two modes $\mathbf{E}_1(x)$ and $\mathbf{E}_2(x)$, real and behaving "symmetrically," if function $\xi(j\omega, x)$ is negative everywhere.

Then the four equations involving real quantities, which are equivalent to (17), are

$$\begin{aligned} 2\omega_1\delta_1 K_1 &= \Gamma K_2 \sin\varphi \\ 2\omega_2\delta_2 K_2 &= -\Gamma K_1 \sin\varphi \\ \gamma(K_1^2 - K_{01}^2)K_1 &= -\Gamma K_1 \cos\varphi \\ \gamma(K_2^2 - K_{02}^2)K_2 &= -\Gamma K_1 \cos\varphi \end{aligned} \tag{19}$$

from which follows

$$-\frac{\delta_1}{\delta_2} = \frac{K_1^2 - K_{01}^2}{K_2^2 - K_{02}^2} = \frac{K_2^2}{K_1^2} \tag{20}$$

$$(K_1^2 - K_2^2)(K_1^2 + K_2^2 - K_0^2) = 0 \tag{21}$$

Equation (21) has two kinds of solutions, which we shall designate A and B, respectively.

Solutions A. $K_1^2 = K_2^2$: synchronized modes with the same amplitude.

From (19) one easily gets

$$\omega = \frac{\omega_{01} + \omega_{02}}{2} \tag{22}$$

$$\sin\varphi = \frac{\omega_1 \Delta\omega}{\Gamma} \tag{23}$$

$$K_1^2 = K_2^2 = K_0^2 \pm \frac{1}{\gamma}[\Gamma^2 - \omega_1^2(\Delta\omega)^2]^{1/2} \tag{24}$$

We see that solutions A can be separated into two categories: A_+ and A_-. A_+ and A_- are oscillations whose amplitudes are, respectively, greater and smaller than those of separate modes.

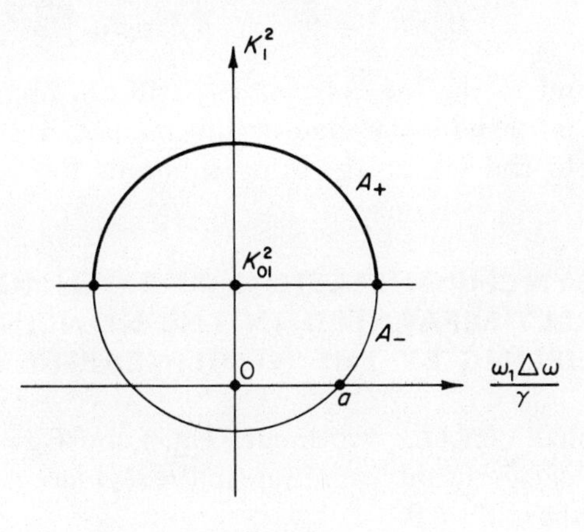

FIG. 1.

These results are shown in Fig. 1. Solutions A, corresponding to a given value of the coupling Γ, are located on a circle, restricted to positive ordinates; solutions A_+ and A_- are found on the upper and lower part of the circle, respectively.

By investigating the stability of the solutions, one finds that solutions A_+, only, are stable. This analysis can be developed from the viewpoint of the classical theory of linear circuits, as outlined by Bertein [2]. Here we shall invoke more heuristic arguments.

Let us consider solution A_-, whose representative point is *a*. It corresponds to no oscillation ($K_1^2 = K_2^2 = 0$), say, to an unstable state of each oscillator. So, one can easily understand that, by continuity, all the solutions which are of the same kind as *a*, namely, solutions A_-, are unstable.

Relations (23) and (24) show that stable solutions A_+ can exist only if

$$\Gamma > \omega_1 |\, \Delta\omega \,| \tag{25}$$

That is the *synchronization condition*; The greater the frequency interval between the two modes, the stronger the coupling must be.

Solutions B. $K_1^2 + K_2^2 = K_0^2$

The solution of (19) can be carried out, and gives, as in the above case concerning A, two categories of solutions which could be written explicitly, and whose limits for $\Gamma = 0$ can be easily anticipated:

$$B_1 : K_1^2 = K_0^2 \qquad K_2^2 = 0$$
$$B_2 : K_1^2 = 0 \qquad K_2^2 = K_0^2$$

They correspond to the separate modes, and are both unstable, since each of them is unstable for one of the modes. Accordingly, similar conclusions hold concerning the other solutions B.

4. SYNCHRONIZATION OF TWO MODES, NONSPATIALLY SEPARATED, IN THE NONLINEAR REGION; COUPLING BY THE NONLINEARITY ONLY

Finally we shall consider two modes $\mathbf{E}_1(x)$ and $\mathbf{E}_2(x)$, both of whose amplitudes are of significant magnitude in the regions where the dielectric is nonlinear, where $\xi \neq 0$.

For the sake of simplification, we shall assume that these modes are real,† and approximately parallel to one another at every point. It follows that

$$\hat{E}^2 = K_1^2 E_1^2 + 2K_1 K_2 E_1 E_2 \cos\varphi + K_2^2 E_2^2$$

and the $\tilde{C}_{ik}$ are given by the following expressions, in practice:

$$\tilde{C}_{ik} = c_{ik} + c_{ik}^{11} K_1^2 + 2c_{ik}^{12} K_1 K_2 \cos\varphi + c_{ik}^{22} E_2^2$$

These expressions will be substituted in (15) for the synchronized oscillation.

Let us introduce the following assumptions:

$$\begin{aligned}
&\text{(a)} \quad C_{12} = C_{21} = 0 \\
&\text{(b)} \quad c_{11}^{11} = c_{22}^{22} = j\gamma \qquad c_{12}^{12} = c_{11}^{-2} = js\gamma \qquad (\gamma, s \text{ real}) \\
&\text{(c)} \quad \text{others } c_{ik}^{ln} = 0 \\
&\text{(d)} \quad K_{01} = K_{02} = K_0
\end{aligned} \tag{26}$$

Assumptions (b) and (d) are similar to (18); (a) means that the two modes are *coupled by the nonlinearity only* (for instance, in the case of homogeneous media; susceptibility independent of x); condition (c) is an approximation which is explained by orthogonality of the two modes.

† They are characterized by an oscillation whose phase angle is well defined, except for a possible phase shift of π.

Equations (15) for the synchronized motion are rewritten

$$\begin{aligned} &[\omega_{01}^2 - \omega^2 + j\gamma(K_1{}^2 - K_0{}^2 + sK_2{}^2)]q_1 + 2js\gamma K_1K_2q_2 \cos\varphi = 0 \\ &2js\gamma K_1K_2q_1 \cos\varphi + [\omega_{02}^2 - \omega^2 + j\gamma(K_2{}^2 - K_0{}^2 + sK_1{}^2)]q_2 = 0 \end{aligned} \tag{27}$$

The resolution of (27) leads to four real equations, which can be obtained by replacing in equations (19):

$$\begin{aligned} &\Gamma \quad \text{by} \quad 2s\gamma K_1K_2 \cos\varphi \\ &K_{01}^2 \quad \text{by} \quad K_0{}^2 - sK_2{}^2 \\ &K_{02} \quad \text{by} \quad K_0{}^2 - sK_1{}^2 \end{aligned}$$

Again we get relation (21), which determines two kinds of solutions, A and B. Here we shall consider only solutions of type A: $K_1{}^2 = K_2{}^2$.

Frequency ω is given by (22), and one gets

$$\begin{aligned} &2s\gamma K_1{}^2 \sin\varphi \cos\varphi = \omega_1 \Delta\omega \\ &(K_1{}^2 - K_0{}^2 + sK_1{}^2 = -2sK_1{}^2 \cos^2\varphi \end{aligned} \tag{28}$$

By eliminating $K_1{}^2$ one gets

$$4s^2(\lambda^2 + 1)u^2 + 4s(1 + s - s\lambda^2)u + (1 + s)^2 = 0 \qquad \left(\lambda = \frac{\gamma K_0{}^2}{\omega_1 \Delta\omega}\right) \tag{29}$$

with $u = \cos^2\varphi$. Thus one can determine $\cos^2\varphi$ and hence $K_1{}^2$ by substituting in (28).

It may easily be seen that (28) has proper solutions, say solutions between 0 and 1, if

$$\gamma K_0{}^2 > \left[\left(\frac{1}{s} + 2\right)^2 - 1\right]^{1/2} \omega_1 |\, \Delta\omega \,|$$

That is the synchronization condition for motions of kind A, independent of any stability investigation. As pointed out earlier, the synchronization is due only to the nonlinearity properties of the medium.

The restrictive assumptions, which we have been led to introduce in the course of this short analysis, show clearly how miscellaneous and complex problems which are raised by nonlinearities can be in the study of modes in electromagnetic cavities.

REFERENCES

1. V. Belevitch, "Theorie des circuits nonlinéaires en régime alternatif", Gauthier-Villars, Paris, 1959.
2. F. Bertein, *J. Phys. Radium* **21**, 137A (1960).

Author Index

Numbers in parentheses are reference numbers and indicate that an author's work is referred to although his name is not cited in the text. Numbers in italic show the page on which the complete reference is listed.

Subject Index